WESTERN EUROPE: CHALLENGE AND CHANGI

Edited by David Pinder

for the Institute of British Geographers

Belhaven Press
(a division of Pinter Publishers)
London and New York

© The Institute of British Geographers 1990

First published in Great Britain in 1990 by
Belhaven Press (a division of Pinter Publishers),
25 Floral Street, London WC2E 9DS

British Library Cataloguing in Publication Data

A CIP catalogue record for this book is available from the
British Library

ISBN 1-85293-032-2 (hbk)
       1-85293-034-9 (ppr)

Typeset by Mayhew Typesetting, Bristol, England
Printed and bound in Great Britain by Richard Clay
(The Chaucer Press) Ltd, Bungay, Suffolk

Contents

List of figures   vii

List of tables   ix

Notes on contributors   xi

Preface and acknowledgements   xv

1   CHALLENGE AND CHANGE IN WESTERN EUROPE: AN OVERVIEW   1
David Pinder

Part I: CHALLENGE, CHANGE AND PRODUCTION   17

2   ENERGY: RESOURCES AND CHOICES   19
Peter R. Odell

3   EUROPEAN INDUSTRY AND GLOBAL COMPETITION   37
Peter Dicken

4   MANUFACTURING TRENDS, CORPORATE RESTRUCTURING AND SPATIAL
CHANGE   56
H.D. Watts

5   SMALL FIRMS: PHOENIX FROM THE ASHES?   72
Colin M. Mason and Richard T. Harrison

6   TECHNOLOGICAL CHANGE AND REGIONAL ECONOMIC ADVANCE   91
Alfred T. Thwaites and Neil Alderman

7   PRODUCER SERVICES AND ECONOMIC DEVELOPMENT   108
P.W. Daniels

Part II: INHERITANCE AND RESPONSE: THE URBAN DIMENSION   123

8   URBAN DECAY AND REJUVENATION   125
Louis Shurmer-Smith and David Burtenshaw

9   FROM SMALL SHOP TO HYPERMARKET: THE DYNAMICS OF RETAILING   142
S.L. Burt and J.A. Dawson

10   THE SOCIAL AND ECONOMIC GEOGRAPHY OF LABOUR MIGRATION: FROM
GUESTWORKERS TO IMMIGRANTS   162
Russell King

11   UNEMPLOYMENT: REGIONAL POLICY DEFEATED?   179
M. Wise and B. Chalkley

12   SUPRANATIONAL ENVIRONMENTAL POLICY AND POLLUTION CONTROL   195
Richard Williams

Part III: CHALLENGE AND CHANGE IN RURAL EUROPE   209

13   THE CHALLENGE OF LAND REDUNDANCY   211
     Brian W. Ilbery

14   COUNTERURBANISATION: THREAT OR BLESSING?   226
     A.J. Fielding

15   TOURISM AND DEVELOPMENT   240
     Gareth Shaw and Allan M. Williams

16   CONSERVATION AND THE RURAL LANDSCAPE   258
     Brian J. Woodruffe

17   CONCLUSION: WESTERN EUROPE APPROACHES THE TWENTY-FIRST
     CENTURY   277
     David Pinder

     Index   283

List of figures

1.1	Post-war energy consumption in Western Europe	4
1.2	Containerisation, selected countries, 1970–86	6
1.3	Sectoral employment and change since the mid-1950s	7
1.4	Unemployment since the late 1970s	8
1.5	Work environments and guestworkers in the Netherlands	13
2.1	The price per barrel of Saudi light crude oil, 1950–88	21
2.2	The North Sea basin: oil- and gasfields and discoveries to December 1988	23
2.3	Western Europe's petroliferous and potentially petroliferous offshore areas	26
2.4	Western Europe's gas production potential in the 1990s and the transmission systems for indigenous and imported gas	28
3.1	Western Europe's position as a manufacturing producer, 1985	39
3.2	Western Europe's manufacturing performance compared with that of the United States and Japan, 1981–88	40
3.3	Western Europe's trade position, 1985	40
3.4	The major elements in Western Europe's international trade network, 1986	42
3.5	US foreign direct investment in manufacturing, 1987	44
3.6	Japanese foreign direct investment, 1987	45
4.1	Trade flows between selected Western European countries, 1986	59
4.2	Passenger car production and assembly, 1978–85	64
4.3	The European Airbus production system	67
5.1	Employment in small firms as a percentage of total manufacturing employment	74
5.2	New business registrations and deregistrations in the UK, 1975–87	75
5.3	New-firm formation in the Netherlands, 1970–80	79
5.4	New-firm formation rates in the UK, all industries, 1980–86	80
5.5	The contribution of small firms to job generation in Western Europe	82
6.1	Regional levels of on-site R & D, Great Britain	96
6.2	The adoption of NC and CNC machine tools, Great Britain, the USA and West Germany	99
6.3	Regional index of CNC machine tool adoption, Great Britain, 1981	100
6.4	Robot adoption rates, Great Britain, 1985	101
7.1	Service employees as a proportion of total employees, 1983	111
7.2	Employment in financial and business services as a percentage of total service employment, 1981	112
7.3	Service structure index, 1983	112
7.4	Number of offices of four leading accountancy conglomerates, by country, 1985	117
8.1	The urban life-cycle model	126
8.2	Typology of locations of inner-urban rejuvenation and change	132
8.3	Processes of inner-urban rejuvenation and change: a suggested framework	133
8.4	East Paris urban regeneration programme	137
8.5	The Waterstad, Rotterdam	138
9.1	Proportion of the population over 65 years of age	144
9.2	The product market matrix	147

9.3	Docks de France: national expansion through acquisition	148
9.4	Hypermarkets in Spain, 1978 and 1988	151
9.5	Sources of competitive advantage	152
9.6	Hypermarkets in Marseille, mid-1980s	156
9.7	Pedestrianisation schemes in France	157
10.1	Major international labour migration flows in post-war Europe	167
10.2	Foreigners in selected European cities	169
11.1	Regional unemployment rates in the European Community, 1987	183
11.2	Changes in the UK map of assisted areas, 1979–84	186
11.3	Regional distribution of ERDF assistance, 1975–87	191
12.1	EC Directives, Regulations and Decisions concerning environment policy and pollution control	202
12.2	European Investment Bank loans for environmental protection, 1986–88	204
13.1	Indicators of agricultural change in European Community countries	212
13.2	Changing regional specialisation in agriculture in the EC9, 1964–77	214
13.3	Schematic flow diagram of the major types of farm diversification	221
14.1	France: annual net migration rates, 1954–82, per thousand population, by settlement size category	228
14.2	Italy and West Germany: annual net migration rates, 1985/6, per thousand population	230
14.3	Norway and the Netherlands: interregional migration rates, 1972–85, per thousand population	232
15.1	Annual tourist flows to selected countries, 1986	242
15.2	Changes in the numbers of foreign tourists visiting Western European countries, 1960–85	243
15.3	Earnings from, and expenditures on, international tourism in Western Europe, 1985	250
15.4	*Center Parcs* holiday complexes, 1989	254
16.1	Protected land in the region of Provence-Alpes-Côte d'Azur, France	259
16.2	The urbanisation of villages within scenic landscapes	260
16.3	The Stelvio National Park, Italy	262
16.4	Val Claret in the Vanoise National Park, France	264
16.5	Landscape protection in Südtirol, Italy	265
16.6	Winter sports and the environment	267
16.7	St Leonhard in Abteital, Südtirol, Italy	268
16.8	Conservation of the built environment	272
16.9	Zoning plan for Evolène, Valais, Switzerland	273
16.10	Niederwald, Valais, Switzerland	274
17.1	Per capita GDP in Western European countries, 1986	278
17.2	Regional variations in per capita GDP in principal EC countries, 1985	278
17.3	Comparative economic indicators for the EC, the USA and Japan, 1980–88	279

List of tables

1.1 Sex, age and unemployment in the European Community 9
2.1 Alternative potential energy supply patterns for Western Europe in 2000 33
3.1 Network of merchandise trade, Western Europe 41
3.2 Inward and outward stock of foreign direct investment by major region 43
3.3 Foreign direct investment ratios, 1975–83 43
3.4 The world's leading automobile producers, 1987 49
3.5 The world's leading electronics companies, 1988 51
3.6 The world's leading pharmaceutical companies, 1982–87 53
3.7 Variations in the importance of domestic sales to the 50 largest pharmaceutical companies, 1982 53
4.1 Manufacturing employment trends by country, 1979 and 1986 57
4.2 Manufacturing employment trends in Western Europe (EC9), 1980 and 1985 58
4.3 The ten largest European-owned manufacturing firms in the European manufacturing system, 1987 60
4.4 Production of crude steel, 1979 and 1986 62
4.5 Aerospace employment in selected countries, 1985 66
5.1 Enterprise size in the private sector economy 73
5.2 Enterprise size in the manufacturing sector 73
5.3 Enterprise size in the service sector 73
5.4 New-firm formation in Northern Ireland and the Republic of Ireland, 1973–86 83
5.5 Components of change in manufacturing employment in Northern Ireland and the Republic of Ireland, 1973–86 83
5.6 Share of innovations by firm size, UK, 1945–80 85
6.1 Gross Domestic Expenditure on R & D: national trends 95
6.2 Business enterprise R & D expenditure: national trends 96
6.3 Trends in UK regional shares of innovations 97
6.4 Regional levels of NC/CNC adoption, Italy 100
6.5 Robot population worldwide 101
7.1 Service employment as a proportion of total economically active population, 1970–82 109
7.2 Producer services' percentage share of total employment (1984) and of service employment growth (1973–84) 110
7.3 Trade in invisibles (services), 1970 and 1984 118
7.4 The top six banking centres in Western Europe, 1986 118
9.1 Rates of population change 143
9.2 Expenditure patterns, France, 1963–90 146
9.3 Index of retail sales volume, 1980–87 146
9.4 Major retailers based in the European Community 146
9.5 Diversification into DIY retailing 149
9.6 Vendex International activities, 1988–89 150
9.7 Hypermarkets and superstores, 1975–87 153
9.8 Targeted small shop formats 154
9.9 Innovation dispersal in store mangement: EAN-system scanning equipment, 1981–88 155

9.10 Major market shares in grocery retailing by major companies, 1987   159
10.1 Distribution of foreign workers by nationality, selected countries, 1976–86   168
10.2 Evolution of stocks of foreign workers and foreign residents, selected countries, 1970–85   168
10.3 Occupational status of foreign employees in French industrial and commercial firms with more than ten employees   172
11.1 Percentage of national population in areas eligible for regional assistance   187
11.2 National allocations of EIB regional development assistance, 1988   189
11.3 EIB Global Loans to promote regional development, 1984–88   190
12.1 The European Community's Action Programmes on the Environment   203
13.1 Regions eligible for EC support under the cessation of farming Regulation 1096/88   215
13.2 Land expected to be set aside from wheat, barley and sugar-beet, EC10   218
13.3 Counties in England with (A) a low and (B) a high competitive ability, by farm type   220
14.1 East Anglia and South-East England: components of social change, 1971–81   234
15.1 Foreign and domestic tourist nights spent in hotels and similar establishments in Western Europe, 1985   244
15.2 Reasons for not taking a holiday during 1985   245
15.3 Changes in the holiday market of British residents, 1976–86   246
15.4 Transnational-associated hotel accommodation in selected Western European countries   248
15.5 International tourism and the national economies of Western Europe, 1984   251
15.6 Employment in tourism in Western European countries, 1985   252
15.7 Farms offering tourist accommodation, 1979–81   151
16.1 National parks in Sweden   261
16.2 National parks, mainland France   263
16.3 Regional nature parks, mainland France   269
16.4 Nature reserves (*Naturschutzgebiete*) in rural West Germany   270
16.5 Regulations for the centre of the old village, Evolène, Valais   274
16.6 Protected land in Südtirol, Italy   275

Notes on contributors

Neil Alderman is a geographer and Research Fellow in the Centre for Urban and Regional Development Studies, University of Newcastle upon Tyne. His research interests are primarily the spatial aspects of technological change in manufacturing industries and the analysis of industrial survey data. His research has involved projects on the location of industrial research and development, the impacts of new technology on skills and training, and modelling the regional patterns of new technology diffusion.

Steve Burt is a Lecturer in the Department of Business and Management, University of Stirling, and is a member of the Institute for Retail Studies. He has conducted research into various aspects of European retailing since 1981. This research has concentrated upon public policy issues, company strategies and structural change within the retail sector, and has been undertaken for a variety of organisations including the Commission of the European Communities and ICL.

David Burtenshaw is a Principal Lecturer at Portsmouth Polytechnic and Deputy Director of the Service Industries Research Centre. He has also taught for short periods at the University of Surrey, Carleton University and Universität Duisburg. His research has been mainly in urban geography, focusing in particular on the cities of German-speaking Europe, and he has written extensively on European urbanisation.

Brian Chalkley took his first degree at Leeds University and completed his doctoral studies at Southampton University. He is currently a Principal Lecturer in Geography at Polytechnic South West (formerly Plymouth Polytechnic). Much of his research has related to small firms, including their role in regional development, and (with Mark Wise and Greg Croxford) he has recently participated in a project analysing the evolution

and impact of the European Regional Development Fund.

P.W. Daniels, formerly Reader in Geography, University of Liverpool, is now Professor of Geography at Portsmouth Polytechnic. He is also Director of the newly-established Service Industries Research Centre at the Polytechnic. His research has largely been concerned with office location, producer services and metropolitan development, with a subsidiary interest in geographical aspects of urban transport. His publications related to these interests include *Office Location* (1975); *Spatial Patterns of Office Growth and Location* (1979); *Movement in Cities* (1980, with A.M. Warnes); and *Service Industries: a geographical appraisal* (1985). He is currently editing a new book on *Services and Metropolitan Development: international perspectives*.

John Dawson is Fraser of Allander Professor of Distributive Studies at the University of Stirling, where he also directs the work of the Institute for Retail Studies. His research into European retailing has been under way for 20 years. This research has involved detailed studies of the food sector; studies of the relationships between structural and spatial change in the sector; and analyses of the effects of government policy on retailing. He is a former Honorary Secretary of the Institute of British Geographers.

Peter Dicken is Professor of Geography at the University of Manchester. His major research interests consist of three interrelated strands: global industrial change; the geography of transnational corporations; and industrial change in East and South-East Asia. He is the author of many articles and several books on these topics including, in particular, *Global Shift: industrial change in a turbulent world* (1986, revised edition in preparation). He is currently Honorary Secretary of the

Institute of British Geographers, a member of the Standing Committee for the Social Sciences of the European Science Foundation and an Editor of *Progress in Human Geography*.

Tony Fielding graduated from the London School of Economics in 1962, stayed there to carry out research for his Ph.D. on French regional planning, moved to Sussex University in 1964 and has taught there almost ever since. He is currently a Lecturer in Geography, and Director of the Centre for Urban and Regional Research. His academic interests are far too broad for his own good and include the history of geographical thought, the urban social geography of Third World cities, computer cartography, and the political economy of urban and regional change in Britain. Plus, of course, migration in Western Europe!

Richard Harrison is a Lecturer in Applied Economics at the University of Ulster at Jordanstown. He previously taught at the Queen's University of Belfast and has worked as a research economist and policy adviser in the public sector. His research interests include regional industrial policy, small business development, local economic development policies and industrial restructuring. Recent publications include studies of the employment effects of regional industrial development policy, innovation processes in small firms, the process of entrepreneurship and new business formation, the regional impact of small-firms policy, and the regional restructuring of the ship-building industry. He is currently working on a study of enterprise development and attitudes in Northern Ireland and on the analysis of the informal risk-capital market in the UK (with C.M. Mason).

Brian W. Ilbery graduated from the University of Wales (Swansea) with a first degree in geography (1972) and a Ph.D. in agricultural geography (1975). His lecturing career began at Dorset Institute of Higher Education, and he then moved to Coventry Polytechnic, where he is now a Senior Lecturer in Geography. His current teaching includes courses in agricultural geography, economic geography and Europe, while his research centres on agricultural change and decision-making in Britain and the European Community. He has authored numerous papers on these topics and his books include *Agricultural Geography: a social and economic analysis*; *Western Europe: a systematic human geography*; and *Agricultural Change: France and the EEC* (with H. Winchester).

Russell King was born in East London in 1945. His undergraduate and postgraduate studies were undertaken at the Joint School of Geography of King's College and the LSE, University of London. In 1970–1 he was Research Fellow in Middle Eastern Studies at the University of Durham, after which he became Lecturer in Geography, and then Reader in Geography, at the University of Leicester. He has held visiting appointments as Lecturer in Mediterranean Studies, University of Malta (1977) and Senior Lecturer in Geography, Ben Gurion University, Beer-Sheba, Israel (1981). Since 1987 he has been Professor of Geography at Trinity College, Dublin. His research interests are population migration, land tenure and rural geography. He is the author of numerous papers on migration, and has written and contributed to several books on Italy and Western Europe.

Colin Mason is a Lecturer in Economic Geography and a Director of the Urban Policy Research Unit at the University of Southampton. He has also taught at the University of Ottawa and the Memorial University of Newfoundland. His research interests include regional industrial development policy, local economic development strategies, the economic impact of small businesses and urban change in Britain. Recent publications include studies of the geography of new-firm formation and small business growth in the UK, the regional distribution of venture capital investments in the UK, the regional impact of schemes to assist the small-firm sector (with R.T. Harrison) and a review of job-creation initiatives sponsored by large firms. He is currently engaged in studies of economic change in Southampton and of the informal risk-capital market in the UK (with R.T. Harrison).

Peter R. Odell is Professor of Energy Studies and Director of the Centre for International Energy Studies at Erasmus University, Rotterdam. He also holds appointments as a Visiting Professor at the London School of Economics and at the College of Europe in Bruges. He is a Fellow of the Institute of Petroleum and of the Royal Institute for International Affairs. He is a member of the Board of the Benelux Association of Energy

Economists, European Editor of *The Energy Journal,* and a member of the editorial boards of *Energy Policy, Geoforum* and the *International Journal of Energy Research.* He is a Fellow of the Royal Society of Arts. His interests in the study of energy resources and their exploitation go back for over 30 years. During this period he has published 12 books (including *Oil and World Power,* now in its eighth edition) and has contributed many articles and papers to a wide range of academic, professional and technical journals. He has advised governments, international organisations and other institutions and companies on international oil and energy matters and has given invited lectures at universities, research institutions and symposia, etc., in many countries in Europe, North and South America, Asia and Australasia.

David Pinder, a graduate of Reading University, is an economic geographer with a particular interest in planning. He is a Senior Lecturer in Geography at Southampton University, and in 1978–9 was a Visiting Lecturer in the Economic Geography Institute, Erasmus University, Rotterdam. From 1982 to 1987 he was Honorary Treasurer of the Institute of British Geographers. Since the early 1970s his research has concentrated on Western European economic and regional development issues. Publications in these fields include *Regional Economic Development and Policy: theory and practice in the European Community* (1983) and (with Hugh Clout, Mark Blacksell and Russell King) *Western Europe: geographical perspectives* (2nd edn, 1989). Recent papers have focused on aspects of EC regional policy and on industrial restructuring, particularly the restructuring of Western European oil refining.

Gareth Shaw is a Senior Lecturer in Geography at the University of Exeter. His research interests include urban historical geography, retail location studies and the geography of tourism. He is a Director of the University's Tourism Research Group and has carried out studies for national, regional and local tourist boards. Much of this work has been concerned with the relationships between tourism and economic development and with longitudinal surveys of tourist behaviour. Together with Allan Williams, he has published numerous reports and papers on tourism, and they have jointly edited *Tourism and Economic Development: Western European experiences.*

Louis Shurmer-Smith is a Principal Lecturer in the Department of Geography, Portsmouth Polytechnic. His main fields of research are in urban and regional development, with particular reference to Western Europe, especially France. In 1975–6 and 1985 he held teaching appointments at the University of Rennes. More recently he has joined a research network spanning a number of university centres in north-western France studying European urban dynamics. Current research interests include political control and the urban planning process, and comparisons between the British and French experiences in the development and problems of medium-sized towns.

Alfred Thwaites is an economist and currently holds the position of Deputy Director of the Centre for Urban and Regional Development Studies, University of Newcastle upon Tyne. His research interests lie principally with the regional dimension of technological change and its impact upon local economic development. He has undertaken many studies on this subject, including a number providing an international, and particularly a Europe-wide, dimension to academic research and policy debates.

H.D. Watts is a Senior Lecturer in Geography at the University of Sheffield with research interests in the impact of the behaviour of large organisations on urban and regional economies. Publications include *The Large Industrial Enterprise* (1980), *The Branch Plant Economy* (1981), and *Industrial Geography* (1987).

Allan M. Williams is a Senior Lecturer in Geography at the University of Exeter. He is also a Director of the Western European Studies Centre and, together with Gareth Shaw, co-founder of the Tourism Research Group. His principal research interests are the economic and social geography of Europe, especially southern Europe and the EC, and the tourism and leisure industries. Amongst the books he has authored or co-authored are *Rural Britain: a social geography* (1984), *The Western European Economy* (1987), *The United Kingdom* (1986), and *Divided Britain* (1989). He is also editor or co-editor of *Southern Europe Transformed* (1984), *Southern European Socialism: parties, elections and the challenge of government* (1989), and *Tourism and Economic Development: Western European experiences* (1988).

Richard Williams holds degrees in geography from the University of Nottingham, and in town planning (Master of Civic Design) from the University of Liverpool. He practised in local government for six years with Lancashire County Council and Nottingham City Council and is a Chartered Town Planner and Member of the Royal Town Planning Institute (MRTPI). Since 1975 he has lectured at the University of Newcastle upon Tyne, specialising in European Community spatial policies, European planning systems, and comparative methodology. He has several publications in these fields; has represented the RTPI in European liaison activities; has advised the Council of Europe and Members of the European Parliament; and is currently Secretary-General of the Association of European Schools of Planning.

Mark Wise took his first degree at the University of Leicester in 1965 and then obtained an M.A. from the University of British Columbia. Whilst studying for a D.Phil. in the Centre for Contemporary European Studies at Sussex University, he spent a year conducting research in Brussels. Since 1973, as a Senior Lecturer at Polytechnic South West (formerly Plymouth Polytechnic), he has maintained his special interest in the European Community. He has written on the EC's regional, social and fisheries policies, and on French attitudes to European unity. At present he is preparing a book on the Community's '1992' project.

Brian J. Woodruffe is a graduate of King's College, London, and is a Lecturer in Geography at Southampton University. His research and teaching are primarily concerned with planning in rural areas and he has undertaken extensive comparisons of approaches to rural planning in Britain and mainland Western Europe. Much of his recent work has focused on landscape evaluation and conservation, including conservation in rural settlements, and has involved close contact with many planners involved in the design, implementation and assessment of conservation strategies.

Preface and acknowledgements

In 1983 the Institute of British Geographers celebrated its fiftieth anniversary, and to mark this occasion John Doornkamp and Ron Johnston edited a volume of essays on *The Changing Geography of the United Kingdom*. Soon afterwards it began to be suggested that companion volumes — in particular, one on Western Europe — might be appropriate. John Dawson and I produced the first sketches for such a volume at the IBG conference in Portsmouth in 1987, and its production since then has reflected the efforts of many individuals.

I am grateful to the authors for their enthusiastic participation from an early stage, particularly as it was proposed that royalties should be assigned to the IBG to further its work on behalf of the profession. No arms needed to be twisted, and a co-operative spirit has prevailed throughout the exercise — even when I despatched my editorial comments on draft chapters! The contributors must also be thanked for undertaking this project at a time of escalating pressure in the academic world. Faced with increasing and conflicting demands on their time, they have fulfilled commitments which would have been far easier to complete ten years ago. Credit is also due to them for dealing with substantial topics within tight word limits. Readers will appreciate that, given such extensive themes, authors could not produce chapters dealing comprehensively with all parts of Western Europe. Indeed, such an unrealistic task was never in their brief. Rather, they have concentrated on selectivity, highlighting important trends within their subject areas, exploring the background to those trends, and extending the discussion to consider their implications. The aims throughout have been to explore the spatial dimensions of Western European development, to communicate the dynamism of economic and social development in the recent past and at the present time, and to provide insights into that dynamism. In pursuing these goals we hope that we have furthered understanding of the complexity and richness of European life in the late twentieth century and that, through this, we have contributed to preparing this major world region to meet new challenges as the twenty-first century approaches.

The production of any book requires the assistance of many people behind the scenes, and this is particularly true of large edited volumes. I am indebted to the many secretaries who dealt so patiently and efficiently with the production and later revision of manuscripts. Many cartographers also deserve thanks, since most contributors have been able to arrange for the preparation of their illustrations in their own Departments. Although this has resulted in some variation in cartographic style, it has not impaired quality, and it has ensured a level of illustration that would have been well beyond the resources of a single Department.

While much support has come from the contributors' Departments, I should like to thank a number of people who have worked closely with me to prepare the final manuscript and artwork. Alison Hind, the Administrator of the Institute of British Geographers, kindly undertook a substantial amount of typing at a time when resources elsewhere were in short supply. Mrs Marilyn Knight and Mrs Shruti Jeevanjee similarly gave very welcome secretarial support. Mrs Jane Rickett, now of the Hartley Library, Southampton University, provided assistance with the manuscript, checking many details and reducing the editorial work necessary on a number of chapters. Although much of the artwork was provided by the contributors, some illustrations required minor modifications, and a number of maps and diagrams needed to be prepared from draft. This work was undertaken by the Cartographic Unit of the Geography Department, Southampton University, under the direction of Mr Alan S. Burn. As always, members of this

Unit worked with great speed, efficiency and unfailing cheerfulness. I must also thank the copy editor of this volume, Richard Leigh, for very welcome support in the form of eagle-eyed attention to detail.

Finally, the authors are grateful to the following copyright holders for permission to reproduce material: the *Financial Times,* Figure 4.3; Croom Helm, Figure 5.3; D.J. Storey and S.G. Johnson, Tables 5.1, 5.2 and 5.3; M. Hart and Taylor and Francis Ltd, Tables 5.4 and 5.5; R. Rothwell and the Chief Editor, *OMEGA: International Journal of Management Science,* Table 5.6; International Institute for Labour Studies, Geneva, quotation p. 84; the Avebury Press, Figures 7.1, 7.2 and 7.3; Gemeente Rotterdam, Figure 8.5; H.I. Ansoff and John Wiley and Sons, New York, Figure 9.2; Free Press, Figure 9.5; INSEE, Table 9.2; Vendex International, Table 9.6; the Geographical Association, Figure 13.2; *Outlook on Agriculture,* Figure 13.3; M. Pacione, Figure 14.1; Office of Population Censuses and Surveys, Table 14.1; *Tourism Management,* Table 15.7.

David Pinder
Department of Geography
University of Southampton
October 1989

1

Challenge and change in Western Europe: an overview

David Pinder

Interest in Western Europe rose rapidly in the late 1980s, a primary reason being the adoption of the Single European Act by the nations of the European Community (Commission of the European Communities, 1986). The timescale of this initiative — to establish by the end of 1992 a genuine common market in products, services and skills — may be optimistic; but the prospect is that progress in the 1990s will be impressive, and the ramifications of this prospect have done much to cultivate a new European awareness in many walks of life. This book naturally reflects this rising interest in Western European, and especially European Community, affairs and interrelationships. However, it is not primarily a volume concerned with 1992 and beyond — others have addressed this theme (Lodge, 1989). Indeed, there are pitfalls in placing too much emphasis on the Single European Act and its implications. Pre-eminent among these is the danger of undervaluing the strength, complexity and importance of forces for change which were deeply entrenched throughout Western Europe long before the Act was conceived by the Community. These forces make the region a highly dynamic environment; they pose numerous challenges on a variety of scales and with respect to many aspects of economic and social life; and an appreciation of their influence, consequences and interrelationships is a prerequisite for deeper understanding of the opportunities and obstacles likely to influence the development of Western Europe into the twenty-first century.

This volume therefore views Western Europe as a dynamic world region and explores that dynamism from a variety of geographical perspectives. To do so it adopts a three-part structure designed to emphasise interrelationships between development forces and to draw out contrasts between the distinct types of challenge that must now be faced. Part I focuses on the production system, exploring the global economic environment within which it operates, considering aspects of its internal restructuring and examining features of the system which — it is frequently argued — are essential to the future well-being of Western European industry. Trends in the production system, and strategies adopted to strengthen it, frequently have major repercussions — for good or ill — in other spheres. Parts II and III therefore consider both the inheritance arising from the short- and long-term evolution of productive activity and responses to that inheritance. In doing so, a division is made between urban and rural Europe. This distinction is to a degree arbitrary, since it is not unusual for key issues to span the urban and rural systems. None the less, the themes covered in Part II are dominantly urban in terms of the origin and impact of the challenges to be faced, while those examined in Part III are primarily concerned with the growing impact of prosperous urban societies in rural Europe.

From this outline it will be evident that this book does not aim to provide exhaustive thematic coverage of the geography of Western Europe. Instead, a selection of themes has been made, a selection relevant to the chosen framework. Equally, although the geographical context is the whole of Western Europe from the Mediterranean to Scandinavia and including the UK and Iceland, the contributors have not attempted to produce comprehensive geographical surveys dealing in equal measure with all areas. Rather, their remit has been to explore the processes governing spatial change; to examine strategies devised to deal with fundamental development problems; and to consider critically the progress of such strategies. The aim, therefore, has been to deepen understanding of development processes and responses to them, and contributors have considered the geographical issues by exemplification at a variety of scales, from the intra-urban to the international.

Because of this approach the volume has an important temporal dimension. Processes of change and development strategies operate through time, and individual chapters are in many ways concerned with the spatial consequences of evolutionary trends. Given this perspective, the remainder of this chapter prepares the ground for later contributions by providing an overview of the evolving economic environment and its relationships with challenge and change in Western Europe today. This overview comprises a survey focusing primarily — but not exclusively — on recent decades, followed by more detailed examination of the orientation of the individual chapters.

Challenge and change in historical context

Western Europe, with its numerous competing economies and its close integration into the world economy, is no stranger to challenge and change in its economic and social systems. Landes (1969), analysing this major world region in the era of industrial take-off, has portrayed it as the 'unbound Prometheus'. As such, Western Europe played a central role in establishing the nature and orientation of global development, but as part of the price its societies were forced to undergo rapid evolution. Unparalleled contrasts emerged, for example, between industrialised and unindustrialised areas, while major urban development and social challenges arose from the rapid pace of urbanisation (White, 1984, pp. 16–22). In the twentieth century, two world wars have dislocated economic and social development throughout much of the continent. Long-term industrialisation trends were disrupted as economies were reoriented to serve the demands of the war machine and, with the close of the Second World War, areas fought over by opposing forces faced

the task of reconstruction. Europe as a whole, and Germany in particular, were divided into East and West. This, in turn, truncated market areas and placed pressure on what is now West Germany as refugees fled Soviet-occupied areas to seek new lives, jobs and homes (Clout and Salt, 1976, pp. 21–2; Kosiński, 1970). So debilitated was the Western European economy that, under the Marshall Plan, in the late 1940s the USA made large-scale dollar advances to enable nations to obtain foodstuffs and rebuild productive capacity. One aim was to prevent the onset of post-war recession, as had happened after the First World War. Equally important, the USA considered rapid recovery an insurance against the spread of Communism west of the Iron Curtain.

War brought devastation and dislocation but, paradoxically, it also had beneficial aspects, particularly opportunities to be grasped in the interests of peacetime recovery and prosperity (Grand'Jean, 1967; Milward, 1970). Damaged production capacity could be replaced by new equipment, raising efficiency and stimulating capital goods industries. Wartime labour shortages drew women into the labour market in large numbers and, although post-war participation rates fell, lasting progress was made towards the principle of the right of women to work. Perhaps most impressively of all, hostilities acted as a forcing-house for technological advances which ultimately proved to be of fundamental importance in post-war economic development. This was true of both wars, but the pace and nature of change were particularly outstanding in the Second World War. To cite only three examples: the development of radar for air, sea and land forces greatly enhanced Western Europe's post-war research and development capacity in the electronics industry; work on military jet propulsion, plus aerodynamic advances and the quest for high-altitude flying capabilities, paved the way for a post-war revolution in civil airliner design; and the need to decipher military codes — particularly the advanced *Geheimschreiber* codes employed by German forces — led to theoretical and technical advances of considerable importance for the development of the modern computer.

Similar conclusions may be drawn with respect to the Great Depression of the 1930s. On the one hand, this inflicted severe hardship as a result of high unemployment at a time when social security systems were in their infancy. Virtually nowhere escaped the scourge of this depression, and its impact was particularly harsh in areas whose economies focused narrowly on traditional, export-oriented industries such as shipbuilding, engineering and textiles. Yet, on the other hand, the hardship experienced in the 1930s did much to encourage the view — ultimately acted on in the post-war years — that governments had a responsibility to protect and enhance the quality of life through improved systems of social security, health care and education. Connected with this, not least to promote the wealth necessary to support large-scale social expenditure, post-war governments were to become deeply involved in economic management, and many were to establish economic planning frameworks to promote industrialisation (Hayward and Watson, 1975; Maunder, 1979). Nationalisation of key activities was to become part of this movement and in some countries would be taken to considerable lengths, the outstanding example being Italy (Shepherd, 1976). In addition, industrial depression before the Second World War coincided with an important period of innovation, along lines predicted by Kondratieff's theory of long-wave development (Balance and Sinclair, 1983, p. 4; Freeman, 1984). In fact many of the technological advances made in the Second World War were not innovations but developments of innovations made in the inter-war period and, most particularly, in the 1930s. This was most often true in industries such as electronics, aviation and road vehicles. Lastly, the 1930s saw a new concern begin to emerge for the physical impact of development. Once again, after 1945 this led almost all Western European governments to establish elaborate physical planning systems to control the nature and location of development at a variety of scales. Surveys of the systems created in a wide range of Western European countries are to be found in Williams (1984).

Taken together, the 1930s and the Second World War may be seen as an era of upheaval in which, however, the foundations were laid for widespread industrial advance and economic prosperity between 1945 and the early 1970s. Compared with earlier experience this century, in the 1950s and 1960s it appeared that — through a combination of industrial regeneration and government policy — sustained long-term growth was reasonably assured (Aldcroft, 1980; Bairoch, 1982; Boltho, 1982; Postan, 1977). Moreover, there was a growing belief that the probability of maintaining sustained expansion could be increased by international co-operation in the economic sphere.[1] In terms of concrete results, early landmarks were the formation of the Benelux economic union in 1948 and the creation of the European Coal and Steel Community under the Treaty of Paris (1951). Progressing from this base, the Paris Treaty signatories (France, West Germany, Italy, the Netherlands, Belgium and Luxembourg) rapidly moved on to establish the European Economic Community under the Treaty of Rome (1957).[2]

One index of the economic upswing that was achieved Europe-wide is the expansion of energy demand. Total energy consumption in Western Europe rose from some 400 million tonnes of oil equivalent (mtoe) in 1950 to 1,200 mtoe in 1973 (Figure 1.1). Central to the satisfaction of this demand was the spectacular growth of the Western European oil-refining industry (Balance and Sinclair, 1983, p. 161). In 1950, annual capacity in this industry was less than 45 million tonnes of crude oil, only 7.6 per cent of the world total. By the mid-1970s capacity had risen to 1,034 million tonnes and amounted to a quarter of the world total. By this time, too, the scale of Western European oil refining exceeded that of North America by almost 20 per cent.

In more general terms, Aldcroft (1980, p. 161) has estimated that per capita GDP rose, on average, by 4.4 per cent per year between 1950 and 1970, well above the world figure (3.0 per cent). Similarly, the annual average rate of increase in industrial production was 7.1 per cent, as opposed to a world figure of 5.9 per cent. Even in this period, however, the rose-tinted spectacles needed to be worn with caution. At the international scale, Aldcroft has also demonstrated that the pace of expansion — although reasonably encouraging in almost all economies — varied considerably from country to country. The complex pattern which resulted still requires full explanation but, even allowing for rapid industrialisation in northern Italy, it broadly underlined the development gap between southern and other parts of Western Europe. Moreover, it also entailed the emergence of a north-western low-performance periphery, primarily comprising large parts of the UK and Ireland.

Meanwhile, within individual countries, traditional industrial areas remained especially vulnerable to competition and to declining

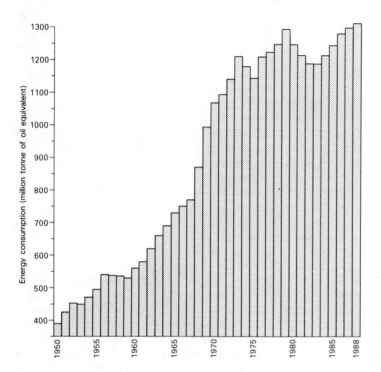

Figure 1.1 *Post-war energy consumption in Western Europe Source:* British Petroleum (1989).

demand for outmoded products. Moreover, numerous factors — such as a revolution in road haulage, motorway development, entrepreneurship patterns and the nature of industrial labour demand — ensured that patterns of industrial decline were not matched by those of growth. In most countries, therefore, regional policies aiming to achieve more efficient use of national resources and a more socially just growth pattern were either launched or revivified, and subsequently underwent substantial expansion. At this time, too, the first steps were taken to try to achieve more even development at the international scale. Fearing that economic contrasts would promote political instability which would frustrate progress towards economic and political goals, the original members of the European Community established the European Investment Bank in 1958 and made its first priority the acceleration of economic growth in major lagging regions (Licari, 1970; Pinder, 1983, pp. 16–18; Pinder, 1986, p. 172).

Crisis and restructuring

Since the early 1970s the economic scene has

changed dramatically. In 1973–4, confidence in the ability of Western Europe to sustain growth was severely shaken by the first oil crisis (Odell, 1986, pp. 22–3). In the aftermath of this shock, European nations were forced to grapple with strong inflationary pressures as the abrupt shift to high energy prices fed through the system. As confidence was starting to recover, the second oil crisis (1979–80) threw industrial economies into turmoil once more. This event again triggered high inflation, with far-reaching implications for the ability of individual countries to compete in the world economy. While it may be a convenient shorthand to explain recession in the 1980s in terms of upheavals in the energy market, however, the reality is much more complex. Shocks to the energy system, particularly short-term price rises, intertwined with and reinforced other trends with long-term implications for the health and nature of Western Europe's production systems.

Outstanding among these was the intensification of global competition. This must be seen as a long-term process, pre-dating the Second World War, but in the post-war period it spread through manufacturing activity, steadily affecting more complex and technologically advanced sectors.

Thus, while industries such as textiles came under considerable pressure in the 1950s, by the 1980s the electronics, automobile and aerospace industries were widely exposed to the new forces. Much attention has been given to the role of transnational corporations in bringing about intensified global competition and, as Dicken demonstrates in Chapter 3, rightly so. Yet, although these corporations were important contributors to the growing difficulties experienced by Western Europe, they were not entirely responsible for creating the new competitive environment.

Government policies in industrialising countries were also of considerable significance, through the attempts that were made to nurture import-substituting industries and, commonly, to develop activities with a strong export orientation. Thus government-inspired development of shipbuilding in South Korea sharply curtailed Western European export opportunities; Saudi Arabia became a major petrochemical exporter, with Western Europe as its main market; and Japanese exporting successes were fostered by government policies designed to protect the home market and favour export-oriented activities. To a degree, therefore, increased competition experienced by Western European producers became — and remains — a consequence of development strategies pursued by competitor countries. Beyond this, however, we may note that Western Europe's growing difficulties have not been confined to relationships between this region and the developing world. Intensified competition from the South has not been negligible but — as the imbalance of trade with Japan demonstrates most graphically — the new competitive climate is also a consequence of shifts in North–North relationships. In addition, the significance of East–West flows must not be overlooked. Thus the decline of shipbuilding has not simply been a consequence of growing exports from first Japan and then South Korea; heavy state subsidisation of this industry in, for example, Poland has also had a major impact on Western Europe's competitive prospects.

Intensified competition faced by Western Europe is also a reflection of technological factors, particularly advances in communications and transportation. Better telecommunications have done much to enable multinational corporations to sustain and improve the efficiency of their operations, but it is improved transportation — and, most importantly, sea transportation — that has facilitated the global dispersion of production capacity. Two types of vessel require particular mention. Container ships, with their associated landward containerisation technologies, have allowed an immense range of manufactures and semi-manufactures to be treated as bulk commodities (Hayuth, 1987). As a result, they have drastically reduced transport-cost penalties associated with globally dispersed production systems. Although comparable data for all parts of Western Europe are not available, Figure 1.2 reveals the strength of the upswing in containerisation in most parts of the region. While the expansion of this transport technology was relatively slow in some countries, such as Greece, it was typical for container traffic to increase five- or sixfold over the 1970–86 period. Meanwhile, ships designed specifically as vehicle transporters became vital to the globalisation of the car industry. Thus, in the Japanese case, efficient sea transportation reinforced production cost advantages to facilitate penetration of the Western European market. Here we may note that advanced countries, including some in Western Europe, were instrumental in developing shipping technologies which subsequently encouraged the global dispersion of industry. Ironically, therefore, Western Europe's search for technological progress has been turned, at least to some extent, against itself.[3]

The impact of these challenges on the production system was severe, but would have been far worse if a number of cushioning factors had not emerged. Social security provision, despite variations from country to country, was everywhere superior to that available in pre-war years, and in most countries was incomparably better (Lawson and Reed, 1975). Equally important, intensified pressure on manufacturing was paralleled by long-term service sector expansion (Ochel and Wegner, 1987). In the mid-1950s, industrial employment accounted for 42 per cent of the Western European labour force and still enjoyed a modest lead over the service sector. By the early 1970s services occupied 44 per cent of the workforce, slightly more than industry. And by the mid-1980s the tertiary sector had forged ahead to account for almost 60 per cent of all workers, compared with industry's 33 per cent. Naturally, the detailed timing of this shift varied from country to country, but the trend was common to all economies

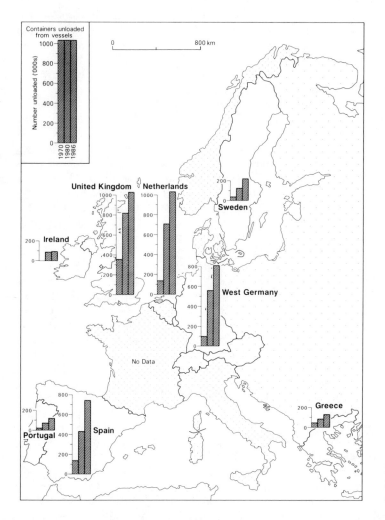

Figure 1.2 *Containerisation, selected countries, 1970–86*
Source: United Nations.

(Figure 1.3). In terms of output, very few parts of this major world region now gain less than half their gross value added from the service sector, and in some areas — particularly capital regions and a number of tourism-dependent regions — the proportion is 70 per cent or more.

To some extent, of course, the shifting balance between industry and services, especially as sources of employment, is a reflection of deindustrialisation. Between the early 1970s and the mid-1980s labour-shedding in the manufacturing sector affected almost all countries,[4] and in Western Europe as a whole industrial employment fell by a fifth. But, beyond this, over the same period service sector employment grew by 37 per cent, creating 21.5 million jobs. This growth was based primarily on the above-discussed increasing

prosperity and rising expectations of Western European societies. In the private sector, prosperity fuelled demand for an ever-widening range of personal services, for leisure-related activities and for full-scale tourism (Williams and Shaw, 1988). In the public sector the rising standards set by nations in the post-war period either created entirely new professions — such as planning or social work — or resulted in the large-scale expansion of existing activities. Thus health care and education became impressive growth 'industries' and, because demand for such services was sustained as the economic climate deteriorated in the 1980s, the economic and social impacts of recession were softened.

None the less, despite this protection, the pressure on the economic system was substantial,

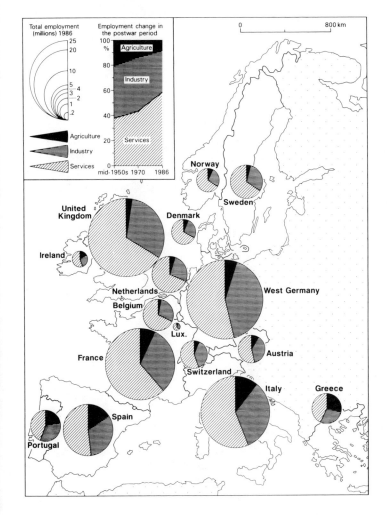

Figure 1.3 *Sectoral employment and change since the mid-1950s*
Source: Commission of the European Communities (1988).

as the trend in unemployment rates reveals (Figure 1.4). In the mid-1980s, 16 million workers were unemployed in the European Community; this was twice the number in the United States, where unemployment improved after 1983. By 1986 the European Community's average unemployment rate, based on harmonised national data, was 10.7 per cent and, above this figure, national rates were as high as 18.3 per cent (Ireland) and 21.2 per cent (Spain). In Western Europe as a whole, only Norway and Sweden controlled the problem at negligible levels. Furthermore, certain groups suffered particularly severely from job shortages (Table 1.1). For example, female unemployment in the European Community was 12.9 per cent in 1986, compared with a male rate of 9.4 per cent. In Spain the female unemployment rate was 25

per cent at that time, and in the worst Spanish region (the Canaries) it was virtually 35 per cent. Even worse was the experience of the young: unemployment among the under-25s in the European Community stood at 22.3 per cent in 1986, and regional breakdowns of this aggregate show that — particularly in the Mediterranean countries, France and the UK — it was not unusual for employment among this group to exceed 25 or even 30 per cent.

As unemployment rose, economic crisis came to be seen as a deep-seated and long-term phenomenon affecting much of Western Europe. In this new climate of opinion, many of the assumptions which had become firmly rooted during the post-war growth era came to be questioned and, not infrequently, cast aside or

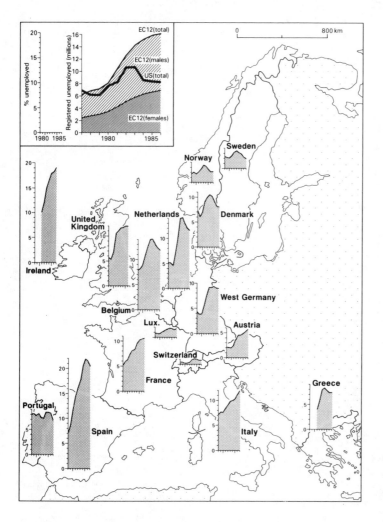

Figure 1.4 *Unemployment since the late 1970s*
Source: Commission of the European Communities (1988).

modified. For example, at a variety of geographical scales, the reputation of economic planning became seriously tarnished because planning systems had not created a development environment capable of resisting severe recessionary pressures. Whether it was realistic that they should be expected to achieve this is, in a sense, an irrelevant issue. The significant point is that success in achieving substantial growth up to the late 1960s was not infrequently attributed to state involvement in economic management and the creation of planning frameworks (Boltho, 1982). These were features that were notably absent in earlier eras, when sustained growth was so elusive. When post-war growth ultimately faltered, state intervention and planning in general were ill placed to avoid criticism.

Similarly, as recession curtailed the supply of interregionally mobile capital, and as deindustrialisation eliminated many regional policy 'successes' achieved up to the early 1970s, regional policy lost much of its support. Also, reduced emphasis on inducing investors to opt for problem regions was paralleled by reappraisal of approaches to physical planning. In this context an influential argument was that planning restrictions — together with their associated 'red tape' — were major disincentives to developers. This viewpoint was most vigorously promoted in the UK, where successive Conservative governments sought to overcome the problems of inner-urban rejuvenation by the introduction of streamlined planning systems, of which Enterprise Zones and Urban Development Corporations are the

Table 1.1 *Sex, age and unemployment in the European Community*

	Female unemployment		Unemployment under age 25	
	National (%)	Worst region* (%)	National (%)	Worst region* (%)
Belgium	12.9	20.2	21.1	30.8
Denmark	7.6	9.5	7.8	9.5
West Germany	8.2	12.3	7.8	15.1
Greece	11.6	13.7	24.1	27.9
Spain	25.0	34.7	46.5	52.6
France	12.2	15.2	23.2	30.3
Ireland	19.4	–†	26.6	–†
Italy	17.0	29.7	33.6	53.5
Luxembourg	4.0	–†	6.1	–†
The Netherlands	12.4	13.7	15.3	18.5
Portugal	11.6	17.4	20.3	32.8
United Kingdom	11.0	15.2	18.7	27.6
EC12	12.9	34.7	22.3	53.5

*Regional data refer to Level 1 regions, the first disaggregation below the national level. Some smaller Level 2 regions exceed the values shown.
†No regional breakdown available.
Source: Commission of the European Communities (1988).

dominant features. Although there is no Europe-wide overview of current thinking with respect to physical planning, Davies *et al.* (1989) have provided a detailed comparative survey of planning control in Denmark, the Netherlands, West Germany, France and the UK.

Thus the 1980s came to be the decade in which overt political attempts were made to limit state activity which, it was increasingly argued, had imposed over-restrictive regimes on private sector development. There were, of course, exceptions to this general rule. Norway, for example, adhered quite closely to Keynsian principles of economic management. Here, rising oil revenues — and loans raised in anticipation of future revenues — were to a great extent employed to dampen down recessional forces and their social consequences (OECD, 1987). But much more pervasive trends were the curtailment of government expenditure; restructuring of legislative and planning frameworks to create more permissive environments for the private sector; and, most recently, the privatisation of state holdings, a strategy in which the UK again took the lead. Hesselman (1983) has charted the onset of this reorientation of policy.

Consideration of the shifting economic environment since the early 1970s should not be limited to its influence on government attitudes and strategies, since it is in this period that the European Community has expanded from a grouping of six nations to one of twelve, in the process increasing its share of Western Europe's population from 63 per cent to 92 per cent. In many ways, the development of the Community can be said to have been in tune with, and to have encouraged, the new emphasis given in the 1980s to revivifying the impetus of the private sector. Although the Community has long-term political aims, in economic and political terms its *raison d'être* is, after all, to achieve an increase in mutual prosperity by creating a common market which will expand the opportunities open to the private sector. In seeking to achieve this, the path has not always been easy. For example, as governments wrestled with the problems of recession in the late 1970s and early 1980s, the Community's policies appeared to be stranded in the doldrums. Overt efforts had to be made to re-establish impetus (Commission of the European Communities, 1981; 1983a; 1983b), efforts which may be traced through to the Single European Act (Commission

9

of the European Communities, 1986). Yet, despite obstacles, the central aim has remained improved prosperity through the liberalisation of intra-Community business opportunities and, beyond this, greater global competitiveness as a consequence of healthier economic development.

Although the role of the Community in promoting the free-market ethos must be recognised, a paradox should not be overlooked. Interventionist approaches have been increasingly adopted by the European Commission in the pursuit of its market-oriented philosophy. National disengagement from control has therefore been paralleled by increased engagement at the international scale. To a great extent this reflects the need for harmonisation of regulations governing the details of production and trade, which naturally vary markedly from member state to member state. But the Commission's interventionist activities have other important motivations. It is striking, for example, that Community regional development initiatives increased after the mid-1970s, in contrast to the reduction of effort that became widespread at the national level. Here the driving force was concern that economic liberalisation would damage, rather than enhance, the economies of the Community's least competitive regions, a possibility that had long been recognised (Mawson *et al.*, 1985, p. 21; Wallace, 1977). Naturally, this concern was magnified as Community expansion in the 1980s greatly increased the scale of lagging regions in the southern periphery (Commission, 1982; 1984; 1987). In addition, some measures are — in greater or lesser degree — reactions to what is perceived as the failure of governments to take adequate regulatory steps with respect to widespread pressing problems. The proliferation of environmental protection measures introduced by the European Commission is a case in point (see Chapter 12, Figure 12.1). So, too, is the proposed European Charter of Social Rights, a primary aim of which is to safeguard workers' employment conditions. First mooted in 1989, and actively promoted by the Commission President and the Commissioner for Social Affairs, at one level the Charter is an attempt at international harmonisation. Yet it also underlines the existence of major discrepancies between national practices in worker protection, and highlights the danger that one price of governmental attempts to stimulate the private sector could be the loss of improved working conditions built up over earlier decades.

Challenge and change in the production process

Recent decades have therefore witnessed many far-reaching developments, including impressive growth, the rise of global competition, economic recession, a major shift in the balance of manufacturing and services, and the rise of international co-operation on an unprecedented scale. Against this background, Part I begins by examining Western Europe in its global context and, in particular, by considering challenge and change in the energy sector. This is now frequently assumed to be a sector in which global forces have brought about development constraints, primarily because of the severe price rises imposed after the early 1970s on economies which had become heavily dependent on the availability of cheap imported energy. In Chapter 2, however, Odell challenges this interpretation, not least on the grounds that the periods of steeply rising energy costs were short-lived. Oil prices soon stabilised in real terms after the first crisis, and after the second they fell almost as quickly as they had risen (Figure 2.1). Consequently, although in real terms crude oil in 1988 cost almost four times as much as in 1970, this was a far cry from the 1970–81 differential (12 times). Odell's alternative interpretation of the Western European energy economy is that of a market served by a group of industries offering economic development opportunities, as opposed to imposing restrictions. The perception that energy resources are scarce is rejected as illusory and counterproductive, and the principle thrust of the chapter is to argue that the real challenge facing Western Europe in the energy sector is to make appropriate choices between competing and plentiful supplies from inside and outside this major world region.

Dicken (Chapter 3) continues the theme of global forces and Western European development trends by considering the internationalisation of national (and regional) economies as industry has spread to new global centres of production. In this survey, Western Europe is set in its global economic context in terms of both production and trade, and particular attention is paid to transnational corporations as agents of change. Much of the analysis is at the aggregate level, but the significance of world-wide trends is emphasised by case studies of the automobile, electronics and pharmaceuticals industries. In selecting these case studies, Dicken has sought to demonstrate the complexity of change occurring in response to global pressures. Challenges to Western Europe's competitive ability vary considerably from industry to industry and, moreover, are in a continual state of flux.

While it is important to view the Western European economy in its global context, the nature of industrial restructuring within the region must also be considered. The four remaining chapters in Part I address this restructuring from a variety of perspectives. In Chapter 4 Watts, following a discussion of the recent evolution of Western Europe's manufacturing sector, considers corporate structures and restructuring in four contrasted industries: steel, aerospace, automobiles and food. Central to the analysis is the identification of contrasting strategies pursued to restore or sustain competitiveness in these strikingly different activities. The influence of Western Europe's political fragmentation on restructuring is also examined, particularly with respect to the extent to which national factors impede or encourage the process of inter-nationalisation within the region. This question is clearly of considerable significance, within the European Community at least, with respect to the 1992 movement.

As major industries have encountered severe challenges, an impressive trend has been to treat small firms — in services as well as manufacturing — as an important potential source of growth. In Chapter 5, therefore, Mason and Harrison switch the spotlight from corporations and corporate restructuring to the small-firm sector. Recent re-expansion of this sector is charted, and competing explanations for the resurgence are examined. In the process, sharp regional contrasts in the dynamism of small firms are demonstrated and an assessment is offered of the contribution small firms may make to urban and regional economic development.

Increasing interest in small firms, and in policies to stimulate them, has been paralleled by an equally impressive recognition of the significance of new technologies. Because these are considered crucial to competitive success in many activities, it is natural that numerous initiatives to promote them have been taken by national governments and, within the Community, by the European Commission (Godet and Ruyssen, 1980). The challenge of new technologies is examined by Thwaites and Alderman in Chapter 6, their particular concern being the likely differential impact of technological change throughout Western Europe. Consequently this chapter not only examines the nature of that change in the late twentieth century, but also questions whether all regions can succeed in exploiting the potential benefits of new technologies. Thwaites and Alderman examine in particular the proposition that inter-regional variations in the generation and uptake of new technologies will seriously exacerbate spatial development problems and will therefore accelerate movement towards a 'two-speed' Europe displaying major economic, social and political difficulties.

Technology, and most particularly information technology, is also central to Daniels's discussion in Chapter 7. The basis of this discussion is that, while the expansion of the service sector as a whole has been impressive, within the sector there have emerged activities which exert an influence in development processes that is clearly disproportionate to their scale. Foremost among these are the producer services — activities providing, for firms throughout the economy, expertise ranging from finance to management consultancy and from legal services to software development. Although they account for only a few per cent of all employees, these producer services are increasingly perceived to be central to the maintenance of growth in modern, information-based economies. Because of this role, their potential influence over spatial inequalities is considerable and, after examining the nature and recent development of these activities, Daniels therefore explores their implications for the geographical distribution of economic growth.

Inheritance and response: the urban dimension

Part II focuses primarily on urban Western Europe and is largely, but not exclusively, concerned with physical and socio-economic problems arising from the evolution of the economy. Disadvantaged sections of Western European society are, through economic forces, generally obliged to live in the poorest, most run-down and least healthy quarters of towns and cities. Their lives are therefore inextricably linked with the widespread challenge of decline in the physical fabric of inner-urban areas. In many cities this decline and its attendant socio-economic problems are closely connected with economic restructuring in the 1980s. Factory closures and the abandonment of outmoded transport infrastructures, such as old port areas, have implanted pockets — and not infrequently tracts — of derelict land around which neglect has spread contagiously (Pinder *et al.*, 1988, pp. 247–52). Yet the need for physical regeneration is not simply the legacy of recent economic crisis. With the chief exception of the bomb-damaged areas of West German cities, most of the areas affected reached their peak in the late nineteenth and early twentieth centuries and were essential to the economic take-off achieved in that era. Since then the demands of economic activity have evolved to make such areas increasingly unsuitable for production, while the housing stock has grown ever more inferior as investment in maintenance has dwindled and as the housing expectations of society have risen.

Long-run forces therefore lie at the root of this major urban challenge; the role of more recent pressures has been to intensify the urban crisis and spotlight the problems of urban areas which were once the building blocks of Western European prosperity. Against this background, urban regeneration is the focus of the study by Shurmer-Smith and Burtenshaw in Chapter 8. Their analysis enlarges on the causes of decline and outlines the emergence of concern for blighted areas, concern expressed in the post-war period in terms of the growth of the urban planning movement. Urban regeneration is seen to present a range of challenges and opportunities, within a societal context that has changed substantially and, in doing so, has encouraged greater conflict over issues such as conservation and social change in the regeneration process.

In contrast, Burt and Dawson (Chapter 9) demonstrate that other challenges have stemmed from growth and prosperity, rather than decline and deprivation. Their concern is Western European retailing, a large, highly successful but neglected activity that has recently changed dramatically in scale and character. The evolution

that has occurred has been generated, above all, by the upswing in prosperity attributable to post-war industrialisation and, subsequently, by the expansion and diversification of the service sector. This long-term influence, Burt and Dawson argue, has simultaneously amplified mass demand and created levels of wealth that have encouraged the customer to seek quality and specialisation in retailing. Thus their chapter concentrates on strategies adopted to satisfy these demands, on changes in corporate structures that have facilitated those strategies and, not least, on the strategies' spatial impacts at a range of scales.

While the spectrum of challenges facing urban Western Europe ranges from those of decline to those of prosperity, the majority relate to problems that are the legacy of past economic and social development. The remainder of Part II therefore concentrates on three major aspects of that legacy: the presence of ethnic minorities, unemployment and pollution.

Despite Western Europe's wealth-creation potential, many groups in society are arguably deprived. Even though social security systems have improved, pension provisions vary markedly from country to country, and everywhere there are substantial numbers of low-paid workers. Europe-wide, such disadvantaged groups are, of course, overwhelmingly European in origin. Yet, in many conurbations and industrial cities, real and potential social problems are closely linked with the presence of 'guestworker' communities, the subject of King's investigation in Chapter 10. Migrant workers have been a feature of the European scene for centuries, but the concentrations which exist today — and which are at the centre of King's analysis — are essentially a creation of the post-war industrialisation era. Drawn into Western Europe's economic system from the 1950s onwards, these workers were hired to undertake menial, repetitive, laborious, dirty and hazardous activities (Figure 1.5). A feature of King's argument is that the migrants (who came from a periphery extending from North Africa to Turkey) were not intended to be a permanent element in Western European society, yet the reality has proved different.[5] Despite the impact of recession on recruitment, their importance for low-order employment has been consolidated and their roots have grown deeper as families have been brought in or have been started in the West. Thus the countries which led the post-war

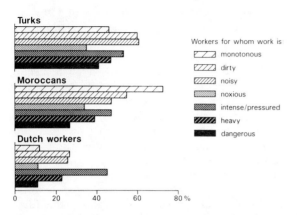

Figure 1.5 *Work environments and guestworkers in the Netherlands*
Source: Central Bureau voor de Statistiek (1986).

economic upswing with the assistance of guest-workers now have permanent migrant communities and therefore face the challenge of integration.

Unemployment has many different faces. Among guestworkers, for example, local male unemployment rates often exceed 25 per cent, and for young people may easily be higher. Historically, however, it is regional unemployment that has been most influential in policy development and, although assisted regions typically cover extensive rural areas, unemployment within them is often concentrated in their towns and cities. Moreover, it is to these towns and cities that the rural unemployed normally look for new job opportunities. Thus regional unemployment has a strong urban component. The background scenario for Wise and Chalkley (Chapter 11) is that contrasts between the 'haves' and the 'have-nots' in economically disadvantaged areas have typically sharpened since the late 1970s as economic restructuring has progressed. At first sight this heightens the argument for regional economic assistance yet, arguably, the failure of regional aid programmes to restrain unemployment may be seen as a policy defeat. Wise and Chalkley examine the recent evolution of regional policy at the national and Community levels in the context of this apparent paradox. Contrasts are drawn between policy trends in individual countries and the European Community, and the chapter questions whether regional policy can be regarded as more than a cosmetic exercise.

Finally, Part II considers pollution and pollution

control, a theme of rapidly growing significance. An impressive feature of the 1980s was the speed with which pollution — in all its various forms — was perceived to be a major European challenge. Like unemployment, the impact of pollution is, of course, geographically widespread, but pollution is primarily urban in terms of the sources which must be eliminated. Industrial conurbations and port cities are the foci for industries created before and after the Second World War, when the emphasis was on economic growth rather than effluent control. Electricity generation, a leading producer of air pollutants, developed at the pace of urban Europe. Furthermore, the populations and industries of urban areas are largely responsible for transport-related pollution and problems of waste disposal. Against this background, the view advanced by Williams (Chapter 12) is that national governments have exhibited contrasting attitudes to pollution control and, despite the rise of 'green' values around Western Europe, much remains to be done if pollution 'havens' are to be avoided. However, pressures for change may now be identified, some of the most powerful being international in nature. The international approach to what is, in terms of impact, an international problem lies at the heart of Williams's survey of environmental policy and pollution control. Several international co-operative efforts — such as those orchestrated by the Council of Europe and the OECD — are identified and discussed, but the primary focus is the European Commission. Above all, this focus reflects the Commission's status as the sole international organisation with regulatory powers in the environmental field.

Challenge and change in rural Europe

Meanwhile, rural Western Europe must respond to its own set of pressures and challenges, most of which originate in urban Europe. This is perhaps least true of the redundant agricultural land problem since, at least within the European Community, the origin of this may be traced to policies to support rural incomes. But even in this context the influence of Western European urbanisation can be identified. Maintenance of a strong agricultural sector, not overdependent on imports, has been seen as a contribution to the security of largely urban populations. Moreover,

13

the general prosperity of those populations has allowed them to accept a regime of high agricultural prices, which in turn has facilitated the large-scale subsidisation of farming. This subsidisation, however, has been positively related to the emergence of surpluses in a range of agricultural products, and it is the cost of supporting these that has led to the conclusion that redundant land must be shed by agriculture in order to control output. In Chapter 13, Ilbery examines the many problems likely to face attempts to bring about significant cutbacks in farming areas. On what scale should land be withdrawn? Which areas are most suitable for cutbacks? How may farmers be encouraged to reduce their cultivated land? And how may land withdrawn from surplus agricultural production best be used?

Quite apart from the problems of agriculture, change in the urban system has placed pressure on rural areas as a consequence of migration trends, a development explored by Fielding in Chapter 14. To a degree, this pressure has entailed the well-recognised processes of sub-urbanisation and decentralisation to villages within commuting range of towns and cities. Yet, more interestingly, counterurbanisation has also halted — and even reversed — downward population trends in many regions well beyond the commuting orbit of major urban centres. Where this has occurred, profound changes have been set in motion with respect to the composition and structures of local society. Much of Fielding's study is concerned with these changes and, most particularly, with the use of cost-benefit concepts to achieve a balanced analysis of the outcome for rural areas.

Far-reaching changes in rural Western Europe have also come about through the contribution of urban prosperity to the dramatic expansion of the leisure industry. More than any other activity, tourism has been seized upon as a long-term growth activity with restructuring potential for the economies of unindustrialised regions. In Chapter 15 Shaw and Williams explore the social and, most particularly, the economic dimensions of this growth industry. Naturally enough, their analysis is partly concerned with the evaluation of mass tourism that is typical of the tourist meccas of the Mediterranean. But, beyond this, the diversity of tourist demand throughout Western Europe is demonstrated, and consideration is

given to the organisational structures and governmental policies which influence each country's ability to exploit successfully the continuing rise in leisure demand.

While tourism can be seen as a phenomenon with potential benefits for many different types of area, however, its rise has serious implications for the environment of rural Western Europe. These implications range from the impact on flora and fauna, through modification of the rural landscape, to pressures for physical and social change in rural settlements. In some regions, of course, the impact of tourism has changed beyond recognition, and beyond redemption, the local landscapes and environments. Much more commonly, the process of change is still accelerating, and the challenge is therefore to seek long-term strategies which, for a given location, will achieve an appropriate relationship between environmental protection and economic progress based on recreation and tourism. This is the challenge examined by Woodruffe (Chapter 16), largely in the context of Alpine Europe. Here the conflict between economic development and environmental protection may be observed particularly acutely as a result of the juxtaposition of outstanding landscapes and economically marginal farming areas. Woodruffe's viewpoint is that lessons learned in this context are of much wider relevance in Western Europe, in areas where environmental conflict emerges less clearly but is none the less significant. Meeting the challenge of this conflict successfully is not simply crucial to the environments in question; it is also fundamental to the long-term quality of life in Western Europe.

Towards the future

Each chapter outlined above examines a major theme or set of issues and in this sense stands in its own right. As the discussion has sought to demonstrate, however, the challenges and changes characteristic of the late twentieth century are consequences of powerful interrelated forces, and an awareness of this can only deepen insights into mainstream economic and social development patterns and trends. This is true if Western Europe is viewed in terms of the present day and the recent past, but it is equally the case if a prospective viewpoint — aiming to identify possible outcomes of current trends — is adopted. In conclusion, therefore, the final chapter offers an integrative and forward-looking analysis of earlier contributions. In this overview, findings are drawn together within a framework which emphasises the triangular interplay between economic, social and environmental forces and which explores the possible consequences of that interplay for the geography of this dynamic world region as the twenty-first century approaches.

Notes

1. The ultimate aims were, of course, much broader than this and included, for some of those involved in the co-operation movement, the goal of political unification. The USA was particularly favourable to more effective political co-operation in Western Europe because of the importance attached to resisting the spread of Communism (Balance and Sinclair, 1983, p. 16).
2. The international grouping of greatest significance outside the European Community was the European Free Trade Association (EFTA), formed in 1959. This comprised a number of countries not participating in the EEC: Austria, Switzerland, Denmark, Norway, Sweden, Portugal and the UK, with Finland becoming an associate member in 1961. However, the importance of EFTA has dwindled as the European Community has expanded.
3. Although these vessels have contributed to the spread of industry away from Western Europe, the original technological breakthroughs from which they were derived — oil supertankers and bulk ore carriers — did not initiate this trend. Because these breakthroughs enabled new sources of raw materials to be tapped at low cost, they encouraged polarisation towards Western Europe, and they do so to this day. Thus the impact of technology has been closely related to the nature of the cargo carried.
4. The exceptions were Greece and Portugal (where industry expanded) and Norway and Austria, where it stabilised.
5. The UK's experience contrasted markedly with that of mainland Western Europe since the dominant sources of its migrant workers were New Commonwealth countries having long-established ties with the host country.

References

Aldcroft, D.H. (1980) *The European Economy, 1914–80*, Croom Helm, London.

Bairoch, P. (1982) International industrialisation levels from 1750 to 1980, *Journal of European Economic History*, 11, 269–333.

Balance, R. and Sinclair, S. (1983) *Collapse and Survival: industry strategies in a changing world*, Allen and Unwin, London.

Boltho, A. (ed.) (1982) *The European Economy: growth and crisis*, Oxford University Press, Oxford.

British Petroleum (1989) *Statistical Review of the World Oil Industry*, British Petroleum, London.

Centraal Bureau voor de Statistiek (1986) *Statistisch Zakboek*, Staatsuitgeverij, The Hague.

Clout, H. and Salt, J. (1976) The demographic background. In J. Salt and H. Clout (eds), *Migration in Post-war Europe*, Oxford University Press, Oxford, pp. 7–29.

Commission of the European Communities (1981) A new impetus for the common policies, *Bulletin of the European Communities, Supplement 4/81*.

Commission of the European Communities (1982) *Commission Communication to Council on a Mediterranean Policy for the Enlarged Community*, COM (82) Final, Commission of the European Communities, Brussels.

Commission of the European Communities (1983a) Increasing the effectiveness of the Community's Structural Funds, *Bulletin of the European Communities, Supplement 3/83*.

Commission of the European Communities (1983b) Prospects for the development of new policies: research and development, energy and new technologies, *Bulletin of the European Communities, Supplement 5/83*.

Commission of the European Communities (1984) *The Regions of Europe: second periodic report on the social and economic situation and development of the regions of the Community*, Commission of the European Communities, Luxembourg.

Commission of the European Communities (1986) The Single European Act, *Bulletin of the European Communities, Supplement 2/86*.

Commission of the European Communities (1987) *Third Periodic Report from the Commission on the Social and Economic Situation and Development of the Regions of the Community*, Commission of the European Communities, Brussels.

Commission of the European Communities (1988) *Basic Statistics of the Communities*, Eurostat, Luxembourg.

Davies, H.W.E., Edwards, D., Hooper, A.J. and Punter, J.V. (1989) *Planning Control in Western Europe*, HMSO, London.

Freeman, C. (ed.) (1984) *Long Waves in the World Economy*, Frances Pinter, London.

Godet, M. and Ruyssen, O. (1980) *The Old World and the New Technologies*, Commission of the European Communities, Brussels.

Grand'Jean, P. (1967) *Guerres, Fluctuations et Croissance*, Paris.

Hayuth, Y. (1987) *Intermodality: concept and practice, structural changes in the ocean freight transport industry*, Lloyd's of London Press, London.

Hayward, J. and Watson, M. (eds) (1975) *Planning, Politics and Public Policy*, Cambridge University Press, London.

Hesselman, L. (1983) Trends in European industrial intervention, *Cambridge Journal of Economics*, 7, 197–208.

Kosiński, L. (1970) *The Population of Europe*, Longman, London.

Landes, D.S. (1969) *The Unbound Prometheus: technological change and industrial development in Western Europe from 1750 to the present*, Cambridge University Press, Cambridge.

Lawson, R. and Reed, B. (1975) *Social Security in the European Community*, Chatham House/PEP, London.

Licari, J. (1970) The European Investment Bank, *Journal of Common Market Studies*, 8, 192–215.

Lodge, J. (ed.) (1989) *The European Community and the Challenge of the Future*, Pinter, London.

Maunder, P. (ed.) (1979) *Government Intervention in the Developed Economy*, Croom Helm, London.

Mawson, J., Martins, M.R. and Gibney, J.T. (1985) The development of the European Community regional policy. In M. Keating and B. Jones (eds), *Regions in the European Community*, Oxford University Press, Oxford, pp. 20–59.

Milward, A.S. (1970) *The Economic Effects of the Two World Wars on Britain*, Macmillan, London and Basingstoke.

Ochel, W. and Wegner, M. (1987) *Service Economies in Europe: opportunities for growth*, Pinter, London.

Odell, P.R. (1986) *Oil and World Power*, 8th edn, Penguin, Harmondsworth.

OECD (1987) *OECD Economic Surveys: Norway*, Organisation for Economic Co-operation and Development, Paris.

Pinder, D.A. (1983) *Regional Economic Development and Policy: theory and practice in the European Community*, Allen and Unwin for the University Association of Contemporary European Studies, London.

Pinder, D.A. (1986) Small firms, regional development and the European Investment Bank, *Journal of Common Market Studies*, 24, 171–86.

Pinder, D.A., Hoyle, B.S. and Husain, M.S. (1988) Retreat, redundancy and revitalisation: forces, trends and a research agenda. In B.S. Hoyle, D.A. Pinder and M.S. Husain (eds), *Revitalising the Waterfront: international dimensions of dockland development*, Belhaven, London, pp. 247–60.

Postan, M. (1977) The European economy since 1945: a retrospect. In R.T. Griffiths (ed.), *Government, Business and Labour in European Capitalism*, Europotentials Press, London, pp. 23–39.

Shepherd, W.G. (ed.) (1976) *Public Enterprise: economic analysis of theory and practice*, D.C. Heath/Lexington Books, Lexington, MA.

United Nations (annually) *UN Annual Bulletin of Transport Statistics for Europe*, United Nations, New York.

Wallace, H. (1977) The establishment of the Regional Development Fund: common policy or pork barrel? In H. Wallace, W. Wallace and C. Webb (eds), *Policy-making in the European Communities*, Wiley, London, pp. 137–63.

White, P. (1984) *The West European City: a social geography*, Longman, Harlow.

Williams, A.M. and Shaw, G. (1988) Western European tourism in perspective. In A.M. Williams and G. Shaw (eds), *Tourism and Economic Development: Western European experiences*, Belhaven Press, London, pp. 12–38.

Williams, R.H. (1984) *Planning in Europe: urban and regional planning in the EEC*, Allen and Unwin, London.

Part I
Challenge, change and production

2

Energy: resources and choices

Peter R. Odell

Antecedents

Pre-1945 components

Current issues and future prospects in Western Europe's energy sector emerge from a background which has evolved over the more than two hundred years of modernisation and development in the region. From the mid-eighteenth to the mid-twentieth century there was a slow evolution in the energy market. Dependence on coal remained of the essence and there was a high, though from the mid-nineteenth century a slowly declining, share of industrial activities concentrated on the coalfields where, of course, the main centres of population growth were also located (Pounds and Parker, 1957; Foley, 1976, pp. 52–7). Even the late nineteenth century discovery and application of electricity made only slow headway in industrial and residential energy-use patterns over most parts of the continent (Schumacher, *et al.*, 1985, pp. 13–14). It was, indeed, only in a few regions, with advantageous physical conditions for the exploitation of hydro-electric potential, that the use of electricity established itself prior to the 1920s as the basis for the location of some of the more energy-intensive industries. These regions, too, were the only ones where electricity was cheap enough to secure widespread application in the commercial and residential sectors.

Elsewhere, much of non-urban Western Europe remained beyond the networks for the generally local distribution of coal-based electricity until the 1930s or even after the Second World War. Most industrial regions remained relatively small users of electricity (for lighting and machine power only), except for a small number of locations in which more sophisticated industrial production became concentrated. Meanwhile, coal continued to dominate the supply of primary energy to the virtual exclusion of oil, let alone natural gas. In contrast with the 27 per cent contribution of oil to the total energy supply of the United States in 1937, the share of oil in Western Europe was less than 8 per cent, with even lower figures of 6 per cent and 2 per cent respectively in the continent's two leading economies, the United Kingdom and Germany.

Continuity into the post-1945 period

Western Europe's recovery after the Second World War and its economic rehabilitation and expansion were predicated on the basis of an 'energy mixture' input as before the war, but with the anticipated eventual addition of nuclear-based electricity (Jensen, 1967, pp. 1–27). It is, indeed, significant that the first two European treaties, pre-dating the Treaty of Rome whereby the Common Market of the Six was formed, were concerned with coal (and steel) and with atomic power. These were the ECSC and the Euratom treaties, respectively (Jensen, 1967, pp. 28–32; Lucas, 1977, pp. 1–8). In 1952 coal still provided 90 and 95 per cent, respectively, of the UK's and West Germany's primary energy, while in Western Europe as a whole oil's share was still only 13 per cent. Among the larger countries, it was only in Italy, poor in coal resources, that oil had made any serious inroads into the industrial and power generation markets by the early 1950s.

In spite, however, of the very considerable efforts — and investments — which were made to stimulate coal production, the industry barely recovered to its pre-1939 size and from the mid-1950s production started to decline. In 1955 Western Europe's own production of 476 million tonnes of coal, plus another 35 million tonnes of imports (largely from the United States), still supplied over 70 per cent of total energy needs. Thereafter, the demise of coal was quick and continuous as it lost its share of the then rapidly expanding Western European energy market increasingly quickly (Jensen, 1967, pp. 40–2, 58–9 and 64–74; OEEC, 1960).

Meanwhile, there had been continuity in the other main pre-1939 thrust in energy sector development — the steady growth of electricity use (Jensen, 1967, pp. 50–4). Networks of supply were expanded and intensified as part of post-war reconstruction, so that virtually the whole of the continent's population — except in the remote areas — secured access to electricity. The range of uses of this energy source also steadily expanded in the residential, industrial, commercial and agricultural sectors. By the early 1960s electricity finally became the norm across the whole of Western Europe (not only for lighting, but also for power and other uses) after three-quarters of a century of slow, but continuous development of the industry (Jensen, 1970, pp. 9–12).

The last thirty years

Since the late 1950s the long and relatively non-traumatic evolution of Western Europe's energy sector has been replaced by a period of highly dynamic and forceful developments in which long-established structures have been undermined not only by changing internal circumstances, but also by the impact of external forces. First, there was a period from 1957 to 1973 in which oil from petroleum-exporting countries elsewhere in the world came to dominate almost the whole of the Western European energy system (Jensen, 1970, pp. 88–91; Odell, 1975). A steadily declining real price for internationally traded oil during this period (Figure 2.1) was the outcome of an increasingly competitive supply system. This enabled oil products to take over residential, industrial and power-generation markets from indigenous coal. Simultaneously, specific markets for oil, particularly in the transport sector and in

petrochemicals, enjoyed rapid expansion. In 1957 Western Europe used only 115 million tonnes of oil (of which 85 per cent was imported). Fifteen years later, in 1972, total oil use was over 700 million tonnes, of which all except some 20 million tonnes were imported. By then oil supplied over 65 per cent of the continent's total energy requirement, while coal had been pushed back into a very poor second place, supplying no more than 22 per cent of the energy market.

Western Europe's ready and willing embrace of cheap oil between 1957 and 1973 was a main factor in stimulating the continent's economy, which grew rapidly at an average annual rate of over 5 per cent for the 17-year period (Prodi and Clô, 1975). At the same time the rising use of oil was influential in changing the geography of industry, particularly as it eliminated the energy-cost disadvantages of coal-less areas (Jensen, 1970, pp. 32–42). It also enabled the continent's cities to explode outwards as the motor-transport revolution allowed population and economic activities to disperse (Odell, 1976).

Until 1973 there seemed to be no reason why these processes should not continue, with the result that most forecasts indicated continued — or even intensified — oil domination of the energy sector. There were confident expectations of 1,000 million tonnes of annual oil use — and of oil imports — by Western Europe by 1985 (OECD, 1974; Brondel and Morton, 1977) and of the need for over 2,000 million tonnes a year by 2000 (Beale, 1972). Such predicted volumes had major implications in terms of the proliferation of import terminals, refineries, petrochemical installations, oil-transport and oil-distribution systems across the face of the continent. The dynamics of a development which had started in the late 1950s seemed to have become powerful enough to make the use of rapidly increasing volumes of oil over the rest of the century a near-certain prospect (Odell, 1976).

The dynamic of Europe's oil developments in particular, and the rapid expansion of the energy economy in general, were, however, halted virtually in their tracks by events in the international oil industry and market in 1973. In what has since become known as the first oil-price shock, the price of internationally traded oil increased in a matter of months by a factor of four (Ray, 1976). This was followed by a second oil-price shock at the end of the decade, when the

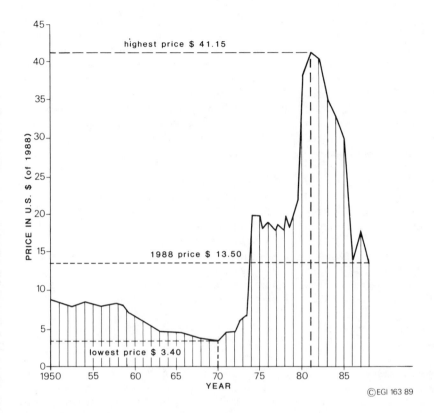

Figure 2.1 *The price per barrel of Saudi light crude oil, 1950–88 (in constant 1988 US dollars)*

oil price again more than doubled. Thus, by 1981 crude oil delivered to Western European refineries was more than an order of magnitude more expensive (in real terms after allowing for inflation) than it had been a decade earlier (Figure 2.1). This traumatic oil-sector development not only helped send the European economic system into recession, but it also created gross uncertainty and a deep lack of confidence concerning future prospects. Moreover, the general perception at the time was of yet worse to come with respect to the price of oil (Manne and Schrattenholzer, 1984). There were, indeed, widely accepted forecasts that oil would in the relatively short-term future rise to $50, $60 or even $100 per barrel (compared with about $35 in 1981) and thus generate further adverse macro-economic effects on the Western European economic and societal systems.

The earlier expectations of a continued rapid growth in oil use were, of course, largely undermined by the massive rise in price of the commodity. Instead, energy sector policies were introduced which deliberately sought to minimise oil use in favour of alternative energy sources (Commission, 1974; Taylor and Davey, 1984). Thus, growth in oil use first slowed down and then ceased. Post-1979 oil demand declined sharply, while energy demand in general also turned down. By the mid-1980s Western Europe's energy use was less than it had been in 1973, and the contribution of oil to total energy supply had fallen back to only about 50 per cent, compared with its contribution of almost two-thirds of the total only a decade earlier. Oil use declined in absolute terms from 750 million tonnes to only a little over 500 million tonnes, instead of continuing to grow as had been expected (Weyman-Jones, 1986, pp. 79–104). Nevertheless, in spite of this large reduction in oil use, there remained a powerful perception of potential supply-side difficulties with oil (Commission, 1974). Thus, there was an expectation of medium-term economic and other problems arising from the continued relatively high degree of dependence on imported oil.

21

Apprehension was heightened because 80 per cent of Europe's oil came from the member countries of OPEC, and of this the Middle East members supplied almost three-quarters. For the longer term there was a widely held perception that global supplies of oil were approaching (or had even passed) their peak, so that competition for future limited supplies would become increasingly intense (British Petroleum, 1979). These powerful perceptions (though not based on a reasonable interpretation of the facts) encouraged energy conservation. Additionally — and increasingly — there was also another motivation for giving greater attention to the efficient use of energy, viz. concern for the environmental impact of both the production and use of energy (European Environment Bureau, 1981; Commission, 1985).

As a result of these various factors, the demand for energy over much of Western Europe now appears to have more or less stabilised at the level it reached in the early 1970s. Indeed, given the rapidly increasing concern for environmental impact questions, declining energy consumption might well emerge as the norm in the 1990s. Thus, contrary to the previous widely held expectations, energy supply-side constraints will be of little or no significance for at least the rest of the twentieth century — and probably well beyond. Choices over supply between energy alternatives will be of the essence over the next 25 years. This has implications not only for the future prospects of both traditional and newly developed indigenous energy resources, but also for Western Europe's geo-economic and geo-political relationships with the regions which supply its energy. Concurrently, under the stimulus of technological changes and of increased competition between energy suppliers (partly as a result of the implementation of the Single European Act), there are likely to be important organisational consequences for the continent's energy industries.

Western Europe's energy resources: an *embarras de richesse*

Indigenous oil and gas

One of the important results of the two oil-price shocks, and the associated deterioration in the security of supply of Western Europe's hitherto almost entirely imported oil needs, has been the stimulus given to the indigenous production of hydrocarbons (oil and gas). In 1967 the region's total production of natural gas and oil was a mere 40 million tonnes of oil equivalent. This came partly from a number of scattered and only locally important production centres in West Germany, Italy and France, partly from the initial exploitation of a new gasfield in the Netherlands (the Slochteren field, later to be renamed the Groningen field) and partly from the first gasfields discovered in the southern sector of the UK's part of the offshore North Sea province (Figure 2.2). By 1972 gas production by the Netherlands and the UK had already grown significantly, but even so total Western European hydrocarbons output was still only 130 million tonnes of oil equivalent. This represented no more than 11 per cent of all energy used, and it was still much less than the declining early 1970s' annual output of coal and lignite which, in 1972, still accounted for 230 million tonnes of oil equivalent.

The North Sea basin

Moreover, even among those who were aware of the geological significance of the Groningen and North Sea finds of gas and oil — in terms, that is, of what they indicated about the potentially petroliferous nature of the North Sea basin stretching northward along the axis of the North Sea (Figure 2.2) — there was a continuing tendency to discount the importance of the early finds (Chapman, 1976, pp. 184–8). The potential for North Sea basin oil and gas was, indeed, generally portrayed as very limited. At best, it was widely argued, it could be the means whereby Western Europe's incremental demand for oil and gas might be met for some years. It was certainly not seen as a phenomenon which would enable large volumes of the then high and growing level of energy imports to be substituted (Odell, 1973).

However, the enhanced motivation for oil and gas exploration that was generated by the international oil-price shocks of the 1970s soon showed that the North Sea basin was a large, complex and potentially highly productive hydrocarbons province, the ultimate resources of which would take decades, rather than years, to be revealed in their entirety, even if there were a continuing exploration effort in all the national sectors. On

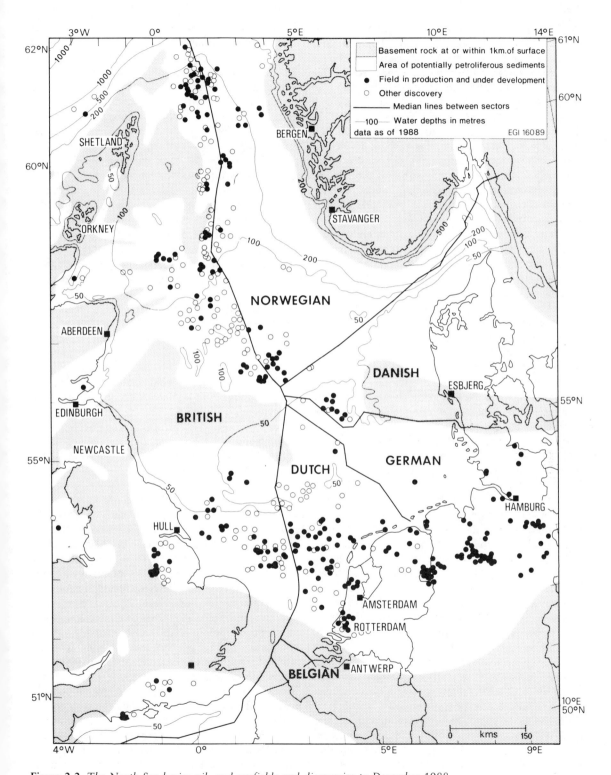

Figure 2.2 *The North Sea basin: oil- and gasfields and discoveries to December 1988*

the basis of the limited data available by the mid-1970s it was then possible, with the help of a simulation study (Odell and Rosing, 1974), to indicate a 90 per cent probability that recoverable reserves would ultimately total almost 80,000 million barrels of oil equivalent (or about 11,000 million tonnes of oil) and a 50 per cent probability that the reserves would exceed 105,000 million barrels (about 15,000 million tonnes of oil). Though the results of that study were generally disbelieved it is, nevertheless, proving to be an understatement rather than an exaggeration of the prospects. By the end of 1988 the oil plus gas reserves already discovered are close to 100,000 million barrels of oil equivalent — and this is only counting the reserves of the fields that are already in production, in development, or for which an indication of development potential has been signalled by the companies concerned. There are additional but still unquantified reserves in the many other fields which have been discovered, but for which development plans have not yet been announced (Nicholson, 1982). Meanwhile, the search for additional fields continues and this continuing exploration effort is still giving a high success rate, when measured against international norms (Department of Energy, 1988; Ministerie van Economische Zaken, 1988; Royal Ministry of Petroleum and Energy, 1988). The average size of new fields is not, moreover, falling away as quickly as had been generally expected. There is thus a high probability that significant additional reserves will continue to be discovered for many years into the future.

Simultaneously, the phenomenon of the 'appreciation' (that is, the upgrading of estimates over time) of the declared reserves of fields which are in production continues, so that production from most of these fields will go on for longer than was anticipated in earlier evaluations of their potential (County NatWest Wood-Mac, 1988). One prime example of this appreciation process is the case of the Dutch Groningen gasfield. Four years after its discovery, at the time it entered production in 1963, its recoverable reserves were declared as 1,125 billion cubic metres of gas (Odell, 1969). Since then more than 1,250 billion cubic metres of gas have been produced from the field, but the reserves which still remain to be recovered are now declared at 1,365 billion cubic metres. Thus today's estimate of the field's original recoverable reserves is almost 2,600

billion cubic metres, an appreciation of 240 per cent over the initial figure — and the appreciation process is not yet complete (Ministerie van Economische Zaken, 1988). Similarly, the first 14 oil fields brought into production in the British sector of the North Sea (prior to 1980), had initial declared reserves totalling 1,050 million barrels. Collectively they have already produced some 835 million barrels, indicating that only 215 million barrels should now remain. The exploiting companies' current expectation of the volume of oil that remains to be recovered is, however, just about twice as much, at some 425 million barrels; and, again, the process of appreciation of these fields has by no means run its full course (Department of Energy, 1988). Elsewhere in the oil world the phenomenon of appreciation of oil reserves has been noted as occurring over a minimum of twenty years (Halbouty *et al.*, 1970; Alberta Energy Resources Conservation Board, 1971; Adelman, 1987). It would thus be surprising if the recoverable reserves of this set of fields do not eventually turn out to be at least double those that were originally declared, so that production levels will be sustained for much longer than originally expected and planned.

The complexity of hydrocarbons-development decisions

Continuity in the discovery and production of oil and gas is, of course, dependent on very much more than the mere presence of resources, though the establishment of the facts of their existence, and recognition of their producibility, are the first essential steps in the development process. Other important influences are government attitudes to the granting of concessions and production licences, and the perceptions of companies whose investment funds have to be committed. For these companies there must be evidence that adequate overall profitability can be secured from the volumes which can be produced (Robinson and Morgan, 1978). This is a prospect which depends on appropriate technology, on markets, on prices and on tax regimes. These variables are individually difficult to evaluate, and collectively they constitute a complex background to the development decision-taking task for the companies concerned, especially as they are all subject to significant changes in value over time.

For the development of a supply of North Sea hydrocarbons, in the heart of an energy-intensive industrial Europe, markets may perhaps be assumed; though even here, in what is generally an open energy market, competition from alternative supplies in a period of easy supply–demand relationships cannot be excluded (see below). The required technological inputs may also be assumed, especially given the high degree to which the technology of offshore oil and gas has been developed (Lewis and McNicol, 1978). And, after a period in which European governments wrongly assumed that oil and gas production activities provided a 'golden goose' phenomenon in respect of guaranteed high tax revenues (Kemp and Rose, 1984), there is now a general recognition that taxes can only be collected when an economic rent (or a supernormal profit) can be earned on a project. The achievement of such rent depends, of course, on the market price of what is produced and thus, in an open economy such as that of Western Europe, on the international price of energy in general, and of oil in particular. Earlier widely held expectations of continuing high, and even still rising, prices for oil have now been replaced by the recognition that prices are more likely to be much lower than those of the late 1970s and early 1980s. This will be true for most, if not all, of the rest of the twentieth century (Adelman, 1987).

Such uncertainty over price does, of course, raise doubts over the economic viability of continued offshore oil and gas exploitation in north-western Europe. But with current levels of costs, which are already declining in real terms as technology improves, and with the continuing adjustment of tax regimes, then, providing international oil prices do not fall below about $15 per barrel for other than short periods, there is not likely to be much price-imposed constraint on potential production levels for the remainder of the twentieth century. Nevertheless, lower prices will slow down the geographical extension of oil and gas exploration and exploitation to other areas which show hydrocarbons potential around Western Europe — notably in Arctic waters to the north of Norway and in the deeper waters to the west of the British Isles (including Ireland) (Figure 2.3).

Uneconomic and unattractive coal resources

In the recent past there has been strong official encouragement for increased use of coal. This was related to the widely held (though mistaken) belief that oil and gas were inherently scarce so that energy policies should be oriented to their replacement by coal (Commission, 1974). It was thus anticipated that imports of coal would be needed to supplement what, it was thought, would only be a slowly increasing volume of indigenous production. However, in contrast to the continuing prospects for further expansion of oil and gas markets, present-day predictions of future coal use are much more pessimistic. The earlier 'belief' in hydrocarbon scarcity has now been largely (though not entirely) discounted and, as it is now generally accepted that oil and gas prices will not return to the early 1980s levels, less emphasis is being given to encouraging coal production in Western Europe. Indeed, little of the rump which remains of the region's formerly widespread and productive deep-mined coal industry can be profitably exploited under existing or prospective costs, prices and technological conditions (Manners, 1981, pp. 85–9 and 99–102). Thus the maintenance of most of 1988's 175 million tonnes of coal production depends on the continuation of formal guarantees of markets and/or on continuing production-cost subsidies (Gordon, 1987, pp. 70–86). Such intervention by national governments and the EC is being increasingly questioned. Moreover, growing awareness of the environmental impact of coal's use is also damaging the prospects for this fuel (Schumacher *et al.*, 1985, pp. 295 and 307). This will not only hit the European coal industry, but also the willingness to import low-cost internationally traded coal from overseas suppliers. Thus, the current prospect is that there will be less encouragement for coal imports and also a much-reduced willingness on the part of governments — and the EC — to subsidise indigenous coal production. The share of coal in Western Europe's energy supply, therefore, seems set to fall below its current contribution of just over 20 per cent.

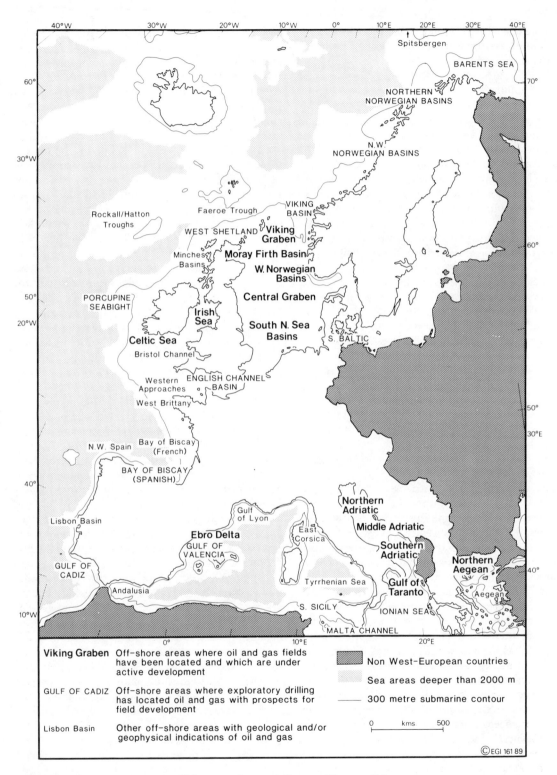

Figure 2.3 *Western Europe's petroliferous and potentially petroliferous offshore areas*

Competition between external suppliers for Western Europe's limited markets

Import options

Within the context of the limited expansion of Europe's energy demand (as argued above), and given the continuing availability of indigenous hydrocarbons in significant volumes (at, say, the 1988 level of about 35 per cent of total energy use), it is evident that potential exporters of oil and gas to Western Europe will be obliged to compete with each other for the markets available. Indeed, the continent has a number of options in respect of its relatively much more limited energy-import needs than is generally indicated in official forecasts (International Energy Agency, 1982). The choice between the options is, moreover, of high-level geo-political significance and thus carries implications for the EC's external policies that go well beyond energy supply and supply security considerations alone. The options involve three potential large-scale oil and/or natural gas supplying regions — North Africa, the Middle East and the Soviet Union — with all three of which Western Europe, for one reason or another, needs to maintain close political and economic relations.

Mediterranean basin prospects

With the expansion of the EC to include Portugal, Spain and Greece, the centre of gravity of the Community's interests is shifting southwards, so heightening the importance of the Mediterranean basin's affairs. These, of course, involve trans-Mediterranean relationships between the EC member countries on the northern side of the Sea and the North African Arab nations on the southern side. Three of the latter (Egypt, Libya and Algeria) depend mainly, or significantly, on exports of energy for their foreign exchange earnings, and a number of formal arrangements relating to energy trade with their close European neighbours on the northern side of the basin are already in place. These include the Algeria to Italy gas pipeline (Figure 2.4) through which supplies have been contracted for 25 years (Stern, 1984; Estrada *et al.*, 1988) and also specific Libyan–Italian oil supply arrangements, including Libyan

interests in the refining industry in Italy. In theory the hydrocarbons resources of North Africa could supply most of the EC's medium-term energy import needs, while Western European markets constitute the only ones in which North African oil and gas have a transport-cost advantage over supplies from elsewhere. In spite of current political difficulties, the potential mutual benefits from the development of EC–North African energy interests seem likely to be increasingly recognised in the coming decade, and will thus lead to attempts to expand Mediterranean basin interdependence in the energy field.

Imports from the Middle East

The oil- and gas-rich countries of the Middle East proper (around the Persian or Arabian Gulf) are geographically more distant but, in terms of general historical, cultural, political and long-standing oil-industry contacts, there exists a powerful relationship with Western Europe (Schurr and Homan, 1971; Shwadran, 1977; Yorke and Turner, 1986). Recent EC energy policy, with its central emphasis on minimising dependence on oil — and especially dependence on oil imports from the Middle East — has been at odds with this background. Nevertheless, EC–Middle East relationships in respect of oil have recovered markedly from their low-point at the time of the Arab–Israeli war of 1973, when there was an Arab embargo on oil supplies to the Netherlands and the threat of reduced supplies for all other countries (Turner, 1978).

EC–GCC (Gulf Co-operation Council) and EC–OAPEC (Organisation of Arab Petroleum Exporting Countries) discussions on matters economic are now well established and have already led to a search for an acceptable solution to a perceived threat to Europe's refining and petrochemical industries. This threat is posed by low-cost oil products and petrochemicals from newly developed facilities in Middle Eastern oil- and gas-producing countries (Commission, 1988a). Meanwhile, all the European-based oil companies have reached agreement over the valuation of their assets expropriated by Iran and other Middle Eastern countries (in contrast to the pursuit of what appear to be unrealistic claims for compensation by many US oil companies). Also, Saudi Arabia and, even more noticeably, Kuwait have

27

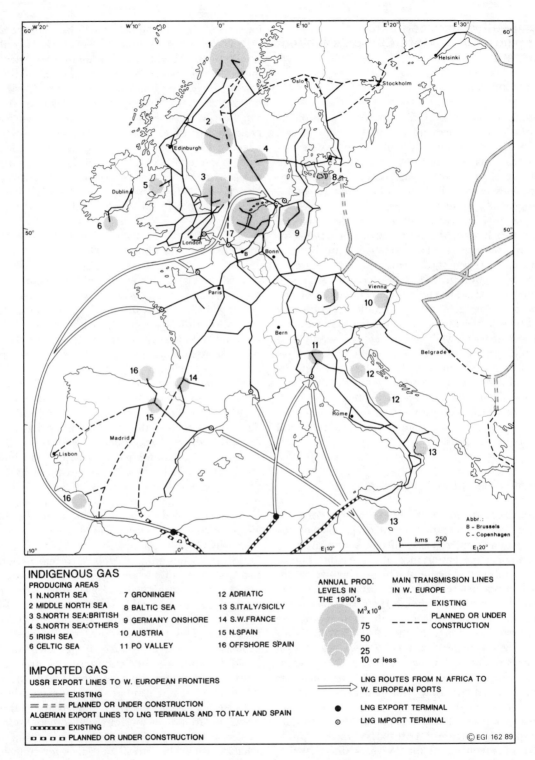

Figure 2.4 *Western Europe's gas production potential in the 1990s and the transmission systems for indigenous and imported gas*

invested in downstream oil activities in a number of Western European countries. An additional important development concerns oil-transport links from the Middle East that are oriented to serving Western European markets. These have been strengthened in recent years by the construction of new pipelines which link the main oil-producing areas of the Gulf with the eastern Mediterranean coast and with newly developed export terminals on the north-eastern coast of the Red Sea (Petroleum Economist, 1987). Such pipelines, specifically dedicated to securing and easing the oil supply routes to Europe, serve to reduce the European importers' perception that there is still danger in dependence on oil shipped from the Gulf.

The key transit countries in this development are Egypt and Turkey, and it is not without significance that both have indicated a wish to be more closely associated with the Common Market. Membership of the EC seems out of the question as far as Egypt is concerned, though its wish for closer relations with the EC can, nevertheless, be appropriately developed in its context as one of the Mediterranean basic countries with hydrocarbons resources (as previously discussed). Turkey, on the other hand, has already submitted an application for membership of the EC and, though its accession will not be a near-future development, it could well become a formal part of the European Community before the end of the century. It thus has a strong and heightening motivation to enhance its role as the country through which Gulf oil is transported to Western Europe, in order to assist its own economic growth and to consolidate its relationship with the EC. Its role in this respect would be even more important — indeed it would be the critical component — in enabling natural gas from the extensive resources of the Middle East (notably from Qatar and Iran) to be transported by pipeline to Western Europe. Turkey could thus make itself central to the potential redevelopment of the hydrocarbons trade between the Middle East and Western Europe. Moreover, from the European importers' viewpoint, Turkey's membership of the EC would take 'Europe' right up to the frontiers of the world's main oil-producing regions — both currently and for the foreseeable future — so reducing the present strong perception of the inherent insecurity of supplies from the Middle East. For the oil exporters, Europe would become a 'near neighbour', with large energy-import demands for the foreseeable future. It would thus offer potential guaranteed purchases of oil and gas. Such a development would, in turn, help undermine the validity of policies in the EC to minimise reliance on Middle Eastern oil, whereby prospects for the oil exporters have been — and currently remain — so adversely affected.

The Soviet option

Finally, the vast (indeed, relative to prospective demand, near infinite) natural gas resources of the Soviet Union (Dienes and Shabad, 1979; Jensen and Shabad, 1983) represent not simply a major new energy resource for long-term use in Western Europe, but also a means whereby relations more generally with the Soviet Union and its Eastern European allies could be fundamentally changed (Odell, 1987). In essence, in order to achieve a more efficient, productive and acceptable economic system, the Soviet Union needs to import large volumes of consumer, capital and investment goods and services over the coming decade. The only means it has to pay for these requirements (apart from borrowing) is by the export of large volumes of natural gas (Estrada *et al.*, 1988, pp. 173–9). The only possible markets for this gas are in Western Europe, where, in return, a wide range of manufacturing industries and suppliers of other goods and services could be offered guaranteed outlets in the Soviet Union — without the difficulties of having to compete with alternative suppliers from other parts of the world. The economic bargain which could be struck between the USSR and the EC is thus a powerful one, with immense potential benefits to both sides. Moreover, it is one that appears to override the hitherto well-rehearsed arguments of strategic dangers arising from reliance on Soviet energy supplies (Maull, 1981; Adamson, 1985; Manne *et al.*, 1986), particularly in the context of the broader geo-political issues which are involved. One of these arises from the fact that the gas has to flow through Eastern European countries whose economies the Soviet Union can no longer afford to sustain. Conceivably, therefore, the gas-for-goods-and-services bargain could, over time, help the loosening of the Eastern European countries' continuing economic ties with the Soviet Union through an agreement that they become more closely involved with the rest of Europe. This political aspect to a massive Soviet–Western European gas deal could prove even more fundamental than the purely economic considerations. The combined economic and political changes which could be set in train by Western Europe's willingness to import several times more Soviet gas than is currently traded (approximately 40 billion cubic metres a year) would clearly have a geo-political importance at least equal to the potential for change from Mediterranean basin or Middle Eastern initiatives arising from oil and gas developments (Odell, 1988).

The choices

Clearly, in the context of a demand for energy in Western Europe which will grow only slowly, and in the light of an indigenous potential to produce large volumes of hydrocarbons for decades into the future, policy-makers may well be obliged to exercise a series of Solomon-type judgements in their search for external energy relationships which make the most long-term sense. Unhappily little thought has been given to date to the alternative options and their significance. This is largely because Europe's energy sector policy-makers, in spite of all the evidence to the contrary, remain generally besotted with the concept of supply scarcity and thus believe it to be more important to continue working and planning mainly for ways in which to protect the EC against future oil-supply crises (Commission, 1988a). This is perhaps a clear example of the way in which policy-makers show that they are a little like the proverbial generals who concentrate their efforts on fighting the last war!

Institutional and technological aspects of Western Europe's energy outlook

The single European market and competition

The changed energy-demand outlook, and the wide range of options which now exist for supplying Western Europe's oil and gas, together represent a significant challenge. The dynamics of the energy sector will, however, be yet further accelerated in the years immediately ahead by both political and legal changes within Western Europe itself and by technological developments. The impact of the latter will, moreover, be heightened in the more competitive environment which the political and legal changes will create. These changes will arise not only from the implementation of the Single European Act (Commission, 1988b) which, in practice, is likely to become gradually effective through the first half of the 1990s, but also from various national moves to denationalise and/or deregulate industries which have traditionally been in the public sector and thus subject to strong political control.

Reduced protection for coal

The coal industry mainly falls into this category. As shown previously, most of the industry which has survived competition from oil and gas has done so as a result of direct or indirect state support, usually in the context of state ownership. Such support emerged, as also shown, out of post-war rehabilitation difficulties and as a consequence of declining oil prices between the mid-1950s and the early 1970s, and it continued to be necessary in spite of high energy prices from 1974 to 1985. Since then there has been a sharp fall in the oil price (Figure 2.1), so that coal's already weak competitive position has deteriorated still further. Thus, most of the small remaining Belgian and French coal industries have been closed, while in West Germany increasing opposition is being expressed to the rising cost of an early 1980s agreement to sustain the use of up to 75 million tonnes per year of indigenous hard coal in power stations until the year 2000. This cost is paid partly by the federal government and partly by electricity consumers through the so-called 'coal pfennig'. In the UK the number of mine closures has been increased since 1985 and this process is expected to continue. Meanwhile, plans are now being made to sell off the remaining coal industry in bits and pieces, so that only a small number of efficient collieries in favoured areas (both geologically and in relation to demand centres) seem likely to remain in production. Unhappily, no technological breakthroughs appear to be in sight for the European coal industry to help it survive the increased competition from gas and from more freely imported and low-cost foreign coal. The absence of progress in, for example, the underground gasification of coal contrasts with important breakthroughs that have increased the efficiency with which gas can be used. And an already bad situation and outlook for Europe's coal industry is made even worse by the developing opposition to the use of coal because of its high contribution (compared with gas and oil) to the atmospheric CO_2 problem and its perceived greenhouse effect on the world's climate.

Gas and electricity and the Single European Act

Forthcoming politico-legal and technological changes are, on the other hand, of great significance for the future of Western Europe's gas and electricity industries. In spite of some trade in these commodities between European countries, the industries have remained essentially nationally organised and controlled. They have, moreover, often enjoyed a high (and sometimes complete) protection from competition. The advocates of the continuation of this situation argue that both the gas and the electricity industries are 'natural monopolies', because of the high levels of fixed investment which go into their production, transmission and distribution systems, and because of the long lead times and stability which are required to enable them to give continuity of service.

The Single European Act is seen to threaten this structure because it casts doubts on the continued legality of the essentially national monopolistic corporations which have been created, and on current restraints on the trading of gas and electricity within the Community. While few opponents of the present situation envisage that competing electricity or gas companies will vie for business among residential and other small users, they do see scope for competition in supplies to large users and in the operation of transmission systems. Such competition is expected to come in ways similar to those emerging from the UK government's privatisation proposals; in these the domination to date of managerial aspirations to achieve technological efficiency is to be substituted by economic efficiency considerations produced by competitive market conditions. In particular, countries will face increasing difficulties in preventing customers from seeking supplies from other than the national suppliers. Thus German industrial users of electricity (who are currently penalised by prices made high by the imposition of the 'coal pfennig') may, for example, contract to buy their requirements from *Electricité de France*. Dutch users south of the Westerschelde could likewise choose to buy their supplies from the Belgian electricity industry; this has surplus capacity that would enable electricity to be sold at short-run marginal prices (just enough to cover running costs and a little bit over to contribute to overheads). Similar developments may be envisaged in the gas sector, in which organisations such as *Nederlandse Gasunie*, *Gaz de France* and *Distrigaz* in Belgium have a *de facto* monopoly. This applies not only to gas sold in the countries in which they operate, but also to natural gas which moves across 'their territories', from a distant supplier on the one hand to a distant consumer on the other. In future the national or regional monopolies may be obliged to accept a requirement that they offer so-called third party access to their pipelines at prices which reflect the costs involved. Alternatively, perhaps other potential suppliers and users of gas will be able to build dedicated lines across the territories of the monopoly organisations which, in effect, currently carve up the European market between them (Estrada *et al.*, 1988).

Technological changes favour gas

Meanwhile, restraints have been imposed on gas demand as a result of the attitudes of the monopolistic and/or the protected gas-transmission and -distribution companies, coupled with government-imposed minimum-price controls and other regulations. These have served to inhibit the implementation of technological breakthroughs in gas-using technology (Rogner, 1988).

The most important of these by far has been the failure to expand gas use in high-efficiency power generation, either through combined-cycle technology, or through combined heat and power systems. In the former, the 50 per cent efficiency at which electricity is produced is almost half as good again as the average 35 per cent efficiency achieved in coal and oil-based power stations; in the latter, heat recovered from the system is used along with the power generated to give overall thermal efficiencies in excess of 65 per cent. In

Japan a 2,000 MW combined-cycle power plant fuelled by natural gas is already operational, together with several other somewhat smaller units. By contrast the largest in operation anywhere in Western Europe is only 230 MW — under the influence of a lack of interest in their development by the gas suppliers, and in the context of highly centralised and usually monopolistic electricity-supply companies whose predilection is to expand their systems by means of large centralised coal-fired and/or nuclear stations.

Freeing the electricity and gas supplies from monopolistic control would lead to a new organisational framework within which the cost, and the environmental advantages, of gas for electricity production could begin to be effective. The ultimate result could be a much more geographically dispersed pattern of power production closer to centres of demand, with the long-distance transport element being that of the gas input, rather than the electricity output. This would give, overall, a much more cost-effective and environmentally acceptable system than the separate gas and electricity systems to which Europe has become used over recent decades (Skea, 1988).

Towards competing systems

Thus, a combination of the politico-legal changes which are under way, and a cluster of technico-economic-environmental advantages, will favour gas over coal, nuclear power and, to some degree, oil (especially fuel oil). This will serve to encourage the completion of comprehensive, but more integrated, energy systems. In this process national frontiers will decrease in significance in determining supply, so that many consumers will achieve a wider choice of suppliers and the prospect of lower prices.

Conclusions

The dominance of fossil fuels

Since the end of the Second World War, Western Europe's energy system has undergone two dramatic changes — and a third is under way. The *first* was in the 1950s and the 1960s, when imported oil replaced indigenous coal and when there was a continuing rapid rate of increase in energy use. The *second* was between 1973 and the mid-1980s, when much higher energy prices curbed demand growth to a near-zero rate and when there was a massive development of Western Europe's own oil and gas resources.

In spite of the traumas associated with the first two changes, Western Europe's dependence today on fossil fuels is very little less than it was in 1945. For thirty years nuclear power has promised much, but its overall contribution to total energy supply is still only about 5 per cent (when the electricity produced in nuclear power stations is appropriately measured in terms of its heat-value equivalent). Indeed, it has only just succeeded in becoming somewhat more important than hydro-electricity, which continues to contribute about 3.5 per cent to total supply (using the same method of measuring its contribution). Hydro-electricity, in spite of efforts over the last decade to develop alternative energy sources, remains the only source of benign (atmospherically non-polluting) energy in large-scale production. The other sources, such as wind, tide and wave power and the direct use of the sun's energy, still contribute much less than 1 per cent of Western Europe's energy supply.

The apparent similarity between yesterday's energy-sector situation and that of today is, however, superficial. Underneath, there have been important changes in the relative contributions of the three fossil fuels, together with changes in their sources of supply. These changes have, moreover, been accompanied by the evolution of energy-sector institutional and organisational structures, reflecting the rapid expansion of energy demand and a 'belief in bigness' and in economies of scale on the supply side. The energy sector has thus become dominated by large public and private corporate entities, with a high degree of centralisation and, in some cases, with monopolistic controls over energy supply — particularly in respect of electricity and gas (Lovins, 1979).

Future choices between fossil fuels

Likewise for the coming decades; the apparent similarity between the Western European energy sector's past and its future, a similarity arising

Table 2.1 *Alternative potential energy supply patterns for Western Europe in 2000 (millions of tonnes of oil equivalent)*

| Energy sources | Estimated 1988 energy supply pattern | | | Alternative patterns in 2000 | | | | | |
| | | | | A | | | B | | |
	Total	Indigenous	Imports	Total	Indig.	Imp.	Total	Indig.	Imp.
Natural gas	215	170	45	230	140	90	335	205	130
%	18.9			18.7			26.7		
Oil	605	200	405	575	125	450	580	200	380
%	53.3			46.7			46.2		
Coal, etc.	215	155	60	285	150	135	220	120	100
%	18.9			23.2			17.5		
Primary elect*	100	100	–	140	140	–	120	120	–
%	8.8			11.4			9.6		
Total energy supply	1135	625	510	1230	555	675	1255	645	610
%	100	55.1	44.9	100	45.1	54.9	100	53.0	47.0

Alternative A assumes continued constraints on natural gas markets.
Alternative B assumes the development of a competitive gas market.
*Primary electricity (including nuclear power) converted to oil equivalent on the basis of the heat value of the electricity produced.

from the continued dominance of fossil fuels in the supply of energy, will be superficial only. On the contrary, changing political ideas and technological developments will, as shown above, jointly create a significant potential for the establishment of a much greater variety of organisational structures and for changed market relationships between suppliers and users. In addition, as already shown in some detail and as set out in Table 2.1, there will be a continuation of the changing fortunes of fossil fuels. Indigenous coal production will certainly continue to decline slowly in importance but, in the event of a reduction in West Germany's high level of support for its coal industry and/or as a consequence of the privatisation of the UK's coal industry, European coal production could be down to little more than half its present scale by 2000. Coal imports will continue to grow, though by no means as rapidly as has been expected and is still widely assumed. Increasing concern for environmental issues (or, put another way, the costs of achieving acceptably low pollution levels from coal use) will slow down the expansion of the market for imported coal. Natural gas will be the major growth element in Western Europe's energy supply for the foreseeable future — for a mixture of easy supply, favourable environmental and new technology

reasons, in the context of a much more open and competitive situation for incorporating gas in a broader range of end uses (Odell, 1988). In particular, natural gas used in high-efficiency generating plants will largely eliminate the present continuing levels of fuel-oil consumption in power-station and industrial use (especially in southern Europe); and it will substitute for what would otherwise have been nuclear and coal-based capacity in most countries. By 2020, natural gas seems likely to be Western Europe's single most important energy source, on the assumption that compressed natural gas (CNG) becomes an important alternative automotive fuel.

Prospects for alternative energy sources

Non-fossil-fuel energy sources will continue to be dominated by hydro-electricity (for which slow capacity growth will continue) and by nuclear power which, barring accidents, will also grow slowly. The downside risk for the latter is, however, high, given that another major accident anywhere would produce an even stronger anti-nuclear response than that over Chernobyl, so leading to the cancellation of the few new facilities planned, as well as the closure of some existing stations. An accident in Western Europe itself, at one of the stations of the pressurised water reactor type that constitutes most of the existing capacity, would seem likely to lead to the near-instant close-down of the industry. This would have severe implications not only for electricity supply, but also for the economies of two countries, France and Belgium, which derive more than 70 and 60 per cent, respectively, of their electricity from nuclear power.

Meanwhile there will be continued, albeit modest, expansion of wind-powered generating capacity, and a few large-scale schemes for harnessing tidal or wave power could well be under way by the turn of the century. Nevertheless, in spite of formal EC support for existing national programmes to promote the development of renewable energy sources (Commission, 1987, pp. 31–2), major contributions from such new systems of power generation will be delayed until well into the twenty-first century. Earlier development will only take place if there is unequivocal and dramatic evidence in the meantime of a real medium-term threat of climatic change from the continued, and increasing, combustion of fossil fuels (Keepin and Kats, 1988). And in the first quarter of the twenty-first century, fusion power and/or power based on the underground gasification of coal may also be competitors for meeting energy demand. This will be especially likely if oil prices — and the prices of fossil fuels more generally — start to rise above the levels to which they have declined since the early 1980s. This may well happen under the impact of increasing 'user costs'; that is, the costs of depleting scarce reserves which might otherwise be saved for the future, when they would fetch higher real prices.

Security of supply?

Political decisions will be required from time to time to try to determine how the choices shall be made. First, Western Europe will need to take into account a broad range of economic and other considerations in its dealings with the other main potential energy supplying regions, viz. North Africa, the Middle East and the Soviet Union. Second, the decisions will need to pay attention to the extent to which it is considered appropriate to give preference to the production and use of

indigenous oil and gas, the continued development of which may otherwise be thwarted by lower-cost imports. Such a continuity of interest in Europe's security of energy supply seems highly likely in a world which will remain replete with problems of international and inter-regional significance, particularly as Western Europe will continue to be second only to Japan as the most energy-intensive part of the world into the first quarter of the twenty-first century. Thus, in the year 2000, a dependence on imported energy to a degree of 'only' 45 per cent, as compared with more than 55 per cent, could be significantly advantageous for Western Europe's economic and political standing in the world. As shown in Table 2.1, this is the possible range of outcomes that could result from the pursuit of contrasting policies for the development over the next few years of Western Europe's indigenous oil and gas.

References

Adamson, D.M. (1985) Soviet gas and European security, *Energy Policy*, 13, 13–26.

Adelman, M.A. (1987) The economics of the international oil industry. In J. Rees and P.R. Odell (eds), *The International Oil Industry: an interdisciplinary perspective*, Macmillan, London, pp. 26–56.

Alberta Resources Conservation Board (1971) *Reserves of Crude Oil, Gas, Natural Gas Liquids and Sulphur, Annual Report*, Calgary, Alberta.

Beale, N. (1972) The energy balance in the year 2000. In *Europe 2000: perspectives for an acceptable future*, Fondation Européenne de la Culture, Amsterdam.

British Petroleum (1979) *Oil Crisis . . . Again?* BP, London.

Brondel, G. and Morton, N. (1977) The European Community: an economic perspective, *Annual Review of Energy*, 2, 343–64.

Chapman, K. (1976) *North Sea Oil and Gas: a geographical perspective*, David and Charles, Newton Abbott.

Commission of the European Community (1974) Towards a new energy policy strategy for the Community, *Bulletin of the European Communities, Supplement* 4/74, Brussels.

Commission of the European Community (1985) *Energy in Europe*, 2 (August), 17–19.

Commission of the European Community (1987) *Energy in Europe*, 9 (September), 31–2 and 68–9.

Commission of the European Community (1988a) *Energy in Europe*, 11 (September), 29–33.

Commission of the European Community (1988b) *Energy in Europe*, 12 (December).

County NatWest Wood-Mac (1988) *North Sea Report*, Edinburgh.

Department of Energy (1988) *Development of the Oil and Gas Resources of the United Kingdom*, HMSO, London.

Dienes, L. and Shabad, T. (1979) *The Soviet Energy System: resource uses policies*, Halsted Press, New York.

Estrada, J., Bergesen, H.O., Moe, A. and Sydnes, A.K. (1988) *Natural Gas in Europe: markets, organisation and politics*, Pinter, London.

European Environment Bureau (1981) *The Milano Declaration on Energy, Economy and Environment*, Brussels.

Foley, G. (1976) *The Energy Question*, Penguin, Harmondsworth.

Gordon, R.L. (1987) *World Coal: economics, policies and prospects*, Cambridge University Press, Cambridge.

Halbouty, M.T. *et al.* (1970) *World's Giant Oil and Gas Fields*, American Association of Petroleum Geologists, Memoir 14.

International Energy Agency (1982) *World Energy Outlook*, OECD, Paris.

Jensen, R.G. and Shabad, T. (eds.) (1983) *Soviet Natural Resources in the World Economy*, University of Chicago Press, Chicago.

Jensen, W.G. (1967) *Energy in Europe 1945–1980*, Foulis, London.

Jensen, W.G. (1970) *Energy and the Economy of Nations*, Foulis, London.

Keepin, B. and Kats, G. (1988) Greenhouse warming: comparative analysis of abatement strategies, *Energy Policy*, 16, 538–61.

Kemp, H.G. and Rose, D. (1984) Investment in oil exploration and production: the comparative influence of taxation. In D.W. Pearce, H. Siebert and I. Walter (eds), *Risk and the Political Economy of Resource Development*, Macmillan, London, pp. 169–96.

Lewis, T.M. and McNicol, I.H. (1978) *North Sea Oil and Scotland's Economic Prospects*, Croom Helm, London.

Lovins, A.B. (1979) Re-examining the nature of the ECE energy problem, *Energy Policy*, 7, 178–98.

Lucas, N.J.D. (1977) *Energy and the European Communities*, Europa Publications, for the David Davies Memorial Institute of International Studies, London.

Manne, A.S., Roland, K. and Stephan, G. (1986) Security of supply in the West European market for natural gas, *Energy Policy*, 14, 52–64.

Manne, A.S. and Schrattenholzer, L. (1984) International energy workshop: a summary of the 1983 poll responses, *The Energy Journal*, 5, 45–64.

Manners, G. (1981) *Coal in Britain: an uncertain future*, Allen & Unwin, London.

Maull, H.W. (1981) Natural gas and economic security, *The Atlantic Papers*, 43, The Atlantic Institute for International Affairs, Paris.

Ministerie van Economische Zaken (1988) *Natural Gas and Oil in the Netherlands*, The Hague.

Nicholson, J. (1982) *Future Oil and Gas Developments in the UK North Sea: reserves of the undeveloped oil and gas discoveries and a commentary on their development options*, Midland Valley Exploration Ltd, Glasgow.

Odell, P.R. (1969) *Natural Gas in Western Europe: a case study in the economic geography of energy resources*, De Erven F. Bohm NV, Haarlem.

Odell, P.R. (1973) Indigenous oil and gas developments and Western Europe's energy policy options, *Energy Policy*, 1, 47–64.

Odell, P.R. (1975) *The Western European Energy Economy: challenge and opportunities*, The Stamp Memorial Lecture delivered before the University of London, Athlone Press, London.

Odell, P.R. (1976) *The Western European Energy Economy: the case for self-sufficiency, 1980–2000*, H.E. Stenfert-Kroese BV, Leiden.

Odell, P.R. (1987) Gorbachev's new economic strategy: the role of gas exports to Western Europe, *The World Today*, 43, 123–5.

Odell, P.R. (1988) The West European gas market: current position and alternative prospects, *Energy Policy*, 16, 480–93.

Odell, P.R. and Rosing, K.E. (1974) *The North Sea Oil Province: an attempt to simulate its development and exploitation, 1969–2029*, Kogan Page, London.

OECD (1974) *Energy Prospects to 1985*, Organisation for Economic Co-operation and Development, Paris.

OEEC (1960) *Towards a New Energy Pattern for Europe*, Organisation for European Economic Co-operation, Paris.

Petroleum Economist (1987) *Map of Middle East Oil and Gas Export Routes*, Petroleum Economist, London.

Pounds, N.J.G. and Parker, N.W. (1957) *Coal and Steel in Western Europe*, Faber & Faber, London.

Prodi, R. and Clô, A. (1975) The oil crisis in perspective: Europe, *Daedalus*, 104, 91–112.

Ray, G.F. (1976) Impact of the oil crisis on the energy situation in Western Europe. In T.M. Rybczynski (ed.) *The Economics of the Oil Crisis*, Macmillan, for the Trade Policy Research Centre, London, pp. 94–130.

Robinson, C. and Morgan, J. (1978) *North Sea Oil in the Future: economic analysis and government policy*, Macmillan, for the Trade Policy Research Centre, London.

Rogner, H.H. (1988) Technology and the prospects of natural gas: results of current gas studies, *Energy Policy*, 16, 9–26.

Royal Ministry of Petroleum and Energy (1988) *The*

Norwegian Continental Shelf, 1988, Oslo.

Schumacher, D., Berkovitch, I., Hesketh, R. and Stammers, J. (1985) *Energy: crisis or opportunity?*, Macmillan, Basingstoke.

Schurr, S.H. and Homan, P.T. (1971) *Middle East Oil and the Western World*, American Elsevier, New York.

Shwadran, B. (1977) *Middle East Oil: issues and problems*, Harvard University Press, Cambridge, MA.

Skea, J. (1988) UK policy on acid rain: European pressures and alternative prospects, *Energy Policy*, 16, 252–69.

Stern, J.P. (1984) *International Gas Trade in Europe: the policies of exporting and importing countries*, Heinemann Educational, London.

Taylor, E. and Davey, W.G. (1984) Energy in Western Europe, *Energy Policy*, 12, 409–24.

Turner L. (1978) *Oil Companies in the International System*, Allen & Unwin, for the Royal Institute of International Affairs, London.

Weyman-Jones, T.G. (1986) *Energy in Europe: issues and policies*, Methuen, London.

Yorke, V. and Turner, L. (1986) *European Interests and Gulf Oil*, Gower, Aldershot, for the Policy Studies Institute and the Royal Institute of International Affairs, London.

3

European industry and global competition

Peter Dicken

Introduction

Less than two hundred years ago Europe was, to all intents and purposes, the world economy. North-west Europe, in particular, was the *core* of a *capitalist world system* which had been evolving since at least the fifteenth century (Bairoch, 1982; Braudel, 1986; Wallerstein, 1979). Today, the position of Europe in the global economy is very different, and it is no longer the dominant economic power. Such a change in relative position has not occurred overnight, of course. For the first 40 years of the twentieth century the United States increasingly threatened Europe's economic status, and after the Second World War that country emerged as the clear and unchallenged world economic leader in the face of an economically devastated Europe. In 1945, the United States was producing more than half the world's manufacturing output and was the hegemonic power of the capitalist nations, not only economically but also politically and militarily (Kennedy, 1988). A major element in the rise to global dominance of the United States was its involvement in Western Europe, initially through trade and later through the direct investments of US companies. By the late 1960s, the French writer Servan-Schreiber (1968, p. 3) was moved to make his classic prediction that 'fifteen years from now it is quite possible that the world's third greatest industrial power, just after the United States and Russia, will not be Europe, but American industry in Europe'. In fact, Servan-Schreiber was wrong on several counts, although not in a way which brought much comfort to Western Europe.

By the early 1980s, Western Europe's competitive position in the global economy (and

that of the United States, too) was under threat from a different direction — Japan. To state that the economic growth of Japan during the past two decades has been spectacular is a massive understatement. Until recently, its major international involvement has been through exports, but this emphasis is changing and overseas investment by Japanese firms is now growing rapidly. Thus, although Western Europe is a huge and affluent market (in aggregate population terms larger than the United States and Japan combined) and continues to be a major centre of industrial production, it is undoubtedly relatively weaker than its two major rivals. Average per capita GNP in Western Europe in 1986 was $9,770, compared with $12,840 in Japan and $17,480 in the United States — although there are, of course, very wide variations between individual European countries. In Ohmae's (1985) conceptualisation of the world economy as a 'triad' of three major economic powers, Western Europe is in third place. However, its competitive position is not only threatened by the economic superpowers of Japan and the United States. Increasingly, its competitive strength in certain industries is being challenged by a small number of newly industrialising countries (NICs), particularly from East and South-East Asia. It is, in part, in recognition of Western Europe's relative economic decline that the drive for a single European market between the 12 members of the EEC is being pursued so vigorously. This, in turn, is prompting fears of the emergence, after 1992, of an economic 'fortress Europe' in which external trade (and investment) barriers may remain and even be strengthened while internal barriers disappear. Hence, the question of Europe's competitive position in the global economy has far more than simply an economic dimension. It is a *political* issue of the greatest international importance.

The aim of this chapter is to explore Western Europe's current competitive position as a region of industrial production and trade in a rapidly changing global environment. The chapter is organised into five parts. The first deals, briefly, with Western Europe's recent industrial performance *vis-à-vis* other parts of the world. The second focuses on the region's position in the international trade network, for it is through international trade that the interdependencies of countries and regions within the global economy are articulated. Because both production and trade are increasingly bound up with the investment and disinvestment decisions of transnational corporations (TNCs), a third section examines inward and outward investment patterns from a European perspective. The first three sections, then, are concerned with aggregate trends and patterns. But aggregates obscure substantial differences of detail. Consequently, the fourth part looks at the competitive picture in three key industries: automobiles, electronics and pharmaceuticals. These are sectors in which global competitive pressures are especially intense and in which Western Europe's position is extremely sensitive. The final part of the chapter draws together the major threads of the argument and looks briefly to the future. Throughout this chapter, the emphasis is on Western Europe as a whole, rather than on individual countries. Inevitably, rather more attention will be devoted to the European Community than to the smaller group of EFTA states.

Western Europe's recent industrial performance in a global context

In global terms, the period since the end of the Second World War can be divided into two broad phases. After the initial post-war hiatus, in which both the political and economic infrastructures of the war-ravaged economies were being rebuilt, the world economy experienced some 20 years of unprecedented economic growth. World industrial production grew at unparalleled rates; world trade grew even faster, a clear indication of the increasing interconnectedness of the emerging global economy. Between 1963 and 1973, world manu-

facturing production increased by 7 per cent per year and world manufactured exports by 11 per cent per year. In 1973, world manufacturing production was 97 per cent greater in volume terms than in 1963; world manufactured exports were 180 per cent greater than in 1963.

The situation changed dramatically after the 1973 oil-price crisis, although the subsequent world recession was not solely the result of the OPEC price rises (Chapter 1). Although the causes were complex, the outcome was unambiguous: economic growth rates throughout the world were drastically reduced. Between 1973 and 1983 they were half those of the 1963–73 period. Since 1980, the cumulative growth of both world output and exports has been modest. In 1987 the volume of world manufacturing production was 22 per cent higher than in 1980; world manufactured exports were 37 per cent higher in 1987 than in 1980. Within this general picture of global economic change, there have been very substantial geographical variations. Some areas, notably Japan and some of the East and South-East Asian economies, have continued to experience very rapid growth. Others, including the United States and a number of Western European countries, have experienced more modest expansion and, in some cases, even decline. Moreover, the economic recovery that has occurred during the second half of the 1980s has not heralded a return to the golden age of economic growth of the 1960s.

Figure 3.1 sets Western Europe's manufacturing status in its global context. The map shows clear variations in the level of manufacturing value added among individual Western European countries. Such internal differentials are the subject of discussion in the next chapter; here we are concerned with the position of the region as a whole. In 1985, Western Europe accounted for a little over one-quarter of world manufacturing output, compared with 32 per cent for the United States and 16 per cent for Japan. Compared with the position in 1970, both Western Europe's and the United States' shares of world manufacturing production had declined (from 30 per cent and 36 per cent, respectively) while Japan's had increased (from 10 per cent). At the same time, the relative industrial importance of a small number of NICs, most notably the four Asian NICs of Hong Kong, Singapore, South Korea and Taiwan, together with Brazil and Mexico, increased dramatically. Overall, the share of the 'middle-income' group of

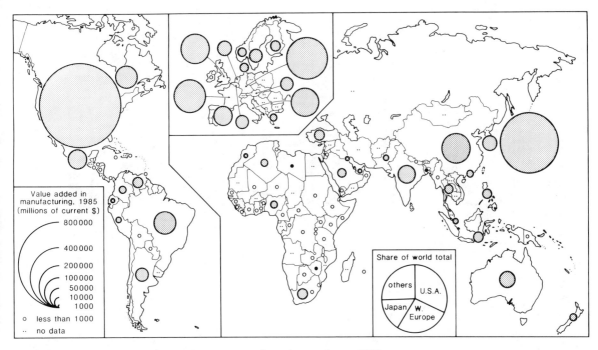

Figure 3.1 *Western Europe's position as a manufacturing producer, 1985*
Source: World Bank (1988), Table 8.

developing countries grew from 7 per cent of total world manufacturing production in 1970 to 11 per cent in 1985.

A rather clearer picture of Western Europe's manufacturing performance during the 1980s, compared with the other two members of the 'triad', is shown in Figure 3.2. Apart from 1986, when in the depths of the global recession Western Europe's manufacturing output grew more rapidly than that of either the United States or Japan, her performance *vis-à-vis* her two major competitors was poor. Even though Japan's growth rate has slackened since 1984, it remains substantial, and in 1988 it exceeded 4 per cent. Similarly, from a trough in 1982 the United States staged a dramatic recovery through the mid-1980s, to expand at nearly 3 per cent in 1988. But in Western Europe growth was only 2 per cent in that year, and throughout the 1980s it scarcely exceeded that figure.

Western Europe and the global trading system

Western Europe is more significant as an exporter

in the world economy than as a producer of manufactured goods. It is, indeed, the world's major trading region. In 1986, Western Europe generated approximately 40 per cent of world merchandise exports, compared with 10 per cent each by the United States and Japan. As Figure 3.3 shows, the leading exporting countries are West Germany, France, the United Kingdom and Italy. But Western Europe is also a major importer. Although in 1986 its trade balance was positive (a surplus of exports over imports) it was only just so, and 10 of the 16 Western European economies now have trade deficits of varying degrees of seriousness. On the one hand, therefore, the region's trade imbalance is much less problematic than the massive deficit afflicting the United States; but, on the other hand, it bears no comparison with the huge trade surplus of Japan.

Table 3.1 and Figure 3.4 summarise the major geographical characteristics of Western Europe's position in the global trading system. Four major features are evident. First, not only is most trade internal to the region, but also this intra-regional trade has increased in importance. Second, North America remains Western Europe's largest

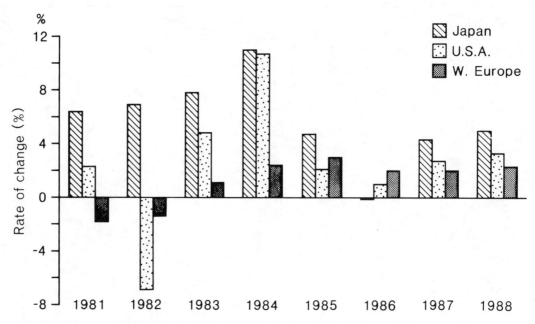

Figure 3.2 *Western Europe's manufacturing performance compared with that of the United States and Japan, 1981–88*
Source: based upon UNIDO (1987) *Industry and Development: Global Report, 1987*, Figures IV, V, VI.)

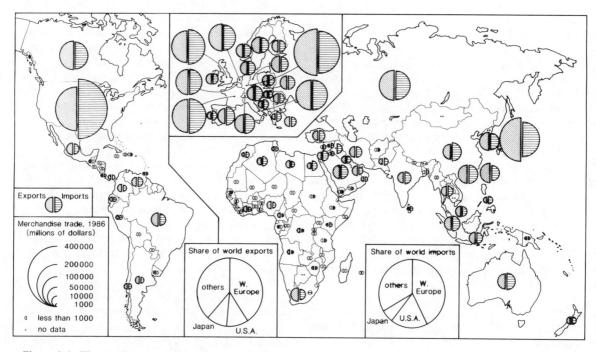

Figure 3.3 *Western Europe's trade position, 1985*
Source: World Bank (1988), Table 11.

Table 3.1 *Network of merchandise trade, Western Europe*

	Export destinations (%)												
	Developed countries					Developing countries					Eastern		
	North America	Japan	Western Europe	Aus, NZ, S. Africa	Total	Latin America	Asia	Middle East	Africa	Total	Trading Area	Unspecified	World
1963	9.0	1.0	63.9	3.9	77.7	4.6	4.1	3.1	5.6	17.6	4.4	0.6	100.0
1973	8.6	1.4	69.2	2.2	81.4	3.3	2.3	2.7	4.2	12.7	4.9	1.1	100.0
1986	10.3	1.5	68.9	1.4	82.0	2.2	3.3	3.9	3.8	13.2	4.4	0.6	100.0
	Import origins (%)												
	Developed countries					Developing countries					Eastern		
	North America	Japan	Western Europe	Aus, NZ, S. Africa	Total	Latin America	Asia	Middle East	Africa	Total	Trading Area	Unspecified	World
1963	14.0	1.0	59.0	2.4	76.3	5.7	3.1	3.8	6.5	19.1	4.6	–	100.0
1973	9.2	2.4	66.7	1.9	80.2	3.1	2.3	4.8	4.9	15.2	4.7	–	100.0
1986	7.1	4.2	71.3	1.0	83.6	2.4	3.3	2.4	3.4	11.6	5.0	–	100.0

Source: GATT (1987) *International Trade 1986–87*, Table A13.

external trading partner, but there has been a substantial change in the flows, particularly of exports. In 1963, almost 14 per cent of Western Europe's imports came from North America; in 1986, the figure was 7 per cent. Third, the biggest single increase in import shares into Western Europe originated from Japan. In 1963, Japanese imports into the region amounted to less than 1 per cent of the total; in 1986 this share had quadrupled, whereas exports to Japan had risen only slightly (from 1 per cent to 1.5 per cent). Fourth, at the more aggregative level, trading relationships with the developing countries have changed substantially in relative terms. In 1963, almost 20 per cent of Western Europe's imports came from developing countries; in 1986 the share was less than 12 per cent. Similarly, whereas almost 18 per cent of the region's exports went to developing countries in 1963, only 13 per cent did so in 1986.

Western Europe as a focus of international direct investment

The traditional view of international trade is that it takes place between independent business organisations at 'arm's length' across national boundaries, and that the kinds of trade figures discussed in the previous section are simply the aggregate totals of such individual transactions. In fact a large — and increasing — share of international trade is not carried out in this way. Rather, much of the world's trade is intra-firm exchange, conducted by transnational corporations (TNCs). For example, in 1980 more than 80 per cent of the UK's total trade was generated by TNCs (both UK- and foreign-owned) and almost one-third of total exports involved flows between 'related concerns' (i.e. intra-firm trade). Increasingly, too, TNCs have come to play a major role in the production of goods and services in virtually all countries, but especially in the developed economies. Indeed, it can be argued that the TNC is, today, the most important force creating global shifts in economic activity. Direct investment in overseas facilities is one major way (although not the only one) through which such international production occurs. For these reasons, therefore, any analysis of Western Europe's position in the global economy must include consideration of direct investment and the activities of TNCs.

Although the statistics are highly imperfect, there is no doubt that Western Europe is both a major destination for, and a major source of, foreign direct investment (FDI). This has been so since at least the turn of the present century, although here our concern is with the current position. Table 3.2 shows Western Europe's share of the world total of inward and outward FDI in 1975 and 1985. The reduction in her share of world inward FDI in this period is a direct

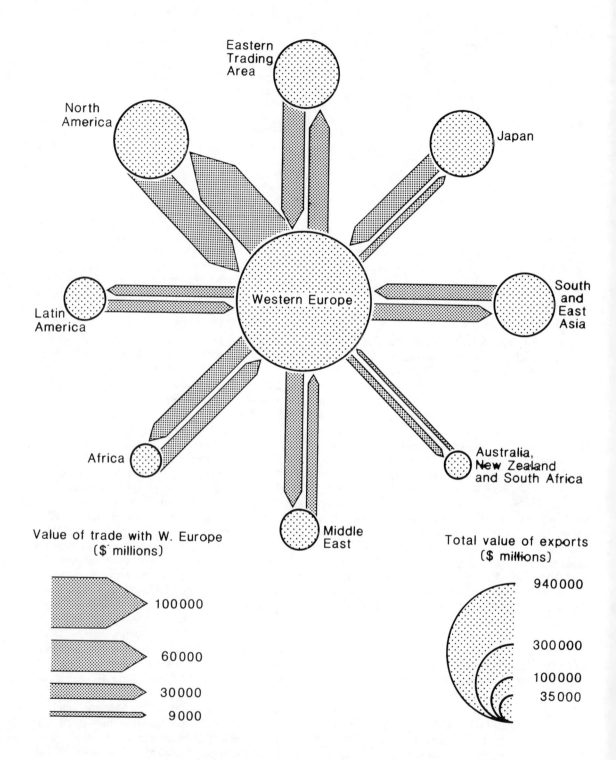

Figure 3.4 *The major elements in Western Europe's international trade network, 1986*
Source: GATT (1987) *International Trade, 1986–87*, Table A10.

Table 3.2 *Inward and outward stock of foreign direct investment by major region (percentage of world total)*

| | Inward | | Outward | |
	1975	1985	1975	1985
Western Europe	40.8	28.9	41.4*	43.7*
United States	11.2	29.0	44.0	35.1
Japan	0.6	1.0	5.7	11.7
	52.6	58.9	91.1	90.5

*Based on data for the United Kingdom, West Germany, Switzerland, the Netherlands, France, Italy and Sweden only.
Source: UNCTC (1988, Tables I.2, I.3).

Table 3.3 *Foreign direct investment ratios, 1975–83*

	1975	1983
Western Europe	1.20	1.56
United States	4.48	1.66
Japan	10.65	12.32

$$\text{FDI ratio} = \frac{\text{Outward stock}}{\text{Inward stock}}$$

Source: calculated from Dunning and Cantwell (1987, Table B1).

reflection of the spectacular change in the position of the United States as a host country. Throughout the post-war period, this country was pre-eminently an outward investor, accounting for the largest share of the world total. In recent years, however, there has been a massive inflow of direct investment into the United States as European and Japanese firms, in particular, have acquired companies and built new facilities there. Hence there are now extremely powerful reciprocal direct investment relationships between Western Europe and the United States, and this has transformed what was formerly an imbalanced situation. As Table 3.2 shows, by 1985 these two major regions had virtually the same share of world inward FDI.

In terms of outward direct investment in the world as a whole, Western Europe's share increased slightly between 1975 and 1985, while that of the United States declined somewhat. The most spectacular change involved Japan, whose share of world outward direct investment more than doubled. Note, however, that Japan's share of *inward* direct investment remained minuscule. If we combine these two measures of outward and inward investment into a single index, the FDI ratio, then we can capture something of the changing balance of international direct investment (Table 3.3). An FDI ratio of 1.0 implies an exact balance of outward and inward investment. In these terms, Western Europe's position remained relatively stable between 1975 and 1983. As a region it is highly significant as both a source of, and as a destination for, FDI. This is now also true of the United States, whose direct investment balance is now much closer to that of Western Europe. The major difference is with Japan, where outward direct investment in 1983 was more than 12 times as great as inward investment. Along with trade frictions, this huge imbalance in the Japanese direct investment account is causing major concern among businesses and policy-makers in the West.

There is, of course, an enormous amount of *cross investment* within Western Europe. At this intra-European scale there have been two very broad tendencies. On the one hand, there has been a substantial concentration of investment in the 'core' of the EC; on the other (in specific industries) there has been a substantial increase in intra-European investment in the 'peripheral' countries of Spain, Portugal and Greece. From the perspective of Western Europe as a whole, however, a particularly important issue is the changing involvement of non-European firms in Western Europe. Hence, it is necessary to look more closely at the two major sources of inward investment: the United States and Japan.

Although Servan-Schreiber's (1968) prediction, referred to in the introduction to this chapter, has not been fulfilled, the fact remains that US firms are an immensely important presence in Western Europe. Such firms have been building new factories and acquiring European firms since the last quarter of the nineteenth century. Some of these US firms are now so long-established that most people probably regard them as domestic, rather than foreign, companies. Although US FDI in general has not grown as rapidly during the 1970s and 1980s as it did up to the late 1960s, Western Europe has actually increased in significance as a destination for such investment. In 1956, around one-quarter of United States

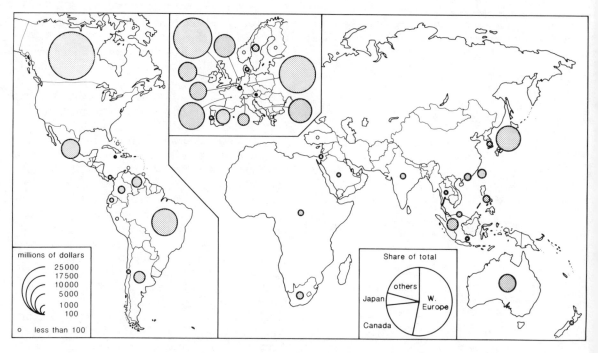

Figure 3.5 *US foreign direct investment in manufacturing, 1987*
Source: US Department of Commerce (1988) *Survey of Current Business* 68, 8, Table 13.

overseas manufacturing investment was located in Europe; by 1987 more than half was located there, 96 per cent of this investment (by book value) being in the EC.

Figure 3.5 demonstrates both Western Europe's position as a host region for US FDI and the differential significance of individual countries as hosts to US manufacturing investment in 1987. Historically, the United Kingdom has been the single most important European host country, not only as a location in itself but primarily as a 'point of entry' for US firms investing in Europe. Over time, as US firms became more experienced and established, and as they developed Europe-wide strategies and structures, this point-of-entry role became less important. But despite such changes the United Kingdom remains the major destination, just ahead of West Germany, and together the two countries accounted for more than half of all US manufacturing investment in Western Europe in 1987.

The nature of Japanese FDI in Europe differs on almost all of these counts. Until around 1970, FDI by Japanese companies was very limited for a whole variety of domestic reasons, both economic and political (Dicken, 1988). Since 1970, and

especially since the early 1980s, it has grown very rapidly indeed, greatly increasing Japan's share of the world total from less than 1 per cent in 1960 to 12 per cent in 1985. Since 1985, when there was a major revaluation of the yen, the rate of growth of Japanese overseas investment has escalated further, and by 1987 Europe accounted for 15.1 per cent of her total FDI, compared with Asia's 19.1 per cent and the United States' 36 per cent (Figure 3.6). Within Europe itself, the leading host country is the United Kingdom, which accounts for more than 30 per cent of the Western European total. In some respects, the United Kingdom is playing a similar point-of-entry role for Japanese investment in Western Europe to that played several decades ago for US investment.

Japanese and US firms account for the bulk of the current flow of FDI into Western Europe although, of course, firms from many other parts of the world are attracted to invest there. Traditionally, these have been firms from such countries as Canada, Australia, South Africa and some Latin American countries. However, a very recent, but potentially highly significant, development is the emergence of investments from firms

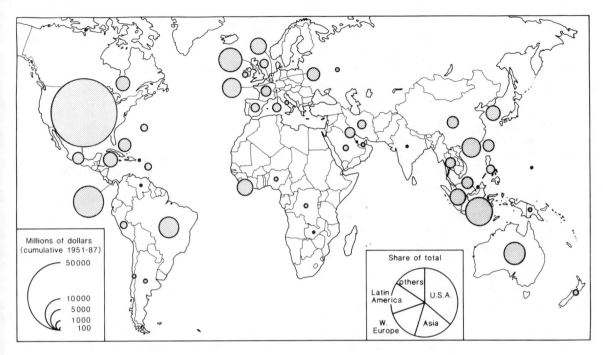

Figure 3.6 *Japanese foreign direct investment, 1987*
Source: MITI (1988), *Direct Overseas Investment from Japanese Companies.*

in the East Asian NICs. As yet such investments are relatively small in terms of numbers and scale, but they may well be harbingers of the shape of things to come. Most prominent have been some of the very large South Korean conglomerates such as Samsung, Daewoo and Goldstar, each of which has recently established new plants in Europe (notably in the United Kingdom and West Germany). But there have also been some direct investments by firms from Taiwan and Hong Kong. For reasons explored below, such investments are likely to increase in the next few years.

These aggregate patterns of FDI in Western Europe are the outcome of complex factors operating, and changing, over a very long period of time. Although we may be able to identify some general reasons for investment, we should always remember that investment requires specific decisions, influenced by the circumstances facing individual firms in particular industries and in particular economic and political conditions. In broad terms we can classify the international investment decisions of firms into two categories: market-oriented investments, and supply- or cost-oriented investments. Most FDI has been, and

remains, primarily market-oriented: firms locate a production facility in a particular overseas location to serve the market directly. In some cases the particular locational pull may be that of overcoming transport costs, but as transport technologies have developed this has become less of an influence. More significant, in economic terms, is the need to be close to markets and customers to provide a better and more direct service, to be sensitive to customer demands and to changes in customer tastes and preferences. In such ways, firms may be better able to exploit their specific advantages (such as superior technology or brand image). A widely accepted view among large corporations is that, to be able to compete in today's intensely competitive global environment, it is necessary to have a physical presence in each of the three major world markets — the United States, Japan and Western Europe.

Supply- or cost-oriented direct investments, other than in the natural-resource-based sectors, are more recent but they have become increasingly significant at the global scale since the 1960s. Most obviously, such investment has taken the form of firms in the industrialised countries moving some of their operations 'offshore' to

Third World countries to take advantage of labour-cost differentials. For Western Europe as a whole this has been a less significant form of FDI than market-oriented investment. But for individual countries and locations *within* Europe it may well be extremely significant. For example, a US firm may decide to invest in Europe to serve the Western European market in general, but the specific location chosen may well be influenced by geographical differences in the cost or availability of labour or of other necessary inputs. Clearly, some types of industry will be more sensitive to such circumstances than others. Equally clearly, the creation of the single Community market after 1992 may increase still further the degree of locational choice enjoyed by inward investors.

International investment decisions are more than simply economic. From the very beginnings of such activity, when both European and American firms began to locate some of their production facilities overseas, political considerations have played an extremely significant role. Indeed, historically, it was the existence of national tariff barriers, protecting specific domestic markets by imposing an import tax, which acted as a major stimulus for firms to switch from serving a particular market through international trade to serving that market through direct investment. They were therefore very important in initiating the pattern of US investment in Western Europe that exists today. However, a major feature of the international trading environment of the post-war years has been the progressive liberalisation of trade, especially the reduction of tariff barriers through the various GATT rounds. For the industrialised countries as a whole, the average tariff on manufactured goods fell from 40 per cent in 1947 to 7 per cent in 1973. In addition, the European Community has adopted a common external tariff with concessionary treatment for specific countries (these are, primarily, EFTA states and some developing countries covered by the Lomé Convention). Consequently, tariff barriers are of relatively little importance today in the European context. Like transport costs, they have become less significant in the process of investment-location decision-making. Instead, non-tariff barriers (NTBs) in the form of so-called 'voluntary export restraints' or 'orderly marketing agreements' have become the dominant consideration. Instead of imports being subject to a tariff they are increasingly subject to restriction in the form of a numerical quota. While the GATT continues to promote multilateral, non-discriminatory trade agreements, many individual countries are introducing bilateral trade agreements with other individual countries. 'Managed' trade is becoming increasingly apparent: international trade friction has grown very rapidly in certain industries (such as automobiles and electronics) and between certain countries (notably between the United States and European countries on the one hand, and Japan and the East and South-East Asian countries on the other). As in the case of tariff barriers, one potential effect of NTBs is to encourage exporters to jump over the barriers to set up production facilities in the protected markets. Hence, in Western Europe the recent proliferation of Japanese investments, and the growing number of South Korean investments, can be explained partly in these terms.

One major influence on international direct investment in Western Europe, therefore, is that of changing trade policies, even though such policies are not specifically directed towards inward investment. In terms of directly targeted policies, some current developments are worthy of brief attention here. In general, the developed market economies (with some exceptions) have adopted a more liberal attitude towards FDI, a reflection of the fact that they are the major source countries of such investment. Within Western Europe the United Kingdom, Ireland and West Germany have probably adopted the most open stance towards potential foreign investors and France the least open, although France's attitude has recently become more favourable. Within the range of possible policies which might be adopted by Western European countries towards investment, there are two sets of measures which are increasingly common but somewhat contradictory.

The first is the set of policies which can be bundled together under the heading of *performance requirements*. In the past, attempts to impose such requirements on foreign firms were predominantly made by developing countries; more recently they have become increasingly common in developed countries. Three types of performance requirement have become most apparent. The first is an insistence on a minimum level of exports from the host-country plant. Second is the requirement that a specified proportion of a plant's inputs should be sourced locally. This 'local content' stipulation has become especially common in assembly industries, such as automobiles and electronics, and is the focus of a great deal of controversy between national governments and private investors, and between national governments themselves, particularly within the EEC. Third, a requirement may be imposed relating to technology transfer, including persuading firms to set up research and development facilities in the host country.

The second set of policies towards FDI runs somewhat counter to the essentially restrictive nature of performance requirements. This is the *competitive bidding* which takes place between nations (and within nations) for internationally mobile investment. Again this is not a new development, but there is no doubt that it has intensified very markedly indeed. It is especially intensive within Western Europe, because here we have a huge contiguous geographical market, in which a direct presence is increasingly believed to be essential, but which offers many potential locations for production. Because this market can be served from a whole range of European locations, national governments (and regional and local agencies within individual countries) are engaged in fierce competition for investment projects. The intensity of such competition verges on the cut-throat, despite the efforts of the European Commission to limit the form and levels of incentive offered. In such circumstances, it is not surprising to find the major foreign TNCs assiduously seeking out the best package and often playing off one potential host country against another. Such intensified international competition for internationally mobile investment has resulted in both a general escalation of incentive levels and efforts to 'differentiate' the national offering through a whole battery of financial and non-financial incentives and through aggressive promotional activity. The problem is, of course, that the upward escalation of incentives may not actually increase a country's share of the investment market if all its competitors behave likewise.

In terms of the changing global competitive environment, one of the biggest political developments facing both national governments and individual business enterprises is, of course, the imminent creation of a single European market after 1992. The stimulus to create a single market was, undoubtedly, the fear that Western Europe was losing out in the global competitive stakes and that the continuing fragmentation of the Community was a major source of competitive weakness both for European firms and for member states. In some respects, whether or not immediate integration actually happens is less significant than the fact that the Single European Act set in motion a whole chain of developments, not least the urgent evaluation of competitive strategies by firms in all industries. As Calingaert (1988, pp. 37–8) observes in a very perceptive view from the outside, '1992 is a process not an event . . . The process will continue beyond 1992. Indeed, it is unlikely that a specific point will be identifiable at which the internal market becomes "complete"'.

The debate on 1992 has been, and continues to be, intense. An enormous amount of hyperbole has been generated and it seems more than likely that some observers will have to eat at least some of their words. Nevertheless, it is quite clear that the prospect of a unified market of some 320 million consumers, set within a single *external* Community customs boundary, has created enormous interest and unease amongst Europe's trading partners and among foreign firms seeking to sell in the Community. For firms not yet established in Western Europe, but anxious to maintain or achieve market access, the overriding concern is the possible emergence of a *fortress Europe* when the various national bilateral trade agreements are replaced by a single Community arrangement. At present, most attention in this respect focuses on those industries in which Japanese market penetration is very high or growing very rapidly. It seems inevitable that the level of Japanese FDI, although still relatively small, will grow and increase the pressures on both European and US companies in Western Europe.

Experiences in three key sectors

Competitive pressures on Western Europe, arising from trends in global production, trade and international investment, take on particular forms in particular economic sectors. Although there are some common features, there are also substantial differences from one sector to another. In this section, therefore, the Western European position in three key global sectors is explored. The focus of the discussion is, again, Western Europe as a whole and, especially, the activities of non-European companies in the European market. The following chapter looks at some of the details of corporate restructuring within Western Europe by European companies.

Automobiles

In some respects, Western Europe is in a very strong position within the world automobile industry. Currently it is the largest world market, with total sales of more than 12 million cars per year, and also the largest region of production. In 1987, some 12.8 million passenger cars were manufactured in Western Europe, compared with 7.1 million in the United States and 7.9 million in Japan. The trade position of Western Europe is also apparently quite evenly balanced. Most of the very high volume of trade is intra-regional (between European countries) and import penetration for the region as a whole is relatively low. The major element in imports is, of course, the flow from Japan, but at present this takes only a little over 11 per cent of the Western European market (although the figure is substantially more in EFTA countries). In addition, car imports are roughly balanced by an outflow, notably to North America, of high-value luxury vehicles.

None the less, the European automobile industry is presently operating with some 20–25 per cent overcapacity, and import penetration is a source of considerable concern (see below). Moreover, the region's position as a centre of automobile production must be set within its global context in functional terms and not just in terms of aggregate production and trade figures. The automobile industry is, by any criterion, increasingly a global industry, dominated by a relatively small number of very large, mostly transnational corporations. In 1987, ten companies produced almost 75 per cent, and 20 companies 90 per cent, of the world total. Table 3.4 summarises the major features of the world's leading automobile producers in 1987. Eight of the leading 20 producers are European, although the largest European company, Volkswagen (VW), is less than half the size of the world's largest producer, General Motors (GM).

As Watts notes in Chapter 4, the European manufacturers have generally adopted production strategies strongly orientated to their home coun-

tries. The major exceptions are VW, Renault and Fiat, which have some operations in Latin America (VW's production plant in the United States was closed in 1988). This tendency contrasts sharply with the strategies pursued in Western Europe by the two truly global players, Ford and GM. For both these companies, Western Europe is the most important focus of their operations outside the United States and both have devised pan-European production systems. Ford set out on this path by integrating its UK and West German operations in the late 1960s, and since then the company has systematically restructured its production geography within Western Europe. For the most part this has involved reallocating models and functions among existing plants in the UK, West Germany and Belgium, but it has also entailed the construction of a major new assembly plant in Spain, a new engine plant in South Wales and a transmission plant in western France. GM, after a later start than Ford, has created a very similar European network, but one in which the role of the UK is less important. Indeed, GM has progressively downgraded its Vauxhall subsidiary to little more than an assembly operation.

For virtually all the major automobile producers in Europe, particularly those selling to the mass market, the current obsession is competition from Japanese producers. Until recently, such competition was expressed entirely through import penetration. Throughout the 1970s, Japanese producers demonstrated the ability to serve world markets from a Japanese production base. Their price competitiveness was founded upon extremely large-scale and highly automated plants; massive technological investment in automobile production during the 1950s and 1960s, aided by government measures and a favourable labour cost structure; and a very high degree of integration with component suppliers. These production cost efficiencies more than offset any transport penalties or import tariffs. Jones (1983) estimated that in 1980 Japanese producers had a landed cost advantage of between 20 and 30 per cent over American and European producers in their own domestic markets. Six years later, a Ford executive suggested that 'Toyota . . . could ship a car from Japan to Europe and still sell it at up to £660 less than a domestically produced vehicle' (*Financial Times*, 12 June 1986).

Hence, if it had not been for the development of

Table 3.4 *The world's leading automobile producers, 1987*

World rank	Company	Home country	Output of private cars	Percentage produced in home country	Percentage produced overseas	Percentage produced in Western Europe
1	General Motors	US	5,605,301	64.3	35.7	25.0
2	Ford	US	4,000,256	45.8	54.2	40.0
3	Toyota	Japan	2,795,964	96.9	3.1	–
4	Volkswagen	WG	2,337,976	77.4	22.6	82.9
5	Nissan	Japan	2,016,869	89.4	10.6	–
6	Peugeot-Citroën	Fr	2,300,647	72.8	27.2	100.0
7	Renault	Fr	1,742,474	79.0	21.0	96.9
8	Fiat	It	1,674,570	89.9	10.1	89.9
9	Honda	Japan	1,361,645	75.0	25.0	–
10	Chrysler	US	1,186,170	93.1	6.9	–
11	Mazda	US	857,509	99.5	0.5	–
12	UAZ	USSR	724,740	100.0	–	–
13	Daimler Benz	WG	597,765	100.0	–	100.0
14	Mitsubishi	Japan	594,654	100.0	–	–
15	Hyundai	S. Korea	544,648	100.0	–	–
16	Rover	UK	471,504	100.0	–	100.0
17	BMW	WG	442,776	100.0	–	100.0
18	Volvo	Sweden	430,794	69.1	30.9	98.2
19	Fuji-Subaru	Japan	206,755	100.0	–	–
20	Isuzu	Japan	203,930	100.0	–	–

Source: MVMA (1989).

increasingly powerful political pressure, there is little doubt that Japanese automobile firms would have continued to operate primarily as domestic producers and to serve the major industrialised country markets through exports. However, the very success of Japanese producers in penetrating these markets has led to very considerable trade friction and to the erection of protectionist barriers in the form of voluntary export restraints. Faced with such restrictions, Japanese automobile producers have begun to establish production plants in Western Europe, although these have not yet had a major impact on production. So far, Europe lags a long way behind the United States in this respect; in the USA there is already huge Japanese automobile investment by Honda (the pioneer), Nissan, Toyota and Mazda, with further plants being built by these and other Japanese firms (including many Japanese components suppliers).

At present there is only one Japanese automobile manufacturer actually producing cars in Europe: the Nissan plant in North-East England, which began production in 1986 and is rapidly expanding its facilities to serve the European market. (The earlier Honda joint venture with BL entailed a much lower level of commitment.) However, by 1992, Nissan will be the United Kingdom's second-largest exporter of cars, while it is clear that other Japanese companies are increasingly keen to establish a production base in the single market. In 1989, Toyota announced its plan to build a major assembly works in Derbyshire and an engine plant in North Wales by the early 1990s. Honda also announced plans to add an assembly plant to its engine factory at Swindon in a complex deal with Rover. Clearly, the intensity of competition from Japanese automobile manufacturers, both through imports and direct investment in Western Europe, is increasing. It has been estimated that the Japanese share of the European car market will increase from 11 per cent to almost 20 per cent by the mid-1990s. There is also a growing likelihood of imports of cars produced in Japanese plants in the United States, as well as a rise in imports from South Korea as Hyundai becomes a major world manufacturer.

Not surprisingly, there has been loud concern voiced at these developments both by some individual European governments (especially France and Italy) and by existing automobile producers in Europe (American and European). As 1992 approaches, the EC has to decide how it will treat Japanese imports, because the existing bilateral voluntary restraint agreements will no longer apply. Within the EC the major arguments have been over the 'local content' of Japanese cars produced in Europe. Both France and Italy, for a time, refused to accept that Nissan's UK-built cars were sufficiently 'European' in content to qualify for free entry to their domestic markets. A figure of 80 per cent local content is being widely used. To achieve this it seems probable that the Japanese automobile producers will encourage some of their existing components suppliers to locate in Europe. Certainly, this is a pattern which has developed in the United States (Rubenstein, 1990).

Electronics

Arguably the electronics industry, particularly the production of semiconductors, integrated circuits and microprocessors, is *the* key industry of today's world. Microelectronics and one of its major derivations, information technology, are at the heart of the radical technological transformations that are occurring with pervasive effect throughout modern economies and societies (Chapter 6). One especially important user of microelectronic components is the consumer electronics industry, whose products now attract a substantial proportion of consumer expenditure. These two industries — microelectronic components and consumer electronics — are but two of a whole spectrum of products which together constitute the electronics industry as a whole. Unlike the automobile industry, the electronics industry is a product of the last 40 years. It came into being following the invention of the transistor in the late 1940s, and its overall growth has been incredibly rapid but very uneven, both in time and space. Temporally, it is an industry subject to very marked fluctuations in demand. The cycle of boom and slump is especially apparent in the microelectronics industry; here huge gluts in the supply of semiconductors push down prices to very low levels, but are invariably followed by equally huge shortages of supply, which force prices up again. Spatially, both of these sectors of the electronics industry are

very unevenly distributed at both the global and the subnational scales.

Western Europe's competitive position in this industry is substantially weaker than it is in automobiles. With some exceptions, the European electronics industry is being squeezed both by the more advanced and sophisticated technology of the United States and Japanese producers, and by the rapidly growing consumer electronics firms of Japan and East Asia. In 1987, Western Europe produced 20 per cent of total world output of electronic components, compared with 30 per cent in the United States and almost 50 per cent in Japan plus the four leading Asian NICs (Hong Kong, Singapore, South Korea and Taiwan). In consumer electronics, Western Europe also produced around 20 per cent of the world total, more than twice that of the United States but less than one-third of Japanese and East Asian production. Almost two-thirds of world consumer electronics production in 1987 was located in Japan and the four Asian NICs. In aggregate, Western Europe has a trade deficit in both microelectronics and consumer electronics.

These two electronics sectors constitute an even more fully developed global industry than automobiles. Indeed, the industry was the first to which the label 'global factory' was applied, a reflection of the complex international division of labour which developed from the 1960s onwards. US firms, in particular, established production facilities in low labour-cost locations in Asia but kept their research and development and more sophisticated operations in the United States and Europe. Table 3.5 lists the leading electronics firms in the world in 1988. If all types of electronics are included (computers, office equipment, telecommunications, defence, and so on, as well as microelectronics and consumer electronics) then the first division consists of three US firms, five from Japan and two from Europe. But in the two sectors which are the particular subject of this discussion the picture is rather different. In semiconductors, only one European firm, Philips, figures in the top ten (in tenth place). The leading semiconductor firms in the world by sales volume are Japanese although, if a rather different measure were used, IBM would be the clear leader (IBM does not appear on the list because all its semiconductor production is consumed in-house). The situation is actually more complex than the table suggests, because the semiconductor industry

Table 3.5 *The world's leading electronics companies, 1988*

Rank	All electronics	Rank	Semiconductors	Rank	Consumer electronics
1	IBM (United States)	1	NEC (Japan)	1	Matsushita (Japan)
2	Matsushita (Japan)	2	Toshiba (Japan)	2	Philips (Neths)
3	Philips (Neths)	3	Hitachi (Japan)	3	Thomson (Fr)
4	NEC (Japan)	4	Motorola (US)	4	Sony (Japan)
5	Siemens (WG)	5	Texas Instr. (US)	5	Hitachi (Japan)
6	Gen Motors (US)	6	Fujitsu (Japan)	6	Toshiba (Japan)
7	Hitachi (Japan)	7	Intel (US)	7	Sanyo (Japan)
8	Toshiba (Japan)	8	Mitsubishi (Japan)	8	JVC (Japan)
9	AT & T (US)	9	Matsushita (Japan)	9	Sharp (Japan)
10	Fujitsu (Japan)	10	Philips (Neths)	10	Mitsubishi (Japan)

Source: BEP Data Services (1989) *World Electronics File 1988/89*; press reports.

produces a wide range of different types of chip, each of which constitutes a separate market. In some market segments the United States is the clear world leader, in others the Japanese dominate. However, it is in consumer electronics that the Japanese are really dominant: eight of the ten leading companies are Japanese, none are American. Significantly, two European companies, Philips of the Netherlands and Thomson of France, are leading world players in this sector, largely because of acquisitions in recent years.

Until the 1960s, Western European microelectronics firms had survived well against what was then overwhelmingly US competition. But US manufacturers had already begun to locate branch plants in Europe for a variety of reasons: to avoid a 17 per cent import tariff; to develop links with defence-industry customers; and to tap supplies of a highly skilled technological and scientific labour force which was substantially cheaper than that in the United States (Morgan and Sayer, 1988). The relative competitive position of US firms, compared with European ones such as Siemens, Philips, GEC, Plessey and Thomson, changed dramatically in the early 1970s. The former surged ahead technologically with the development of the integrated circuit, and their strong basic position enabled them to respond to the first of the industry's major slumps by aggressively cutting prices and increasing market share at the expense of European producers. The latter, in response, tended to pull out of the standardised chip markets and retreat into more easily defended specialist niches (Sharp, 1989). European firms subsequently have found it very difficult to re-enter the larger markets. Meanwhile, the leading

US semiconductor producers developed pan-European strategies, operating wafer fabrication plants in some locations (notably Scotland and West Germany) and further assembly plants elsewhere.

In turn, however, the US producers began to be threatened at the global scale by growing Japanese competition in certain sectors of the semiconductor market, notably the very large-scale integration (VLSI) segment. As Sharp (1989) points out, the major Japanese electronics firms (such as NEC, Toshiba, Hitachi and Fujitsu) adopted the same aggressive marketing strategies as those used by US companies ten years earlier. The Japanese firms thus moved from being virtually 'nowhere' in the world semiconductor industry in the early 1970s to a dominant position in the late 1980s. In Western Europe, from the late 1970s onwards, increasing numbers of direct investments were made by Japanese semiconductor firms, for reasons very similar to those which earlier caused investment from the United States, but reinforced by trade frictions which now exist between Japan and Europe over levels of import penetration and local content. Most recently, as in the automobile industry, a further increase in Japanese inward investment is occurring in anticipation of 1992. The major problem facing the European producers is what appears to be a large and, in some cases, growing technological gap; this has prompted attempts at collaboration between leading European electronics companies, both privately and with government support.

In consumer electronics, the European (and global) position has been almost totally dominated by the Japanese. Through a combination of productive efficiency and a continuing wave of product innovations and modifications, virtually all branches of the consumer electronics industry have been captured by Japanese companies. The domestic situation in Western Europe has not deteriorated to the same extent as in the United States, but it is still problematical. The sequence of events is quite clear: early development of a consumer electronics industry in Japan encouraged by government support; a rapid increase in Japanese exports and a consequently high level of import penetration in European countries; a response by individual European governments in the form of 'voluntary export restraint' agreements to limit the inflow; direct investment by Japanese companies to get round these trade barriers. As a result, all the major Japanese consumer electronics firms now have production plants in Western Europe, with a particular concentration in the United Kingdom and West Germany. Indeed, the United Kingdom no longer has any domestically-owned colour television manufacturer since the Ferguson brand was sold to the French company Thomson in 1988. Philips and Thomson remain the only world-scale European players in this industry. But the Japanese position is now being challenged by other Asian producers, notably from South Korea and Taiwan. Imports into Europe from these sources have grown rapidly, and a few firms from both countries have established European production plants in recent years. Again, these tendencies seem likely to intensify as 1992 approaches.

Pharmaceuticals

Our discussion of automobiles and electronics — together with the experiences of some other industries — might suggest that Western Europe's position in the global economy is uniformly weak, at least compared with the United States and Japan. In fact that is not the case. Even in the two industries we have discussed above, there are clearly some major European players and the competitive performance of some European countries is strong. The third case study in this chapter illustrates this point more explicitly. Western Europe's position in the pharmaceuticals industry, another clearly global industry, is very strong, and the position of some companies and countries is very strong indeed. In contrast, Japanese competi-

tion at both the global and regional scales is not as marked, although it is undoubtedly growing.

To an even greater extent than either automobiles or electronics, the pharmaceuticals industry is extremely research-intensive. Competition is especially severe within a specific category of drug:

> The name of the game ... [is] ... to launch a successful new drug and to muscle in as quickly as possible on your competitors. This makes the pharmaceuticals industry rather like a game of roulette. R and D is essential, both to the development and launching of new drugs and to the process of imitation. The development of new drugs has been a relatively random process. It has meant screening many thousands of chemical compounds and it involves constant testing and evaluation; in the process many fall by the wayside because they prove inappropriate, ineffective or harmful. Those which survive (which may be as few as one in 10,000, and may take as long as ten years to develop) risk being pre-empted by another firm which has been working along similar lines and has succeeded in launching its product first. The outcome of this process of competition is that, like a continuing game of roulette, success can be transitory. The firm that strikes lucky and makes high profits on one drug is unlikely to strike lucky on each successive throw. It is not surprising, therefore, that there are considerable changes in the ranking of pharmaceuticals companies over time (Brech and Sharp, 1984, pp. 43–4).

The second important characteristic of the pharmaceuticals industry is that it exists within a very stringent and complex political and regulatory environment. Before a drug can be put on the market it has to be subjected to very thorough, expensive and time-consuming pre-clinical trials. The time taken for this process varies from one country to another but can be up to ten years. In addition, in most countries the major customer for drugs, in the sense of paying the bill, is either the state or a state-regulated private agency. Differentials in health-care policies between countries are therefore important to the activities of pharmaceuticals companies. These regulatory devices act as a kind of non-tariff barrier to trade in pharmaceuticals, creating quite distinctive national markets.

Western Europe as a whole is the second-largest consumer of pharmaceutical products, accounting for 28 per cent of the world market in the mid-1980s, compared with 34 per cent in the United States and 24 per cent in Japan. However, it is the largest centre of pharmaceuticals production, with

Table 3.6 *The world's leading pharmaceutical companies, 1982–87*

	1982			1985			1987	
Rank	Company	HQ	Rank	Company	HQ	Rank	Company	HQ
1	Bayer	WG	1	Merck	US	1	Merck	US
2	Merck	US	2	AMH	US	2	Hoechst	WG
3	AMH	US	3	Hoechst	WG	3	Glaxo	UK
4	Hoechst	WG	4	Ciba-Geigy	Switz	4	Ciba-Geigy	Swtz
5	Ciba-Geigy	Swtz	5	Bayer	WG	5	Bayer	WG
6	Pfizer	US	6	Pfizer	US	6	AMH	US
7	Eli Lilly	US	7	Warner Lambert	US	7	Takeda	Jap
8	Hoffman La Roche	Swtz	8	Abbott	US	8	Sandoz	Swtz
9	Sandoz	Swtz	9	Eli Lilly	US	9	Eli Lilly	US
10	Bristol Meyers	US	10	Bristol Meyers	US	10	Abbott	US

Source: SCRIP Pharmaceutical Company League Tables, various issues.

Table 3.7 *Variations in the importance of domestic sales to the 50 largest pharmaceutical companies, 1982*

	Percentage in home market	Range
Average, 36 companies	45.4	5–99
United States companies (10)	52.8	40–67
German companies (4)	15.3	9–20
French companies (3)	47.3	35–60
United Kingdom companies (5)	15.2	10–20
Dutch companies (1)	11.0	–
Swiss companies (3)	5.0	5.0
Swedish companies (1)	25.0	–
Italian companies (1)	31.0	–
Japanese companies (8)	93.4	86–99

Source: calculated from Commission of the European Communities (1985, Table 3.3).

44 per cent of the world total, compared with 30 per cent in the United States and 23 per cent in Japan. A most striking characteristic of the global structure of the pharmaceuticals industry is the trading position of Japan. In the early 1980s, this country accounted for only slightly more than 2 per cent of world pharmaceuticals exports, as opposed to Europe's 72 per cent and the United States' 17 per cent. Only around 2 per cent of Japanese pharmaceuticals production was exported, whereas for automobiles the figure was well over 50 per cent. In contrast, one-third of Europe's and 12 per cent of United States' pharmaceuticals production was exported. So far, at least, the Japanese pharmaceuticals market is relatively isolated, although there are clear signs of change.

The general production characteristics are reflected in the league table of the world's leading pharmaceuticals companies (Table 3.6). Three points are worthy of emphasis. First, Brech and Sharp's comment on the volatility of the ranking of pharmaceuticals companies is clearly evident. Second, European companies are very strongly represented in the lists, especially Swiss and West German concerns. Third, only one Japanese firm (Takeda) appears, and then only on the 1987 list. Table 3.7 confirms the strength of Western European manufacturers, in terms of orientation away from the home market, and also underlines the heavy reliance of Japanese companies on domestic sales. Although some of the very big Japanese producers, such as Takeda, Yamanouchi and Fujisawa are aiming to increase substantially their overseas operations, at present the competitive balance between Japan and Western Europe is very different from that in either automobiles or electronics.

The spatial structure of these successful Western European and American corporations has a number of common features. The strongly regulated, nationally based nature of the pharmaceuticals market has led the major companies to disperse their basic production processes very widely indeed. A national presence is virtually mandatory. Hence,

it is common for drugs to be made up into dosage forms in many places. The manufacture of active ingredients, however, is confined to a few locations. Development work, especially clinical testing, is done in a number of countries, but serious innovative research is usually carried out in the nation of origin (Commission of the European Communities, 1985, p. 19).

European pharmaceuticals companies, therefore, are active in most Western European countries as well as overseas, including both the United States and Japan. At the same time, US companies have a very extensive and long-established presence in Europe. In the EC, for example, US subsidiaries accounted for some 12 per cent of total pharmaceuticals production in the early 1980s. They also perform a substantial amount of R and D in Europe, primarily in the United Kingdom, West Germany and France.

These various corporate strategies by the major companies, both European and non-European, are being re-evaluated in the light of the imminent arrival of the single European market. The study by the Commission of the European Communities (1988, p. 125) of the possible effects of 1992 on the pharmaceuticals industry concludes that 'in the longer term, the effect of unifying the European market will be to make the strong stronger and the weak weaker. The steps so far will benefit all firms . . . those of the United Kingdom, the United States and Switzerland will benefit the most'. It might be added that it seems inevitable that the level of Japanese direct investment will increase in response to the 1992 initiative, possibly taking on a variety of forms, including acquisitions and joint ventures. If this should happen it will further intensify competition and increase the pressures on both Western European and American pharmaceuticals companies.

Conclusion

The aim of this chapter has been to set Western Europe as a whole in its global competitive context, particularly in comparison with its two major rivals, the United States and Japan which, together with Europe, form Ohmae's (1985) triad of world economic powers. In aggregate terms, Western Europe appears to be the weakest of the three although, as the three industry case studies demonstrate, the picture is far from uniform across all economic sectors.

A major theme throughout this chapter has been the strategies of transnational corporations.

Such organisations are not the only force shaping the global economy, or of regions within it, but they are probably the most important. A very large proportion of the economic activity of all Western European economies is generated, either directly or indirectly, by TNCs, both European and non-European. It is the strategies of TNCs in investing or disinvesting in particular locations, in switching and reswitching their activities from place to place, in channelling the flows of materials, products and technology within, and outside, their corporate networks which contribute most to spatial economic change. Western Europe plays a key role in the investment decisions of TNCs, and its competitiveness depends a good deal on them. On the one hand, Europe is a major source region of outward investment; on the other, it is a major destination for the investments of non-European firms. A European presence is regarded as essential to firms in globally competitive industries. Much depends, therefore, on the relative efficiency and competitiveness of European and non-European firms.

Although TNCs are the most important single shaper of today's global economy, their behaviour is influenced and modified to a substantial, but varying, degree by political forces. In particular, the ability of TNCs to pursue strategies of global integration is often constrained by the pressures to conform with national and local requirements. Historically, at the geographical scale considered in this chapter, trade policy in the form of tariff barriers has greatly influenced the international location of much investment by TNCs. Today, tariff barriers are less important and non-tariff barriers more so. The interaction between TNCs and nation-states is seen especially clearly in the current drive towards a single European market. The proponents of the single market argue that the best way for the EC to strengthen its global competitive position is through the removal of all internal barriers (physical, fiscal and legislative) which still exist some three decades after the signing of the Treaty of Rome. While many would agree with this, there is much disagreement as to whether the removal of such barriers will create a single community in a real sense. Certainly, the process of change set in motion by the Single European Act is generating enormous waves in the global economy as both governments and businesses re-evaluate their strategies towards

Europe. Internally there is likely to be substantial reorganisation and rationalisation of economic activity. This is one of the subjects for discussion in the next chapter.

References

Bairoch, P. (1982) International industrialisation levels from 1750 to 1980, *Journal of European Economic History*, 11, 269–333.

Braudel, F. (1986) *Civilization and Capitalism, 15th–18th Century, Volume 3: The Perspective of the World*, Collins, London.

Brech, M. and Sharp, M. (1984) *Inward Investment: policy options for the United Kingdom*, Routledge and Kegan Paul, London.

Calingaert, M. (1988) *The 1992 Challenge from Europe: development of the European Community's internal market*, National Planning Association, Washington, DC.

Cantwell, J. (1987) The reorganisation of European industries after integration, *Journal of Common Market Studies*, 26, 127–52.

Cecchini, P. (1988) *The European Challenge 1992: the benefits of a single market*, Wildwood House, Aldershot.

Commission of the European Communities (1985) *The Community's Pharmaceutical Industry*, CEC, Brussels.

Commission of the European Communities (1988) *The 'Cost of Non-Europe' in the Pharmaceutical Industry*, CEC, Brussels.

Dicken, P. (1986) *Global Shift: industrial change in a turbulent world*, Paul Chapman Publishing, London.

Dicken, P. (1987) Japanese penetration of the European automobile industry: the arrival of Nissan in the United Kingdom, *Tijdschrift voor Economische en Sociale Geografie*, 78, 59–72.

Dicken, P. (1988) The changing geography of Japanese foreign direct investment in manufacturing industry: a global perspective, *Environment and Planning, A*, 20, 633–53.

Dunning, J.H. (1988) *Explaining International Production*, Unwin Hyman, London.

Dunning, J.H. and Cantwell, J. (1987) *IRM Directory of Statistics of International Investment and Production*, Macmillan, London.

Dunning, J.H. and Robson, P. (eds) (1987) Multinational corporations and European integration, *Journal of Common Market Studies*, 26, special issue.

Jones, D.T. (1983) Motor cars: a maturing industry? In G. Shepherd, F. Duchène and C. Saunders (eds), *Europe's Industries: public and private strategies for change*, Frances Pinter, London.

Kennedy, P.M. (1988) *The Rise and Fall of the Great Powers*, Unwin Hyman, London.

McCalman, J. (1988) *The Electronics Industry in Britain: coping with change*, Routledge, London.

Morgan, K. and Sayer, A. (1988) *Microcircuits of Capital*, Polity Press, Cambridge.

MVMA (1989) *World Motor Vehicle Data*, Motor Vehicle Manufacturer's Association, Chicago.

Ohmae, K. (1985) *Triad Power: the coming shape of global competition*, The Free Press, New York.

Rubenstein, J. (1990) The impact of Japanese investment in the United States. In C.M. Law (ed.), *Motor of Change*, Routledge, London.

Servan-Schreiber, J.-J. (1968) *The American Challenge*, Hamish Hamilton, London.

Sharp, M. (1989) European technology: does 1992 matter? *Science Policy Research Unit, University of Sussex Papers in Science, Technology and Public Policy*, 20.

UNCTC (1988) *Transnational Corporations in World Development: trends and prospects*, United Nations, New York.

Wallerstein, I. (1979) *The Capitalist World Economy*, Cambridge University Press, Cambridge.

World Bank (1988) *World Development Report, 1988*, Oxford University Press, New York.

4

Manufacturing trends, corporate restructuring and spatial change

H.D. Watts

Introduction

In 1986 Western European manufacturing employment totalled 35 million workers, some 10 million more than the United States.[1] Yet the European economy in the mid-1980s was less mature than that of the United States. Eight per cent of EC workers were still employed in agriculture (compared with only 3 per cent in the USA), while only 59 per cent were in services (69 per cent in the USA). Equally, with a third of its employees in industry, Europe's dependence on this sector was greater than that of the United States (28 per cent).

In many ways the current situation in Western Europe reflects a steady movement towards the North American pattern. As recently as 1980, employment in industry represented 38 per cent of total European (EC10) employment. A fall in the relative importance of manufacturing employment need not be an indication of absolute decline, but the reality is that Western Europe has lost 5 million manufacturing jobs over the last decade. What is most interesting geographically is that these rates of manufacturing employment decline have varied from country to country. Although Europe is facing a number of external pressures arising from changes taking place in the global economy (see Chapter 3), the response to those pressures has varied from place to place, and from company to company. It is the variations in corporate responses, and the spatial outcomes arising from them, which form the principal concern of this chapter.

To understand corporate restructuring and spatial change in Western Europe it is important to have an overview of the manufacturing system and the changes it has experienced in recent years. This study therefore begins by describing briefly some of the major features of contemporary manufacturing industry. It then explores in detail corporate structures and the processes of corporate restructuring, with reference to four broad industrial groups: iron and steel, vehicles, aerospace, and food and drink. Reasons for the selection of these groups are discussed below.

The geography of manufacturing employment

Although Western Europe has lost several million manufacturing jobs over the last decade, such losses are not necessarily associated with loss of output. Indeed, the EC12 data for industrial production (excluding construction) show that output was 5 per cent higher in 1986 than in 1980, indicating labour-productivity gains, without which Western Europe's global competitiveness would have been very seriously undermined. Analysis of the manufacturing system in terms of changes in output would be an interesting exercise, but output data disaggregated by region, country and industry are not readily available. Consequently, in Europe, as elsewhere, descriptions of manufacturing activity must, of necessity, be based on employment data. However, reliance on these data does enable a direct link to be made between industrial change and its effect on employment, unemployment and regional inequalities (Chapter 11).

The international distribution of manufacturing employment in 1986 is shown in Table 4.1. The dominant centre of manufacturing employment

Table 4.1 *Manufacturing employment trends, by country, 1979 and 1986**

| | Paid employment in manufacturing | | |
	1979 (thousands)	1986 (thousands)	change (%)
Austria	912	842	− 8
Belgium	888	741	− 17
Denmark	568	515†	− 9
West Germany	8,370	7,716	− 8
Greece	485**	500	+ 3
France	5,291	4,589†	− 13
Ireland	228	187	− 18
Italy	4,715	4,038	− 14
Luxembourg	43	38	− 12
Netherlands	1,037	959	− 8
Norway	370	337†	− 9
Portugal	641	626†	− 2
Spain	2,737	2,163	− 21
Sweden	608	535†	− 12
Switzerland	678	679	0
UK	7,253	5,243	− 28
Total	34,824	29,708	− 15

*Because the definition of paid manufacturing employment varies between countries, the data provide only broad indications of the scale of the manufacturing workforce and must therefore be treated with caution.
**1981.
†1985.
Source: International Labour Office (1987) *Yearbook of Labour Statistics*, Geneva, United Nations.

was West Germany (with 7.7 million employees) while France, Italy and the UK shared second place, with 4–5 million employees each. Outside these leading four countries, Austria, Belgium, Denmark, Greece, the Netherlands, Portugal, Sweden and Switzerland all had between 0.5 and 1 million industrial employees. Spain's 2.1 million workers placed it in an intermediate position.

The relative importance of industry does, of course, vary from country to country. Above average dependence is characteristic of West Germany (with 41 per cent of its civilian employment in industry) and Switzerland and Austria (38 per cent each). The lowest dependence occurs in Denmark, the Netherlands and Norway, where industry accounts for only 27 per cent.

Between 1979 and 1986 Switzerland was the only economy in Western Europe to maintain its employment in manufacturing. The average decline was 14 per cent, with most countries lying

in a range from 8 per cent (the Netherlands, Austria and West Germany) to 18 per cent (Ireland). The exceptions to this general pattern were the United Kingdom, which experienced a dramatic fall of 28 per cent, and Portugal and Greece where employment fluctuated around a static level. It is, of course, tempting to interpret these contrasts as arising from the early industrialisation of the UK (which is now adapting to a new European, rather than a colonial, role) and from the late industrialisation of the strongly agricultural communities on the Western European periphery. However, this rather neat interpretation might be questioned by the fact that Spain lost 21 per cent of its paid employment in manufacturing.

It will be recalled that employment loss cannot be equated necessarily with loss of output, and it may be that countries showing the highest losses are also showing the most rapid productivity gains and are thus adapting most rapidly to changes in Europe's position in the world economy. Nevertheless, it is clear from these contrasting employment records that the impact of global competition and recession has varied markedly throughout Europe. One possible explanation for these contrasts is that employment loss may be related closely to differences between national employment structures. Brief consideration of these structures is therefore necessary.

Four activities account for 42 per cent of the EC workforce in manufacturing: electrical engineering, mechanical engineering, the manufacture of metal articles, and food and drink production. Chemicals, motor vehicles, footwear and clothing, and paper and printing account for a further 27 per cent. An overview of this industrial structure reveals that the percentage of 'the industrial base devoted to goods with a high technological content remains relatively narrow' (Emerson, 1988, p. 25). More detailed spatial analysis shows that the food, drink and tobacco industry is the largest source of employment in many countries (Belgium, Denmark, France, Ireland and the Netherlands). But each of the other countries has its own main source of manufacturing employment, such as electrical engineering in West Germany, mechanical engineering in the UK, and the manufacture of metal articles in Italy.

Each of these activities had its own distinctive employment trend between 1980 and 1985 (Table 4.2). Only one sector, office machinery, increased

Table 4.2 *Manufacturing employment trends in Western Europe (EC9), 1980 and 1985, ranked by change*

	1980 (thousands)	1985 (thousands)	change (%)
Office machinery	253	283	+ 12
Mechanical engineering	2,751	2,646	− 4
Food, drink, tobacco	2,707	2,570	− 5
Paper and printing	1,855	1,740	− 6
Electrical engineering	2,864	2,684	− 6
Instrument engineering	497	449	− 10
Rubber and plastics	1,061	955	− 10
Chemical industry	1,875	1,683	− 10
All manufacturing	*27,850*	*24,373*	*− 12*
Other transport equipment	1,052	900	− 14
Metal articles	2,844	2,429	− 15
Timber and wood articles	1,414	1,193	− 16
Motor vehicles and parts	1,968	1,646	− 16
Non-metallic mineral products	1,257	1,006	− 20
Footwear, clothing and leather	1,967	1,558	− 21
Textiles	1,634	1,253	− 23
Other manufacturing	420	317	− 25
Production of metals	1,431	1,008	− 30

Source: Eurostat (1983) *Basic statistics of the Community*, Luxembourg; Eurostat (1988) *Basic statistics of the Community*, Luxembourg.

its employment, but a further four industries (mechanical engineering, food and drink, paper products and printing, and electrical engineering) lost less than 10 per cent of their employment. Marked employment decline was associated with textiles (23 per cent), footwear, clothing and leather (21 per cent) and non-metallic mineral products (20 per cent). Most marked of all was the contraction of the metals workforce (30 per cent). Despite these contrasting trends, and despite the initial variations in industrial mix, shift-share analysis suggests that the effects of industrial structure on employment decline between 1980 and 1985 were small. In France, for example, the industrial structure increased job losses by no more than 20,000. In West Germany the mix slowed down the contraction, but only by 50,000 jobs. These shifts arising from industrial mix represent only a minor part of total employment decline in each country over this period; thus explanations

for the different employment trends in each country must be extended to include other factors.

Interdependence in the manufacturing system

Although the responses of individual countries may be measured in isolation from one another, this ignores the high level of interdependence within the manufacturing system. This interdependence has three important forms. First, the trading links between the individual countries; second, the powers of the EC; and third, the influence of international multi-plant manufacturing organisations.

The countries and sectors in Western Europe are tied into a complex internal trading system. Detailed data on manufactured goods are not readily available and trade data available for analysis relate to all imports and exports. Fifty-eight per cent of the exports are sent to other members of the EC. Naturally, these overall figures disguise considerable international variations. Ireland and Belgium/Luxembourg, for example, purchase 73 per cent and 70 per cent of their imports from other EC countries, whereas the UK and Spain import only about half from Community sources. Export links to other EC countries are highest (over 70 per cent) in the Netherlands, Ireland and Belgium/Luxembourg and lowest (50 per cent or less) in the UK and Denmark. Some of the important trading links between countries are illustrated in Figure 4.1, which depicts the flows from and to the five main manufacturing countries of Western Europe. An increasing proportion of this trade comprises intra-firm flows between the sites of the multinational firms discussed below (Greenaway, 1987, p. 161).

The existence of the EC has encouraged a higher degree of integration among member countries, but barriers have remained in place throughout the 1980s. In the mid-1980s the cost of border formalities alone equalled 2 per cent of the value of total intra-Community trade (Emerson, 1988, p. 49). Further, most countries have continued to maintain closed markets for public procurement, and non-tariff barriers have posed problems (see p. 46). Attempts to remove the former have often been blocked on grounds of national security (see the discussion of aerospace below),

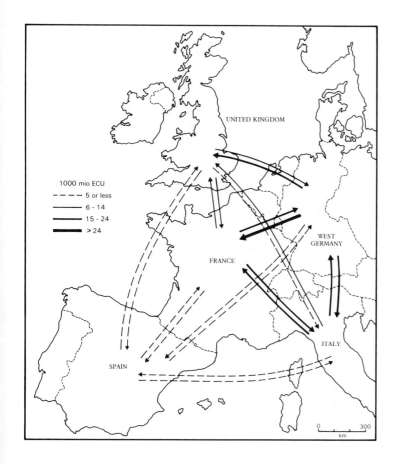

Figure 4.1 *Trade flows between selected Western European countries, 1986*
Source: Eurostat (1988) *Basic Statistics of the Community,* Table 6.10

while non-tariff-barrier reform was blocked partly because common standards would make it easier for North American and Japanese firms to penetrate the Western European market. In addition, common standards have been opposed in some quarters because they may give advantage to European countries whose technical specifications form the basis of the European standards. While the EC has been allowed to intervene in certain sectors, the general pattern is that 'member states preferred to pursue their own policies and economic markets in the growth sectors while allowing the Community a say in the crisis sectors' (Hall, 1986, pp. 11–12). For the most part, the corporate restructuring of the 1980s has taken place within a fragmented market. Restructuring within a market constrained, for example, by EC-imposed quotas is unusual.

While European-wide planning of capacity changes under the guidance of the EC is infre-

quent, large multinational organisations have undertaken both intra-corporate and inter-corporate restructuring of their production facilities. The term 'corporate restructuring' was once confined mainly to financial restructuring involving changes in the ownership of firms and plants (Massey and Meegan, 1979, p. 160), but it is used here in the wider sense to cover all forms of corporate reorganisation. It is this corporate restructuring which is the main concern of this chapter but, before moving to a detailed examination of corporate responses to external pressures, a brief sketch of the corporate structure of Western European industry is provided.

Corporate structure

The importance of individual countries as head-quarters of major firms varies as one moves down

59

Table 4.3 *The ten largest European-owned firms in the European manufacturing system, 1987**

Firm	Base	Products	Turnover (£m)	Employees (thousands)
1 Daimler Benz	Germany	Automobiles, electronics	38,401	327
2 Volkswagen	Germany	Automobiles	31,094	260
3 Fiat	Italy	Automobiles	29,460	271
4 Unilever	UK/Netherlands	Processed foods	29,396	294
5 Siemens	Germany	Electrical equipment	29,270	359
6 Philips	Netherlands	Electrical equipment	26,678	337
7 Nestlé	Switzerland	Processed foods	24,106	163
8 BASF	Germany	Chemicals	22,900	134
9 Bayer	Germany	Chemicals	21,138	164
10 Hoechst	Germany	Chemicals	21,032	168

*Excludes oil companies and diversified holding companies.
Source: Financial Times, 21 November 1988.

the corporate hierarchy. When the ten largest Western European-owned firms are considered, West Germany appears dominant (Table 4.3). Six of the ten are based in that country, compared with one each in Italy, Switzerland and the Netherlands plus one joint UK–Netherlands operation. Consideration of the 100 largest European-owned firms, however, reveals a different picture. Geroski and Jacquemin (1984, p. 359) found that, of the top 100, 39 were based in the UK, 27 in West Germany and 20 in France. Overall, the top 100 firms of *all* nationalities accounted for 26 per cent of EC output in 1982, compared with 24 per cent in 1975 (Emerson, 1988, p. 138). This increasing concentration ratio is itself an indication of the extent of inter-corporate restructuring. Among the 1,000 largest industrial firms in the EC there has been a steady increase in acquisition activity in the 1980s, from 117 acquisitions in 1982–3 to 226 in 1985–6. High takeover rates were associated with France and West Germany and with the food and chemicals industries, these two industries being particularly notable for large-scale acquisitions. In terms of increasing European integration, whereas the number of national acquisitions was static between 1984–5 and 1985–6, the number of intra-Community but extranational takeovers rose by 18 per cent (European Commission, 1986a, pp. 217–18).

Of course, among the largest firms there is an increasingly significant number of overseas investors, notably from North America and

Japan. The reasons for their interest in Western Europe was discussed in the last chapter and they are considered here mainly in terms of the way they have restructured their European activities over the last decade.

The range of factors involved in the European-wide corporate planning strategies of major European companies can be exemplified by ICI's expansion of production capacity, particularly with respect to its Wilhelmshaven polymer plant opened in 1980. Although a need for lower energy costs was cited as being of importance in the decision to expand in mainland Europe (electricity costs were 50 per cent lower in West Germany than in the UK[2]) this investment was also a deliberate attempt to move closer to the successful mainland car market and away from the dying market in the UK. The Wilhelmshaven location was selected only after consideration of other European sites, notably at Dunkerque. It was reported at the time that what tipped the balance in favour of the West German site was that the politicians at both the city and regional levels were the most helpful ICI had met (Hudson, 1983, pp. 105–22; Clarke, 1985, pp. 133–5). Significantly, therefore, in this major location decision both economic and behavioural variables played a role.

While this example is concerned with the construction of a new plant, Creasey (1987, p. 1) was right to observe that in many cases 'growth is now coming from re-organisation within firms and from a more abrupt closing of uncompetitive plant'. The remainder of this chapter focuses on this corporate restructuring, with its far-reaching implications for manufacturing employment patterns throughout Western Europe. Before proceeding, a word of caution is necessary. It is dangerous to see restructuring as arising solely from the outcome of these organisations' decisions, for similar patterns might have emerged under different organisational forms. Yet major corporations are very visible features of the European scene and there is little doubt that the information and production systems which many of them possess accelerate significantly the process of change.

Corporate restructuring

There are few published studies of the restructuring of industrial systems in Western Europe (for an exception see Pinder and Husain, 1987b, on oil refining) and this discussion is therefore based on firms operating in four carefully selected sectors whose restructuring is considered with particular emphasis on European spatial outcomes. Three of the sectors were selected because each one is at a different level of maturity. The steel industry is typically associated with the second Kondratieff long-wave, vehicles with the third, and aerospace with the fourth (Watts, 1987, p. 36). The fourth sector, food and drink, is included because it is an industry which allows observation of the early stages of restructuring in the European market.

Although this chapter adopts a European perspective, it will be evident from the case studies that many of the European-owned firms operate mainly within their home country. This is particularly true of the steel and aerospace industries. Multiple-country operations are more common in the automobiles sector, but such operations are much more marked in the US-owned firms than those originating in Europe. This distinction is also apparent in the food sector. Consequently, much of the discussion must focus on spatial change at the national scale. Nevertheless, as has been noted above, European-owned firms are now taking a greater interest in developing European-wide production systems.

The steel industry

Table 4.4 outlines the recent international production pattern of this declining industry. In 1986, West Germany was the leading steel producer in Western Europe, with production exceeding that of France and the UK combined. Italy was the second largest producer (but with output equal to only 61 per cent of West German production), while the output of the Benelux countries was roughly equivalent to that of France. Economies of scale in this industry are so marked that even countries the size of West Germany can support only a few major steel producers. Ownership of the industry in each country is therefore highly concentrated: the major steel producers are Finsider (Italy), British Steel (UK), Thyssen (West Germany), Arbed (Luxembourg), Sacilor (France) and Usinor (France). In 1982, early in the current restructuring phase, all these firms produced over 8 million tonnes of steel.

The industry has long been associated with state ownership, and the recent crisis has encouraged further state intervention in Cockerill-Sambre in Belgium and Sacilor and Usinor in France (Parris, *et al.*, 1987, p. 40). In the 1980s, therefore, more than half the EC's crude steel production came from state-owned firms (Tsoukalis and Strauss, 1985). Recently, this trend has weakened with the privatisation of British Steel, yet this firm is one of only three large corporations in private ownership, Thyssen and Arbed being the other two. State participation has accentuated the national production systems of most steel-producing firms, and has had many other influences on their activities. One of the most fundamental of these has been the provision of contributions to operating costs and loan charges ('contributions' is now a less emotive term than 'subsidies'). One UK study (House of Commons, 1983, p. 110) suggested that state contributions amounted to £11 per tonne in the UK, £15 per tonne in Germany, and £18 per tonne in France. Clearly, such aid has had an important effect on cost differences between Western European countries.

The main pressures on firms in the 1980s arose from a fall in demand associated with recession, loss of export markets and increased import penetration of both steel and products utilising steel (such as the automotive industry). The problems were accentuated by the fact that steel is a homogeneous product, so that the outputs of

Table 4.4 *Production of crude steel, 1979 and 1986**

	1979 (kilo-tonnes)	1986 (kilo-tonnes)	Change (kilo-tonnes)	(%)
West Germany	46,040	37,134	− 8906	− 19
Italy	25,250	22,883	− 2367	− 9
France	23,360	17,670	− 5690	− 24
UK	21,472	14,769	− 6703	− 31
Spain	12,254	11,882	− 372	− 3
Belgium	13,442	9,713	− 3729	− 28
Netherlands	5,805	5,283	− 522	− 9
Austria	4,917	4,660[†]	− 257	− 5
Sweden	4,731	4,813[†]	+ 82	+ 2
Luxembourg	4,950	3,705	− 1245	− 25
	162,221	132,512	− 29,709	− 18

*Countries producing over 3 million tonnes in 1986.
[†]1985.
Source: Eurostat (1988) *Basic Statistics of the Community*, Luxembourg; Eurostat (1981) *Basic Statistics of the Community*, Luxembourg.

different plants and countries are often inter-changeable. One significant indication of the declining nature of the industry is the lack of interest in the sector by multinationals based outside Europe but, beyond this, the production figures are stark. Between 1979 and 1986 only Sweden achieved a modest production increase, and output contraction ranged from 3 per cent in Spain to 31 per cent in the UK. Overall, the production decline among the countries covered by Table 4.4 was 29.7 million tonnes (18 per cent).

Restructuring strategies associated with this contraction have been dominated by the large public sector and private firms. One response that might have been anticipated of them is diversifica-tion into new products. Significantly, by the 1980s Thyssen, Europe's largest private sector steel-maker, had reduced its steel-production activities to only 24 per cent of its turnover. But state ownership generally blocked the adoption of this diversification strategy, and a more popular route for state-owned firms was to adjust their product ranges towards higher-quality steels.[3] In France, for example, both Sacilor and Usinor pursued this strategy. Despite the possibilities of diversification and upgrading, however, these solutions were overshadowed as pressures forced all firms to slim down or close some of their steel-making operations. A particularly interesting

feature of this contraction process is that both a planning framework, and increased emphasis on the use of market forces, have been utilised to stimulate progress.

Planning in the EC commenced with a system of voluntary quotas, introduced in the Davignon plan of 1977 (Price, 1981, p. 88). This voluntary approach quickly failed through the reluctance of firms to cut output without the co-ordinated agreement of competitors, a failure which quickly led the European Commission to declare a 'state of manifest crisis' in the industry. Linked with this declaration was the introduction of compulsory quotas in 1980. This new quota system allowed firms to pursue their restructuring and rationalisa-tion efforts within an orderly commercial climate (Emerson, 1988, p. 82), the aim being to cut capacity by 27 million tonnes between 1980 and 1986 (European Commission, 1984, p. 133). Capacity would then be used, it was anticipated, at the 70 per cent level compared with only 50 per cent in 1982–3.

Simultaneously, greater responsiveness to market forces was encouraged by close monitoring of state aids to the firms. The aim of this, and indeed of the quota system, was 'the restoration in the medium term of normal market conditions in which undertakings are profitable, i.e. they cover from the proceeds of production, without state aid, the cost of the factors of production, including a normal level of depreciation and a reasonable rate of financing costs' (European Commission, 1984, p. 116). Broadly, state aid was to be acceptable only if the firm had a specific restructuring plan, if there was to be no increase in capacity (unless there was a growth market), if aid was to be progressively reduced, and if it did not distort competition. Overall, the volume and intensity of government assistance had to be proportional to the restructuring effort.

Within this framework firms such as Thyssen reduced capacity from 16 to 11 million tonnes, and Krupp from 5.4 to 4.4 million tonnes (Hudson and Sadler, 1987, pp. 12–16). In Italy Finsider virtually shut down the Bagnoli plant in Naples, while in France Usinor planned to cut 7,200 jobs in Longwy and 5,600 in Denain between 1979 and 1982. Such closures inevitably, and quickly, led to protest, notably in the steel-making districts of north-east France where in July 1982 the head-quarters of Usinor's special steels division were burnt down (Hudson and Sadler, 1983, p. 416).

During the mid-1980s the quota system was gradually reduced. From the first quarter of 1986 certain steel products were excluded from it. Then, in June 1988, despite the lack of an agreement on further cuts to reduce overcapacity, the quota system was abolished. However, aid for redundancies continued to be offered and negotiations for further cuts through EUROFER (the European integrated steel producers' organisation) took place. By mid-1987 they had offered 15 million tonnes of cuts, but not in the hot-rolled area where state-aid is still most marked and over-capacity is greatest (Creasey, 1988, p. 75). Clearly, despite attempts at co-ordinated restructuring throughout the EC, such co-ordination was only partially successful.

The automobile industry

The automobile industry, like the steel industry, operates with relatively homogeneous products throughout Western Europe. There are detailed differences between countries, but these differences can be met by minor modification of the basic vehicle. Like the steel industry, the automobile industry is seen as a priority sector by European states and state involvement in the industry is high. None the less, private ownership remains more important than in the case of steel, while the presence in the automobile industry of major multinational corporations causes an even sharper contrast between the two industries.

The industry is perhaps the most highly concentrated in Western Europe: ten major producers account for 99 per cent of automobile production. These producers can be divided into three groups: the US global corporations (Ford and General Motors), the state corporations (Renault, Volkswagen) and the private firms (Peugeot, Fiat, Daimler-Benz, British Aerospace[4] and BMW). Volkswagen is included in the state corporations since the West German Federal government holds 20 per cent of the voting stock and the state of Lower Saxony a further 20 per cent (Marfels, 1983).

With the exception of the US corporations, most firms have mainly national production systems in their European operations, although Peugeot operates plants in Spain (Simca) and the UK (Peugeot-Talbot). Volkswagen has a number of plants in Spain. The persistence of national production systems among European firms reflects the fragmented European political system and makes it difficult for them to achieve economies of scale in automobile production. This is clearly a handicap in an increasingly competitive world. Recent estimates suggest that doubling the current output of a typical plant making car bodies would reduce car body costs by 20 per cent (Ludvigsen Associates, 1988). Fiat is a firm which places great emphasis on the economies of scale that can be achieved in motor-vehicle production. Based mainly at the massive Turin plant, the firm is 'against scattering production plants too far apart' (Cornelius, 1988, p. 1), a philosophy perhaps seen in its sale of its Barcelona factory to the Spanish government. In contrast to the European firms, Ford and General Motors operate integrated European production systems. These were established mainly in the 1960s and the 1970s and include components and assembly plants in different parts of Western Europe.

In national production terms, the sector is dominated by West Germany, which produced over 4 million passenger cars in 1985, compared with just under 3 million in France. The UK, Belgium and Spain each produced or assembled around 1 million vehicles, twice the output of the next largest producer, Sweden.

Corporate restructuring within fairly static output levels has resulted in some marked geographical shifts in the automobile industry (Figure 4.2). Between 1978 and 1985 the most striking proportional increase in output (58 per cent) was in Sweden, while within the European Community output fell in Italy, the UK and, most particularly, France (22 per cent). The 'technologically strong' West German industry was the only long-established branch of the sector to expand in the Community (Cantwell, 1987, p. 143), but growth also occurred in Spain, the latter's success resulting from a shift to take advantage of lower labour costs. For example, Volkswagen claims its new assembly plant at Matorell (near Barcelona) reflects the fact that labour costs in the West German motor industry are 86 per cent above those of Spain. In addition to this major production shift, a number of other changes have taken place in response to external pressures.

Firstly, diversification has been adopted as a defensive strategy by a small number of firms. The outstanding example is Daimler-Benz, which

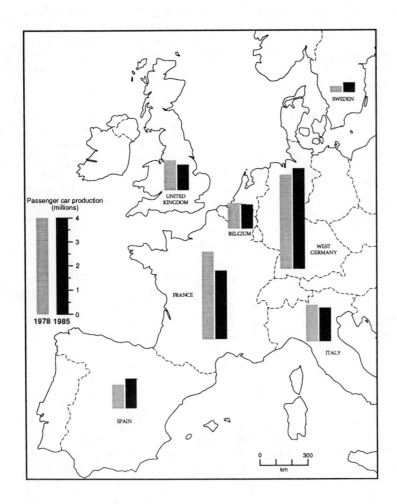

Figure 4.2 *Passenger car production and assembly, 1978–85*
Source: Eurostat (1980 and 1988) *Basic Statistics of the Communities*, Luxembourg. Data relate to countries producing or assembling over 200,000 passenger cars a year.

in the mid-1980s bought MTU (an engine producer), Dornier (aircraft) and AEG (electronics). It was these acquisitions which gave the firm its leading European role noted earlier in this chapter. This trend, however, has not been strong and has in some respects been countered by the acquisition of car manufacturers by firms operating in other industries. Thus, in the UK, the Rover Group was acquired by British Aerospace.

Secondly, within the industry, there have been a number of changes in the ownership of production facilities. To a degree, this reflects interest on the part of major producers in the highly profitable activities of small specialist companies. In the UK, for example, Aston Martin and Lotus were purchased by Ford and General Motors, respectively. However, the takeover movement is also a

consequence of the advantages of acquiring large-scale producers with the potential for achieving economies of scale through the concentration of production activities on fewer sites. Ford has made attempts to buy both Alfa Romeo and the Rover Group, but the most substantial acquisition in recent years has been Peugeot's purchase of Chrysler's European activities.[5] While this acquisition led to the closure of Chrysler's Linwood assembly plant in Scotland it also led to major investment in the former Chrysler plant at Coventry. Such adjustments following ownership change are not unusual, and in the late 1980s Peugeot was still reported to be 'digesting' Chrysler's European operations (Cornelius, 1988).

Thirdly, automation plus (in some countries) reduced demand has seen major job losses associated with restructuring. Fiat has adopted robot-assisted production methods more extensively than any other Italian producer (van Tulder and Junne, 1988, p. 29), and this high level of automation has been a leading factor in the decline of the workforce from 169,000 in 1979 to 99,000 in 1986 (Brookes, 1987, p. 26). From 1984 to 1986 employment in motor vehicles and motor-vehicle parts fell in France from 429,000 to 362,000. Clearly, labour-shedding on these scales, and in restricted time periods, has substantial implications for the sub-regional economies that are affected (see Chapter 11).

These declining employment figures relating to vehicles *and* parts hide the significance of a fourth strategy. This is the externalisation of production activities and increased links between motor-vehicle assemblers and their suppliers. Increasingly, Fiat is working towards a system of producing in-house only suspension systems and technically important components. Externalisation of supplies enables some of the risks associated with market fluctuations to be passed on to suppliers and, because the suppliers are less constrained by corporate labour policies, the inputs can be supplied at lower cost to the dominant firm. Linked to this change is the development of just-in-time systems which enable car-assemblers to reduce significantly their costs through the reduction of stocks of components. Just-in-time, and the externalisation of many key activities, create a need for stronger links between assembler and supplier because of the necessity to monitor component quality and supply flows. The requirement for closer contacts has, in turn, led to a fall in the number of component suppliers. Mouhot (1987) describes the way in which the reduction in the number of suppliers has strengthened the relationships between the Peugeot assembly plant at Sochaux and its suppliers in north-east France. Similarly, Renault's 1,415 suppliers of the early 1980s had been reduced to 900 by 1986 (Emerson, 1988, p. 74). Also, this firm's component stocks were cut from 11 days' supply in 1984 to just under three days' in the late 1980s (Baker, 1988, p. 63). At Peugeot, too, 2,000 suppliers were reduced to 1,229 between 1981 and 1986, and to only 950 by the end of 1988.

These four strategies were in some cases supplemented by state intervention, especially in the case of the Rover Group in the UK and Renault in France. The state sold off the Rover Group's successful high-value car-making operation (Jaguar) and announced the closure of plants at Llanelli and Oxford before selling the slimmed-down volume car activities to British Aerospace. Renault made major losses in 1984–5, in part related to its attempts to penetrate the North American market. Through its stake in American Motors it sold 125,000 Alliances in North America in 1985; the following year it sold only 73,000. Early in 1987 it announced it was selling its stake in the North American firm. Renault still has a huge debt and depends on major capital injections from the French government (Brookes, 1987, pp. 25–6). These contributions were challenged by the EC, as was the British government's decision to write off many of the losses incurred by the Rover Group in earlier years.

Despite these varied strategies and adjustments, appraisal of their success must be cautious. It was estimated that in late 1986 there was 20 per cent overcapacity in the motor-vehicle industry (European Commission, 1986b, p. 18). Although some of this slack may be taken up by exports, the possibility that any improvement in exports might be offset by an increased Japanese challenge cannot be discounted. Indeed, this challenge has taken a new dimension through the introduction of European-made Japanese cars at the Nissan plant in north-east England.[6] Against this background, medium-term cutbacks in assembly capacity now seem likely, and a recent study suggests that the number of 'platforms' (the basic chassis on which different models can be built) may well fall from 30 to 21 (Ludvigsen Associates, 1988).

The aerospace industry

The aerospace industry includes firms involved in the manufacture of airframes, engines, aero-equipment, missiles and space vehicles. To many, the aerospace sector is a key sector in the Western European economy. Like steel, it is still very much a European-controlled industry with national firms owning national production facilities. This is not because of a lack of external interest, but because of a determination to keep outsiders away from a sector with strong defence links. Hartley (1986, p. 245) describes the industry as one characterised by fragmented and protected

Table 4.5 *Aerospace employment in selected countries, 1985 (thousands)*

	1985
UK	204
France	127
West Germany	77
Italy	43
Spain	13
Netherlands	10
Belgium	5

Source: EC Information Memo P-68, 1 June 1988.

markets, wasteful duplication and short production runs. Unlike the automobile industry, in which multinationals have at least helped to lessen the problems caused by production systems focused on individual countries, the national firms in the aerospace industries have accentuated them.

In employment terms, the workforce has fluctuated between 400,000 and 500,000 since 1972. This compares with a US employment figure of 1.3 million. The major Western European employer is the UK, followed by France (Table 4.5). These two countries between them account for 70 per cent of the labour force, while West Germany accounts for a further 15 per cent (European Commission, 1988).

The leading firms in the industry are Aerospatiale (France) and British Aerospace (UK). Although state ownership is a feature of only some firms, intervention in the industry is high. This arises partly from the fact the state is a major customer (particularly for military equipment) and partly because of the high profile of the sector at the leading edge of new technology. The high-risk projects undertaken by these firms also encourage state backing for some activities. Export activities are, of course, constrained by the fact that military sales can be made only to politically acceptable regimes.

The peculiarly fragmented structure of the aerospace industry arises because political considerations have outweighed the cost savings which arise from economies of scale. Such economies are most marked in research and development (R & D) and in production runs. Some idea of the differences in scale between

European and North American operations is given by the fact that the average production run for a European-made combat aircraft in the 1970s and 1980s was 685, compared with a US average of 2,172. Similarly, the average for a European airliner was 142, against a US average of 655 (Hartley, 1986, p. 256). Clearly, there is considerable potential for corporate restructuring to achieve economies of scale within Western Europe.

The dominant response to the competitive pressures has been to maintain national interest by the development of joint projects. Estimates based on current joint programmes suggest that a nation can save around 35 per cent of its R & D costs, and some 5 per cent of unit production costs, through such programmes. Many aircraft producers now subcontract work on key components to firms based in other European countries. For example, the Fokker 100 jet is in many ways more British than Dutch. UK firms make the undercarriage (Dowty), the engines (Rolls-Royce) and the wings (Shorts). Even this level of international co-operation is modest compared with that developed for the European Airbus programme.

The first Airbus flew in 1972, while the latest version (the A320) left the assembly plant in 1987. The aircraft had a market share of 34 per cent of jet airline orders in 1987, and in this essentially global market it ranks second to the market leader Boeing (47 per cent) and ahead of McDonnell Douglas (17 per cent). France's Aerospatiale and West Germany's Messerschmitt-Bolkow-Blohm (MBB) each hold a 37.9 per cent stake in the Airbus, British Aerospace 20 per cent and the Spanish partner (CASA) 4.2 per cent. The work is spread between the partners in relation to their share of the overall project. British Aerospace produces the wings, MBB the fuselage and CASA the tailplane. The resultant intra-European flows are shown in Figure 4.3, the parts being transported by air in specially converted transport aircraft for final assembly in Toulouse.

There is little doubt that national aerospace corporations can do much to overcome scale handicaps by joint international ventures. The Airbus has significantly reduced Boeing's share of jet airline orders, an achievement probably beyond the scope of any single European company (Santini, 1987; *The Economist*, 1988). However, several observations on these co-operative programmes are appropriate. One is that the

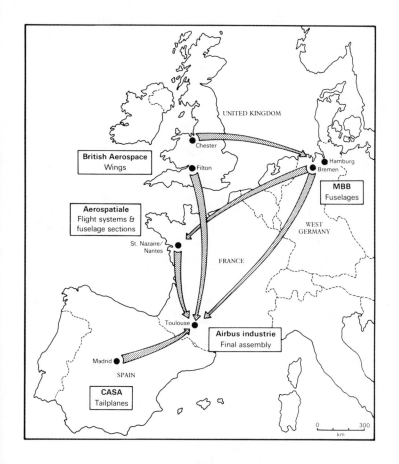

Figure 4.3 *The European Airbus production system*
Source: The Economist, (1988, p. 10).

success of the Airbus project has been achieved with significant financial support from various national governments, and this is not unusual. Also, while international ventures are given considerable publicity, in reality competition between European firms is still strong. For example, although British Aerospace, MBB, Aeritalia and CASA have set up a consortium to build a European fighter aircraft for NATO, in their non-European markets they will be challenged by the French-designed Rafale aircraft. Lastly, although joint projects provide an alternative to major Europe-wide corporate restructuring blocked by national considerations, they present problems which might be less obvious if integrated European operations were established. Joint ventures require considerable organisation, they need to ensure equality of treatment for each of the partners, and the mix of languages and managerial styles is a recipe for poor co-ordination. It was observed in the case of Concorde (the joint Anglo-French supersonic airliner project) that the inherent difficulties of the multiplicity of contracts, the number of interested parties and the participation of two sovereign states meant that the project 'acquired a life of its own and was out of control' (House of Commons, 1981, p. xx). Even more telling evidence comes from a detailed examination of six joint European ventures undertaken by Hartley (1986). This indicated that, compared with US manufacturers, European civil and military programmes suffered a production delay of about a year. A year is a long time in the aerospace industry, and is certainly sufficient to increase the risk of losing major contracts to more efficient competitors.

Food and beverages

Until quite recently, the food sector has avoided extensive Europe-wide restructuring. Competition

67

from imported processed foods has been low and the heterogeneity of tastes within Europe, often accentuated by national differences in food-related laws, has also prevented the extensive development of Europe-wide corporate strategies. This influence is more evident in traditional foods, such as pastas and beers, than in new food lines such as breakfast cereals and soft drinks.

Given the predominance of bulky, low-value goods, it is perhaps not surprising that the industry is spread across Western Europe approximately in proportion to each country's population. In general, there are around 20,000 food employees per million persons, the two marked exceptions being Italy with only 11,000 and Denmark with 32,000. The extent to which production is concentrated in a limited number of firms is difficult to measure, as detailed concentration ratios are not available for the entire industry. Certainly concentration is still low in some activities (such as brewing in West Germany), but in others it is high. Again in West Germany, the top three firms take 58 per cent of the cake market, while in France three firms control 90 per cent of mineral water production.

The most recent study of the corporate structure of the EC food industry (Group MAC, 1988a; 1988b), indicates that the five largest European firms (in terms of EEC food sales) are Unilever (based in the UK and the Netherlands), Nestlé (Switzerland), Allied Lyons (UK), BSN (France) and Suchard (Switzerland). Indeed, Nestlé is the world's largest food group with food sales some 50 per cent larger than its nearest rival, Unilever. Despite the presence of these giants, European firms based on the mainland in general have only a small presence in world food sales. Only Nestlé, Unilever, BSN, Suchard and Heineken figure in the top 30 in terms of world sales. The remaining firms are either of US or UK origin.

Until recently, there have been great differences in the European strategies of the European and US firms. US firms account for 19 (28 per cent) of the 67 major food firms in Western Europe. Their main geographical feature is that they tend to have wide coverage of the European market, albeit for a limited range of products. These firms have specialised in products in which they have the largest market share (on which they can earn the best return) and they have achieved production economies of scale through the geographical diver-

sification of their markets across Europe.

Indigenous firms tend to have a more limited geographical spread. Among the 45 major food firms with an EC head office, only 9 per cent have a presence in four or more West European countries. In other words, the major European groups are very largely oriented towards their national markets alone. This pattern has arisen partly because of the traditional views of many of the food firms, reinforced by the technological characteristics of the industry. Using traditional technologies it has been calculated that Amsterdam beer would have to be sold at a 16 per cent cost disadvantage in Hamburg; London beer in Hamburg would bear a 35 per cent cost penalty. Even with respect to products for which transport disadvantages may be less marked, spatially distinct markets can still be identified at the regional scale. In the French chocolate industry, for example, Suchard-Lindt is strong in the east, Côte d'Or in the north, Poulain in the west, and Nestlé in the south (Group MAC, 1988a, p. 318). In contrast to the US firms (such as Kellogg) which have promoted European brands, EC firms have tended to diversify into new products *within* their own country rather than diversifying *across* countries in a limited number of foodstuffs.

Today, these patterns are changing as the leading food groups respond to external pressures, mainly from North American and, increasingly, Australian food groups. The latter are entering the mainland European market through UK links. The European firms which have recognised this challenge have responded to it primarily by a strategy of acquisition to increase the geographical spread of their European operations. In so doing they are, of course, moving towards the model established by the US food firms operating in the European environment.

BSN provides an example of the recent expansion of a major European firm. In 1988 it was the second largest pasta producer in Europe, the third largest biscuit maker in the world and the world's largest mineral water producer. In 1986–7 alone it acquired Sunnen-Bassermann, a German manufacturer of pasta and soups, plus a minority or majority interest in five Italian pasta-makers; in General Biscuits in France; in Ferrelle, an Italian mineral water producer; and in Aquas Fort Vella, a Spanish mineral water company. This latter holding supplemented its control of Evian water in France. BSN's frenetic activity, working towards a pan-European operation, continued through 1988 with the acquisition of the UK activities of Imperial Foods.

The other large food firms have followed similar paths. Nestlé acquired Buitoni, the Italian pasta group, in 1988. Not only did this strengthen Nestlé's activities in Italy through control of Perugina (chocolate and confectionery), Vismara (meat products) and Olio Sassa (olive oil), it also gave control of Darigel in France which makes frozen foods. Nestlé's acquisition activities continued with the purchase of Rowntree, giving the Swiss firm major production sites in the UK as well as access to Rowntree's activities in France. Rowntree had acquired three French firms in the 1970s (Chocolat Meurier, Chocolat Ibled, and Lanvin) and operated production plants at Noisiel (near Paris) and Dijon. On a more modest scale, Cadbury Schweppes, the other major UK producer, moved into France in late 1987 by acquiring Chocolat Poulain and its two plants in and near Blois.

This recent corporate restructuring has yet to be reflected in the spatial reorganisation of production facilities, although such a development is an almost inevitable repercussion of the restructuring process. Indeed the spatial reorganisation may be more dramatic than in the other industries considered in this chapter because the food and beverage industry is not fettered by national strategic considerations or national prestige. Modified production patterns are likely to reflect new investment and technical change as the dominant firms concentrate increased production capacity on a limited number of sites. Recent estimates from Cantalou's plant at Perpignan suggest that new chocolate-bar plants need a minimum capacity of 200 tonnes per day to reach optimal scale. Some idea of the scale of future changes can be gained from the fact that, whereas 95 per cent of Italian beer comes from breweries with an annual capacity of over 50 million litres, in West Germany only 55 per cent is produced by such plants.

Towards the unified European market

The firms operating in each of the four sectors considered above have restructured, or are restructuring, the spatial pattern of their production activities. In each case, although there are distinct spatial outcomes, the constraints within which the restructuring decisions are made vary quite markedly. It has been shown there are different market trends, different levels of non-European ownership, differing degrees of state intervention and different levels of influence by supranational organisations such as European multinationals and the EC. All these factors are themselves intertwined and interact with one another and with questions of national pride and prestige. Indeed, in three of the industries, national considerations have been shown to be a powerful hindrance to corporate restructuring. What has been demonstrated is that the complexity of these forces makes it essential to focus upon corporate activities within a particular industrial sector in order to understand the changes which are taking place. Viewing manufacturing as a whole does not provide significantly deep insights.

Examining manufacturing at the sectoral level demonstrates vividly that in no case is restructuring entirely satisfactory and that, despite the corporate restructuring of European manufacturing sites over the last decade, 'the process of the reorganization of European industry to European integration is far from complete' (Dunning and Robson, 1987, p. 123). Further challenges therefore remain. The most dramatic of these is likely to be the introduction of the unified European market from 1992. When the internal economic barriers are removed, firms will have access to a home market of 320 million consumers, although it is likely that some of the non-tariff barriers arising from national policies and languages will remain for some time. The fundamental aim of the larger market is to contribute to European development by stimulating new entrants to the markets and by encouraging increased competition among existing firms. Further, it is hoped that European industry as a whole will benefit from substantial gains in economies of scale (Cecchini, 1988). In these ways, costs will be controlled and competitiveness improved. What must always be remembered, however, is that the new market does not simply offer new opportunities for European firms; it may also be served and exploited by Japanese, Australian and North American multinationals. This clearly underlines the importance of European progress towards more efficient production systems.

Adjustments stimulated by both global pressures

and increased market size in Europe will, like the adjustments described in this chapter, have distinct geographical consequences. As the industries and firms adjust to the new pressures, so those adjustments will create new problems for the localities in which the changes are implemented. Not only may there be pressures on land and labour supplies in areas where new investments take place, but the specific towns and regions in which redundancies and plant closures occur will be faced with the problems of adjusting to higher unemployment. They will also have to seek new uses for the redundant spaces vacated by the restructuring organisations (Pinder and Husain, 1988). In these ways corporate restructuring and the spatial changes arising from it will continue to impinge on communities and societies throughout Western Europe.

Further reading

Material used in this chapter can, in many cases, be updated from *Basic Statistics of the Community*. Issued in July each year, this series reports primarily on the EC, but also provides some data for Austria, Norway, Sweden and Switzerland. More detailed information can be obtained from the main Eurostat statistical series and official publications of the EC. All major cities in the UK have an official collection of EC documents in one of the educational institutions. Unfortunately there is no one book which takes the geography of Western European industry as its central theme, and contemporary patterns have to be pieced together from more general studies. A useful review of post-war changes up to the early 1980s is provided in Williams (1987), while Clout *et al.* (1989) includes a comparison of the spatial restructuring strategies of three major twentieth-century industries. A special issue of the *Journal of Common Market Studies* (December 1987) on multinational corporations and European integration provides some important insights into contemporary changes. A useful, though increasingly dated, review of changes in a number of different sectors is provided by Shepherd *et al.* (1983). This has been supplemented by the more recent material in de Jong (1988).

Notes

1. Western Europe is defined here as the EC plus Austria, Norway, Sweden and Switzerland. Comparison with the United States is only approximate because of differences in the way in which industry is defined.
2. The decision was taken before West German electricity prices reflected the introduction of the 'coal pfennig' to support the country's coal industry. See Chapter 2.
3. The strategy of combating industrial crisis by upgrading the product range is not confined to the steel industry. It frequently requires heavy investment and has been applied particularly widely in European oil refining in the 1980s (Pinder and Husain, 1987a).
4. British Aerospace entered the industry by acquiring the Rover Group (see below).
5. Chrysler's financial problems in the US market resulted in a decision to pull out of its European operations.
6. A second example, again involving Japanese investment in the UK, was Toyota's decision to locate a car plant at Derby. This plan was announced in April 1989. Similarly, in June 1989, Honda announced plans to expand its engine plant at Swindon into a full car plant. The UK is thus becoming the focus of Japanese car production in the Community.

References

Baker, S. (1988) Renault renaissance, *The Observer*, 17 July, p. 63.

Brookes, S. (1987) Car Wars: Europe's industry gears up, *Europe*, 269, 24–6.

Cantwell, J. (1987) The reorganisation of European industries after integration: selected evidence on the role of multinational enterprise activities, *Journal of Common Market Studies*, 26, 127–51.

Cecchini, P. (1988) *The European Challenge 1992 — the benefits of a single market*, Gower, Aldershot.

Clarke, I.M. (1985) *The Spatial Organization of Multinational Corporations*, Croom Helm, London.

Clout, H., Blacksell, M., King, R. and Pinder, D. (1989) *Western Europe: geographical perspectives*, 2nd edn, Longman, London.

Cornelius, A. (1988) Battle of Europe moves into top gear, *The Guardian*, 5 July, p. 11.

Creasey, P. (1988) *Structural Adjustment in Europe*, Parker, London.

de Jong, H.W. (ed.) (1988) *The Structure of European Industry*, Kluwer, Dordrecht.

Dicken, P. (1986) *Global Shift: industrial change in a*

turbulent world, Harper and Row, London.

Dunning, J.H. and Robson, P. (1987) Multinational corporate integration and regional economic integration, *Journal of Common Market Studies*, 26, 103–25.

The Economist (1988) Civil Aerospace Survey, 3 September, pp. 9–10.

Emerson, M. (1988) The economics of 1992, *European Economy*, 35, 1–218.

European Commission (1981) *Eleventh Report on Competition Policy*, Office for Official Publications of the European Communities, Luxembourg.

European Commission (1984) *Fourteenth Report on Competition Policy*, Office for Official Publications of the European Communities, Luxembourg.

European Commission (1986a) *Sixteenth Report on Competition Policy*, Office for Official Publications of the European Communities, Luxembourg.

European Commission (1986b) *Report on the EC Automobile Industry*, European Parliament Report, DOC 171/86.

European Commission (1988) *The EC Aircraft Industry*, EC Information Memo, P-68.

Geroski, P.A. and Jacquemin, A. (1984) Large firms in the European corporate economy and industrial policy in the 1980s. In A. Jacquemin (ed.), *European Industry, Public Policy and Corporate Strategy*, Clarendon Press, Oxford, pp. 343–68.

Greenaway, D. (1987) Intra-industry trade, intra-firm trade and European integration: evidence, gains and policy aspects, *Journal of Common Market Studies*, 26, 154–72.

Group MAC (1988a) *Research on the 'Cost of Non-Europe' — basic findings — Volume 12, Part A — the 'cost of non-Europe' in the foodstuffs industry*, Office for Official Publications of the European Communities, Luxembourg.

Group MAC (1988b) *Research on the 'Cost of Non-Europe' — basic findings — Volume 12, Part B — the 'cost of non-Europe' in the foodstuffs industry*, Office for Official Publications of the European Communities, Luxembourg.

Hall, G. (ed.) (1986) *European Industrial Policy*, Croom Helm, London.

Hartley, K. (1986) Defence industry and technology: problems and possibilities for European collaboration. In G. Hall (ed.), *European Industrial Policy*, Croom Helm, London, pp. 245–60.

House of Commons (1981) *Concorde*, HC 265, Session 1980-1, HMSO, London.

House of Commons (1983) *The British Steel Corporation's Prospects*, HC 212, Session 1982-3, HMSO, London.

Hudson, R. (1983) Capital accumulation and chemicals production in Western Europe in the postwar period, *Environment and Planning A*, 15, 105–22.

Hudson, R. and Sadler, D. (1983) Region, class and the politics of steel closures in the European Community, *Environment and Planning D*, 1, 405–28.

Hudson, R. and Sadler, D. (1987) *The Uncertain Future of Special Steels*, Sheffield City Council, Sheffield.

Ludvigsen Associates (1988) *Research on the 'Cost of Non-Europe' — basic findings — Volume 11 — the EC92 automobile sector*, Office for Official Publications of the European Communities, Luxembourg.

Marfels, E. (1983) *Concentration, Competition and Competitiveness in the Automobile and the Automotive Component Industries of the European Community*, Office for Official Publications of the European Communities, Luxembourg.

Massey, D. and Meegan, R.A. (1979) The geography of industrial reorganisation, *Progress in Planning*, 10, 155–237.

Mouhot, P. (1987) L'évalution des rapports entre le groupe automobile PSA et ses fournisseurs franc-comtois, *Revue Géographie de l'Est*, 27, 171–84.

Parris, H., Pestian, P. and Saynor, P. (1987) *Public Enterprise in Europe*, Croom Helm, London.

Philip, A.B. (1986) Europe's industrial policies: an overview. In G. Hall (ed.), *European Industrial Policy: an overview*, Croom Helm, London, pp. 1–20.

Pinder, D.A. and Husain, M.S. (1987a) Innovation, adaptation and survival in the West European oil refining industry. In K. Chapman and G. Humphrys (eds), *Technical Change and Industrial Policy*, Blackwell, Oxford, pp. 100–20.

Pinder, D.A. and Husain, M.S. (1987b) Oil industry restructuring in the Netherlands and its European context, *Geography*, 72, 300–8.

Pinder, D.A. and Husain, M.S. (1988) Deindustrialisation and forgotten fallow: lessons from Western European oil refining. In B.S. Hoyle, D.A. Pinder and M.S. Husain, (eds), *Revitalising the Waterfront: international dimensions of dockland redevelopment*, Belhaven, London, pp. 232–46.

Price, V.C. (1981) *Industrial Policies in the European Community*, Macmillan, London.

Santini, J.L. (1987) Strife over the Airbus, *Europe*, July–August, 8–9.

Shepherd, G., Duchène, T. and Saunders, C. (1983) *Europe's Industries: public and private strategies for change*, Pinter, London.

Tulder, R. van and Junne, G. (1988) *European Multinationals in Core Technologies*, John Wiley, Chichester.

Tsoukalis, L. and Strauss R. (1985) Crisis and adjustment in European steel: beyond laissez-faire, *Journal of Common Market Studies*, 23, 207–28.

Watts, H.D. (1987) *Industrial Geography*, Longman, London.

Williams, A. (1987) *The West European Economy*, Hutchinson Education, London.

5

Small firms: phoenix from the ashes?

Colin M. Mason and Richard T. Harrison

Introduction

For much of the post-war period, small firms attracted much less attention from social scientists and policy-makers than large firms. This was based on the widespread belief that small firms, due to inferior organisation, poor management and backward technologies, would play an increasingly more residual role in advanced industrial economies as the economic development process intensified integration and concentration processes (Sengenberger and Loveman, 1988). However, since the late 1970s the picture has changed. The impact of recession on employment prospects in the large-firm sector, largely as a result of factors discussed in Chapters 1–4, combined with new claims that small firms created the majority of new jobs in the United States (Birch, 1979), have supported the conclusion that a new dynamic of small-firm growth may provide a solution to the unemployment problem. In the policy domain, active encouragement and financial support for the small-firm sector became the norm (Burns and Dewhurst, 1986). In the academic domain, small firms and the new entrepreneurship were identified with characteristics such as competitiveness, innovativeness and flexibility, which gave them an inherent advantage over large organisations. As a result research on the sector expanded rapidly (Curran, 1986).

In 1971 the survival of the small firm as an organisational form was at issue (Bolton, 1971), yet small firms are now placed in the vanguard of the restructuring of advanced industrial economies at the national, regional and local levels (Giaoutzi *et al.*, 1988). In this chapter we examine the nature and significance of the industrial restructuring in favour of small firms in Western Europe. Following a summary of spatial variations and temporal trends in the small-firm sector, we examine some of the main explanations of its recent revival. In the remainder of the chapter we discuss the nature of the shift of employment towards smaller enterprises and its impact on urban and regional development. Our approach throughout has been to examine small-firm development and policies at the national level; we do not consider policy developments at the European Community level, which are discussed elsewhere (see Chapter 11; Pinder, 1986).

One final introductory point must be noted. International comparisons of the small-firm sector are made problematic by substantial international variations in both the coverage of official data and the basis of their compilation in terms of minimum size cut-off points, time-periods and classification intervals. These variations particularly affect the availability of data on the service sector which contains the majority of small firms (Ganguly, 1985, pp. 3–21). None the less, valid conclusions may be drawn provided the data are treated with caution.

Recent trends in the small-firm sector

The most comprehensive information on the significance of small firms in Western European economies is provided by Storey and Johnson (1987a; 1987b). If small firms are defined as those with less than 100 employees, their importance is demonstrated by the fact that — in five of the six EC countries for which information on the whole

Table 5.1 *Enterprise size in the private sector economy (percentage of population employed)*

		Enterprise size (number of employees)			
		< 20	20–99	100–499	500 +
France	(1986)	29.7	25.4	[44.9]	
Netherlands	(1980)	26.6*	30.9**	[42.5]	
Belgium	(1983)	25.0	20.9	21.5	32.6
Spain	(1986)	24.3*	34.3**	20.0	21.3
Greece	(1978)	51.7*	17.0†	[31.3]	
Portugal	(1985)	[57.6]		25.5	16.9

Notes: * 1–9
    ** 10–99
    † 10–49

Source: Storey and Johnson (1987a, Table 1.1).

Table 5.2 *Enterprise size in the manufacturing sector (percentage of population employed)*

		Enterprise size (number of employees)			
		< 20	20–99	100–499	500 +
United Kingdom	(1983)	[22.0]		14.4	63.6
Italy	(1981)	22.9*	36.0**	21.3	19.8
West Germany	(1983)	–	16.0	24.8	59.2
France	(1980)	18.8	25.3	28.8	27.1
Netherlands	(1980)	10.7*	27.1**	[62.2]	
Belgium	(1983)	12.1	20.7	25.8	41.3
Luxembourg	(1980)	7.7	11.5	25.8	55.0
Spain	(1978)	20.2	23.2	21.8	34.8
Portugal	(1985)	[43.8]		33.7	22.5
Ireland	(1980)	9.5	28.6	30.6	20.4
Denmark	(1982)	10.1‡	29.7	34.6	25.6
Greece	(1978)	39.3*		[60.7]	

Notes: * 1–9
    ** 10–99
    † 20 +
    ‡ 6–19

Source: Storey and Johnson (1987a, Table 1.2).

Table 5.3 *Enterprise size in the service sector (percentage of population employed)*

		Enterprise size (number of employees)			
		< 20	20–99	100–499	500 +
France	(1986)	41.8	30.3	18.8	9.1
Netherlands	(1980)	35.9*	29.1†	[35.0]	
Belgium	(1983)	33.8	21.9	20.2	24.2
Spain	(1986)	[59.7]		18.0	22.3
Greece	(1978)	63.7*		[36.3]	
Portugal	(1985)	[78.8]		14.8	6.3

Notes: * 1–9
    † 10–99
Source: Storey and Johnson (1987a, Table 1.3).

economy is available — they employ the majority of the working population (Table 5.1). This overview, however, disguises a contrast in the importance of small firms in the manufacturing and service sectors. Small firms account for around 20 per cent of manufacturing jobs in the UK and Luxembourg; between 30 and 45 per cent in France, the Netherlands, Belgium, Spain, Portugal, Ireland and Denmark; and over half in Greece and Italy (Table 5.2). In the service sector, meanwhile, such firms provide a substantial majority of all jobs in the six countries for which information is available (Table 5.3).

Furthermore, the relative importance of small businesses, measured in terms of their shares of both national employment and net output, has been increasing in virtually all EC countries since the mid-1970s. This is, in part, a reflection of the shift in most countries from manufacturing to services, in which most small firms are concentrated. But, even within the manufacturing sector, small firms have become more important in all countries except Greece and the Netherlands (Figure 5.1).

Clearly, a variety of processes could have contributed to this increase in the relative size of the small-firm sector. In part it is simply an outcome of plant contraction or closure by large enterprises since the mid-1970s, especially in 'mature' manufacturing industries such as steel, shipbuilding, heavy engineering, textiles and motor vehicles, but also in modern ones such as electronics. In addition, the contraction of medium-sized firms has brought many of them into the small-firm category. Here we note that in the UK, France, Belgium and Denmark the small-firm sector in manufacturing has also been losing jobs, but at a slower rate than large firms. Only in West Germany, Italy and Ireland have small manufacturing firms increased employment (Storey and Johnson, 1987a).

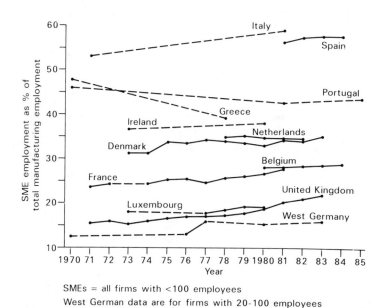

Figure 5.1 *Employment in small firms as a percentage of total manufacturing employment Source:* Storey and Johnson (1987a, Table 1.4).

SMEs = all firms with <100 employees
West German data are for firms with 20-100 employees

However, the relative rise of the small-firm sector is also a result of more positive trends, notably an excess of firm births over deaths, although the scale of this increase, and its timing, have varied between countries (Keeble and Wever, 1986). Figure 5.2 details the UK trend for start-ups and reveals an excess of births over deaths since 1979. While there is a lack of comparable statistics for other Western European countries, it is clear that the UK's experience has been paralleled in West Germany (Hull, 1987), France (Aydalot, 1986) and the Netherlands (Wever, 1986), although not in Scandinavia (Madsen, 1986; Lindmark, 1983). Most new businesses have been established in the service sector. For example, in Belgium 80 per cent were in service industries in 1984 (Donckels and Bert, 1986), and 73 per cent of new businesses registered for VAT in the UK in the early and mid-1980s were in services, with almost 30 per cent being in wholesale and retail distribution (*British Business*, 1988).

Three further aspects of the increase in new-business formation are also of note (Mason and Harrison, 1985; Curran, 1986). First, women are forming new businesses in increasing numbers. In Great Britain the number of females classified as self-employed increased by 42 per cent between 1981 and 1984, by which time women comprised 24 per cent of the self-employed population (Creigh *et al.*, 1986). Second, there has been a growth in the number of small businesses owned by members of ethnic minority groups, although this trend is confined to certain groups and is clearest in the UK. Here Asians have the highest level of self-employment, but some smaller groups (e.g., Cypriots) also exhibit high rates of self-employment (Ward, 1987). Ethnic businesses are less prominent in other European Community countries because of their different pattern of immigration (the 'guestworker' system, see Chapter 10), yet there are signs of change. Examples include the emergence of Turkish entrepreneurs in West Germany and Surinamese entrepreneurs in the Netherlands (Ward, 1987). Third, there has been an increase in new forms of small business, notably franchised businesses (Stern and Stanworth, 1988) and workers' co-operatives (Lowe, 1989).

Explaining the revival of the small-firm sector

Consideration of why the small-firm sector has recently increased requires discussion of various

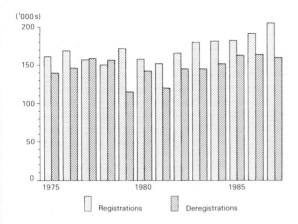

Figure 5.2 *New business registrations and deregistrations in the UK, 1975–87*
Source: British Business (1988).

factors, the relative importance of which varies between sectors and regions and also over time. These factors are of two kinds — causal and facilitating. Causal ones can be categorised under three broad headings: recession-related influences such as rising unemployment and corporate restructuring; technological change, which has created opportunities for new innovative enterprises in emergent industries and has encouraged smaller-scale production in some established industries; and structural changes, of which the emergence of the information economy and increasing consumer affluence are the most significant. Facilitating factors include the availability of resources (e.g., finance, markets, premises, labour, and advice) which must be present for business formation and growth to occur, and societal attitudes towards entrepreneurship.

RECESSION-RELATED EXPLANATIONS

There is much support for the view that the revival of the small-firm sector reflects the impact of the deepening global recession since the mid-1970s. It is argued that redundancy, job insecurity and the effects of recession on promotion prospects have 'pushed' many individuals into establishing their own businesses. At the macro-scale, Harrison and Hart (1983) and Hamilton (1986) have identified a statistical association between new company registrations and unemployment in

the UK. Although there is a strong secular upward trend in new business start-ups (Binks and Jennings, 1986), rising unemployment is one of the main influences on cyclical variations in company births (Hudson, 1987). At the micro-level the evidence is less clear-cut. Research in the late 1970s suggested that between 10 per cent and 25 per cent of founders in the UK had been unemployed immediately prior to establishing their businesses (Cross, 1981; Storey, 1982). Recent evidence suggests, however, that unemployment and redundancy have become increasingly significant factors in new-firm formation in the 1980s (Mason, 1989).

Rising unemployment is only one dimension of the recession-push interpretation of new-firm formation trends; recession may also have contributed to the increase in business start-ups in Western Europe in more positive ways. Plant closures, liquidations and bankruptcies since the onset of recession have increased the availability of cheap second-hand plant and machinery, thereby reducing the start-up costs for new businesses, especially in manufacturing (Binks and Jennings, 1986). In addition, restructuring by large firms has created opportunities for small businesses. Shutt and Whittington (1987) argue that large firms have pursued various fragmentation strategies, including devolution of production to independent small firms with which revenue links are retained (e.g., by licences), and disintegration by subcontracting support services (such as catering, cleaning, security and printing) to independently owned firms. Increases since the late 1970s in franchised businesses, management buyouts, corporate venturing, networking schemes and subcontracting all confirm that fragmentation strategies have led to increased opportunities for small businesses, although some of these developments are still on a small scale (Mason and Harrison, 1985). Opportunities for small firms have also been created by the withdrawal of large firms from their least profitable activities, allowing entry by small firms which have succeeded in these market niches because of their lower overheads and greater flexibility (Carlsson, 1989).

TECHNOLOGICAL CHANGE

The post-1979 recession has undoubtedly been

influential, but other factors have also created opportunities for small firms — one of the most important being technological change (Keeble and Wever, 1986). Central to this explanation is the notion of long waves of economic development, or Kondratieff cycles (Freeman, 1986). According to this theory, the Europe-wide recession of the 1970s and 1980s was a slump which inevitably followed the post-war boom of growth industries (such as electronics, petrochemicals, synthetic materials, semiconductors and pharmaceuticals) which were based on intensive innovation in the 1930s and 1940s. The model predicts the emergence, during the slump, of industries utilising radical new technologies which will form the basis of the next economic upswing — the fifth Kondratieff cycle. Many commentators suggest that these technologies are already identifiable: computers, telecommunications and information technologies, linked by the common theme of microelectronics (Freeman, 1986). Industry case studies (such as Rothwell, 1983) show that risk-taking entrepreneurs are often the first to grasp the technoeconomic opportunities offered by new technologies; consequently, the recent rise in business start-ups in Western Europe may partly reflect the exploitation of the technological opportunities of the fifth Kondratieff cycle by these entrepreneurs. The significant increase in new-firm formation in the computer industry since the mid-1970s provides one illustration of this phenomenon (Keeble and Kelly, 1986). New technologies have also created significant opportunities for small firms in established industrial sectors. For example, the adoption of micro-processor electronics in process technologies has produced a new generation of flexible industrial equipment especially suited to small-batch production; this has reduced barriers to entry for new businesses (Carlsson, 1989).

Despite this, technological change is of only secondary importance in the revival of Western Europe's small-firm sector. Relatively few small firms are found in high-technology industries and the vast majority are not innovative. Moreover, Western Europe lags well behind the USA in the formation of new-technology-based firms (NTBFs). In biotechnology, Feldman (1985) has identified 113 firms in the USA but only 27 in the rest of the world. In semiconductors, small firms played a key development role in the USA, whereas in Western Europe this technology was exploited by large companies such as Philips, GEC and Siemens. By the time the industry developed in Europe, the technology had become relatively standardised, and scale and technology barriers largely precluded entry by NTBFs (Rothwell, 1983). Other reasons suggested for the larger numbers of NTBFs in the USA than in Western Europe include differences in cultural attitudes, the size of domestic markets, capital market structures and government procurement strategies. Consequently, Rothwell and Zegveld (1982, p. 40) suggest that in Western Europe existing firms might once again play the major role in the exploitation of the technologies of the fifth Kondratieff cycle.

STRUCTURAL CHANGE

In contrast to recession-push and technical-change explanations, the structural change thesis interprets the recent increase in the small business sector as an outcome of long-run shifts in the Western European economy. Possibly the most significant structural change in the more industrialised countries has been their transition into 'post-industrial societies' (see also Chapter 7). As aggregate incomes have risen, expenditure on basic items such as food, clothes and durable goods has increased less rapidly than the amounts spent on more discretionary — and often intangible — items such as leisure and entertainment services. In addition, increased emphasis has been placed on various 'natural' products (e.g., wholefoods, medication and cosmetics). Rising real incomes have also contributed to the break-up of the mass market as demand has shifted in favour of more varied, customised and more sophisticated products and services. Associated lifestyle changes — urbanisation, increased leisure time, the decline of the nuclear family, smaller family sizes and rising female employment — have reinforced this trend by increasing demand for a wider range of services. The result has been the appearance of numerous market niches which small firms have been better able to exploit than large ones (see Chapter 9). Finally, the pursuit of fragmentation strategies — noted above — is not simply a reaction to recession. As the economies of industrialised countries have become more complex, expertise has commonly been bought in from smaller specialist firms (Chapter 7, Daniels, 1985).

Many small firms have been able to take advantage of the opportunities created by these trends. Service industries often have low barriers to entry and lack significant economies of scale, especially in knowledge-based sectors. Their complexities make standardised impersonal procedures difficult to evolve, and potential economies of scale are also limited by the fact that the main assets required are largely intangible — human creativity, knowledge and person-to-person skills (Curran and Stanworth, 1986). It therefore appears that high rates of new-firm formation throughout the non-manufacturing sector may be interpreted, in large measure, as a function of Western Europe's transition towards 'post-industrial' economic structures. Similarly in manufacturing, the rise of small businesses in such industries as food and drink, clothing and furniture reflects consumer resistance to mass-produced, often synthetic, items and the resultant appearance of numerous market niches. In these niches, short production runs and more flexible production processes are often required, and small firms have been better able to meet such demands than large enterprises (Brusco, 1982; Mazzonis, 1989).

A 'structural change' explanation for the revival of small firms in Western Europe therefore has much to commend it. Although quantification of its precise impact is impossible, a causal relationship is suggested by the fact that industries which have experienced a large increase in new business start-ups — notably business services, personal services, leisure services and fashion-oriented and customised manufacturing industries — are precisely those in which the effects of structural change have been most significant. Such an explanation is consistent with the results of research in the UK which established the existence of a long-term relationship between company formation and rising disposable incomes (Hudson, 1987). However, the structural change theory — with its implication that the revival of small firms in Western Europe conforms to a long-term secular trend — does not specifically account for the sharp rise in new-firm formation rates since the mid-1970s and especially since around 1980. This rise is the latest manifestation of a well-established cyclical pattern (Harrison and Hart, 1983) and too heavy reliance on the structural change explanation has consequently been sharply criticised by Shutt and Whittington (1987). It is,

therefore, evident that other factors have also contributed to the revival of the Western European small-firm sector. Recession and technological change are two of these, but no explanation should overlook the role of the facilitating factors discussed below.

Facilitating factors — the enterprise climate

The improvement in the climate for enterprise during the 1980s has had two main dimensions: societal attitudes to enterprise and the availability of resources.

SOCIETAL ATTITUDES

In much of post-war Western Europe professions and administration have traditionally been the most socially acceptable occupations, and the small business owner-manager has been held in low esteem. Yet in the UK, only a decade after an official report noted the low social standing of small business owners (Bolton, 1971), Bannock (1981, p. 21) was able to claim that it is now 'no longer a disgrace for a clever person to set up in business instead of going into the civil service, teaching or a large company'. A similar shift in societal attitudes towards entrepreneurship can be observed in other countries (*Fortune*, 1987). One indication of this change in social values is the marked increase in the number of small-business, entrepreneurship and investment-related magazines that have been launched since the late 1970s (Bannock, 1987). It seems undeniable — although impossible to prove — that this shift in public opinion has contributed in no small measure to the rise in business start-ups by removing an important intangible constraint on firm formation.

AVAILABILITY OF RESOURCES

Because of improved resources — notably finance, information, advice, training and premises — fewer potential entrepreneurs have been discouraged from putting their business ideas into practice, and many constraints on the growth of existing small firms have been reduced. In this context, governments at all levels — local, provincial and national — have played a significant role in the revival of the small-firm sector. This they have done either through the direct provision of resources, or through encouragement to the private sector to play a greater role in assisting the formation and growth of small businesses. Almost all Western European countries (Switzerland is an exception) have specific policies designed to encourage the growth and development of small firms, although there are differences between countries in the 'delivery' of this assistance (Burns and Dewhurst, 1986).

The difficulties that small firms encounter in obtaining external finance on acceptable terms have been reduced in most, if not all, countries. In terms of equity finance, tax incentives — such as the UK's Business Expansion Scheme (Mason and Harrison, 1989) — have increased the flow of equity into small companies by offering substantial tax 'breaks' to private investors. A major improvement in the availability of venture capital, felt initially in the UK (Martin, 1989) is now evident in several other countries — notably the Netherlands, West Germany and France. And secondary financial markets have developed to provide small companies with a source of equity capital. Meanwhile the supply of loan finance has also been improved in key respects. Banks — which have traditionally taken an unsympathetic attitude towards new and recently established businesses — are now introducing a variety of special lending schemes (NEDO, 1986). Some countries, including West Germany, have introduced subsidised loans for small businesses (Kayser and Ibrielski, 1986). Most countries now have schemes to underwrite commercial loans to small firms which are high-risk or lack security to offer as collateral, and some have provided financial incentives to encourage unemployed people to start their own businesses (Harrison and Mason, 1986; Burns and Dewhurst, 1986).

Better training and specialist business advice represent a further important improvement in the resource base. Because owner-managers often lack the necessary commercial, financial and management skills, many start-ups are condemned to early failure. Similarly many potential founders — conscious that they lack such skills — may be deterred from pursuing self-employment. But short training courses offered by higher education institutions, business-support organisations and government agencies are now burgeoning (Curran and Stanworth, 1989). Longer entrepreneurship courses have also been established in numerous higher education institutions, while advice on such topics as preparing a business plan, financial management, product development, exporting and marketing is also widely available in most countries (Burns and Dewhurst, 1986).

Two property-related resource developments are of note. One is the conversion and subdivision of large, usually redundant, industrial buildings into small incubator workshop complexes. These offer 'easy-in, easy-out' lease terms and provide a range of common services, such as secretarial facilities, bookkeeping, photocopying, cleaning, and conference/meeting rooms. Most of these 'incubators' have been developed by local authorities and other public sector organisations, and some are run by joint public/private sector companies. Science parks, meanwhile, are university-linked high-quality industrial estates for technology-based enterprises engaged in R & D, but not in mass production. They aim to encourage the formation and growth of high-technology enterprises by providing access to university facilities and resources (e.g., libraries and laboratories); by facilitating interaction between technical entrepreneurs and academics; and by commercialising university research through, for example, licensing agreements and the formation of university-owned companies. Since the early 1980s the popularity of the science park concept has grown rapidly: 36 were in operation in the UK in 1988; developments either planned or initiated elsewhere in Western Europe include those at Louvain, Utrecht, West Berlin, Wilhelmshaven, Stuttgart, Karlsruhe, Gothenburg, Malmö, Bari, Venice and Trieste (Monck *et al.*, 1988).

These significant improvements in Western Europe's enterprise climate have reduced — or even eliminated — many constraints on small business. However, we must stress that neither the increased availability of resources outlined above nor government policies have been a *fundamental* cause of the revival of the small-firm sector; that revival largely pre-dated these developments. Instead, recovery has primarily been based on shifts in Western Europe's economic and social structures, supported by recessional and technological trends; these have created displacements and opportunities which have either pushed or pulled individuals into forming businesses.

Geographical variations in new-firm formation

By no means all regions and sub-regions within Western Europe have shared equally in this revival of the small-firm sector, and three broad generalisations can be made about the spatial pattern of new-firm formation (Keeble and Wever, 1986). First, economically diversified metropolitan regions exhibit high rates and large volumes of new-firm creation in the manufacturing and service sectors. Examples include the Paris region, the Amsterdam region (Figure 5.3), the London region (Figure 5.4) and the Munich area. Because of the clustering of corporate headquarters and R & D establishments, these regions contain above-average proportions of managerial, professional and technical employees — occupational groups with the highest propensities for entrepreneurship. In addition, the concentration of corporate decision-making, national or regional government functions, information-intensive activities (such as producer services) and information-generating organisations (such as research institutes) has created in these areas an information-rich environment. Potential entrepreneurs therefore have access to a large, and continually enriched, stock of knowledge (Sweeney, 1987). Large metropolitan regions also provide better access to external capital (Mason and Harrison, 1989; Martin, 1989), while concentrated consumer and commercial market opportunities exist for many products and services, including high-technology and producer service industries.

Second, high rates, but lower volumes, of business start-ups are found in a number of relatively unindustrialised rural regions. Some of these have a tradition of small-scale, often agriculture-related, industrial development. Examples include the Midi in southern France; parts of the southern and eastern Netherlands (Figure 5.3) and East Anglia, the South-West and mid-Wales in the UK (Figure 5.4). North-central Ireland is another such region (O'Farrell, 1986), while others have been identified in Jutland (Illeris, 1986), central Italy, southern Germany bordering the Alps, various parts of Spain (Cuadrado Roura, 1988) and rural areas of central Portugal (Lewis and Williams, 1987). In northern Europe new firms in many of these less-industrialised and rural regions are often in

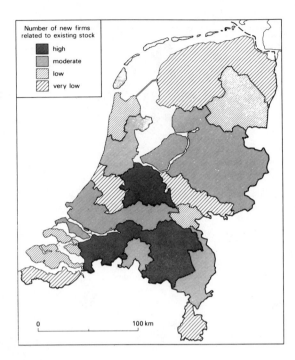

Figure 5.3 *New-firm formation in the Netherlands, 1970–80*
Source: Wever (1986).

technologically advanced sectors such as electronics and computers. Examples include the Côte d'Azur, Toulouse and Grenoble regions of France; Bavaria and Baden-Württemberg in West Germany; and the Cambridge region in the UK. This bias is partly related to the nature of industry in such regions, and to the presence in small and medium-sized cities of 'leading-edge' activities owned by multi-plant companies (Sweeney, 1987). These factors have combined to create market opportunities for small firms, which have been reinforced by occupationally selective migration to scenically attractive small towns and rural areas with appealing climates (Keeble and Wever, 1986).

In contrast, in rural parts of the Mediterranean region and Ireland, new firms are typically associated with traditional industrial sectors such as textiles and clothing, furniture, pottery and metal products (Brusco, 1982; Lewis and Williams, 1987). In these regions, high rates of new-firm formation naturally reflect different factors (Keeble, 1989a). Using evidence from Italy, two models of rural industrialisation based

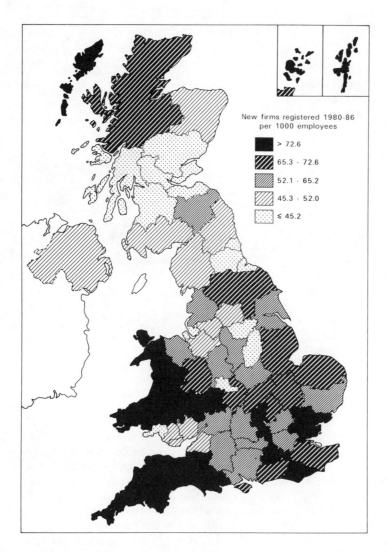

Figure 5.4 *New-firm formation rates in the UK, all industries, 1980–86* *Source:* Department of Employment; VAT registration data.

on small firms have been proposed. The 'productive decentralisation' model envisages that much of the increase in small-firm activity in the north-east and central regions has resulted from the vertical disintegration of production by large firms in cities such as Milan and Turin. These firms, it is argued, seek links with small subcontracting firms in rural areas where labour is cheaper and non-unionised. The alternative 'flexible specialisation model' suggests that industrialisation is an endogeneous, rather than a dependent, process. The economy of north-east and central Italy is characterised by various industrial districts, each exhibiting a high level of specialisation and displaying a low level of vertical integration

(Mazzonis, 1989). In any district, only a limited number of small firms market finished goods; the others work as subcontractors, and the formation of subcontracting firms is not difficult for workers with specialist knowledge (Brusco, 1982). However, further research is required into the general applicability of both models (Amin, 1989; Lewis and Williams, 1987).

With respect to rural areas, it is necessary to note that not all unindustrialised regions have exhibited high rates of new-firm formation. For example, in Greece small firms remain firmly concentrated in Greater Athens and Thessaloniki, the two main urban centres, reflecting conditions in peripheral regions: limited market opportunities, a lack of skilled labour, inefficient communication networks and limited access to information (Giaoutzi *et al.*, 1988, pp. 264–81). Other remote rural regions, such as the northern provinces of the Netherlands, south-west Ireland and southern Italy, also have low rates of new enterprise creation.

Third, the lowest rates and smallest volumes of new-firm formation are in older urban-industrial regions such as industrial Belgium; the Nord and Lorraine in France; and North-East England, central Scotland and south Wales (Figure 5.4). Poorer performance is directly related to the traditional economies of these regions, which are generally based on dominant industries such as coal, steel, shipbuilding, textiles and motor vehicles — a legacy of nineteenth-century industrialisation and the establishment of large branch plants attracted by regional policy initiatives in the 1960s and 1970s. Such industries have high barriers to entry which reduce spin-off opportunities. External ownership and truncated management functions result in an occupational structure that is biased towards blue-collar semi- and unskilled jobs; workers in these have the lowest propensity to establish their own businesses (Storey, 1982). Large plants provide employees with work experience which is of limited usefulness as a training for independent business, and the nature of the production process often emphasises strength, endurance and manual skills rather than problem-solving abilities (Segal, 1979). Rates of new-firm formation are further depressed by the limited purchasing autonomy of branch plants and their high level of vertical integration, which constrain local market opportunities. New-firm creation may also be hindered by high local unemployment (which restricts consumer demand), an impoverished information environment (Sweeney, 1987), and lack of access to external finance (Mason and Harrison, 1989).

Small firms and urban and regional development

The revival of small firms, and the patterns outlined above, raise an important question: how significant are small businesses in urban and regional economic development? This section examines this question from four important perspectives: employment creation; employment quality; innovation; and economic impact.

Employment creation

The contribution of small firms to employment is normally assessed by means of 'job-generation' studies which classify enterprises or establishments according to whether they have opened or closed, or have increased or decreased their employment, over the period of time studied. This methodology requires access to micro-level longitudinal data on individual enterprises (or establishments) and, unfortunately, in Western Europe differences in data sources, in sectoral and temporal coverage and in presentation make international comparisons problematical. Nevertheless, most studies undertaken in different countries and regions indicate that small and medium-sized enterprises created jobs during the 1970s and early 1980s, while large firms simultaneously reduced their employment (Figure 5.5).

This conclusion is well illustrated by a recent study based on specially constructed establishment-level data banks which now make it possible to draw comparisons between regions on a fully standardised basis (Gudgin *et al.*, 1989). This new research provides comparative analyses for three regions: Northern Ireland, the Republic of Ireland, and Leicestershire in the English Midlands. The fullest comparison is available for the two parts of Ireland, and it is on these areas that the present discussion will concentrate. The data available for them cover the period 1973–86 and relate to almost 6,000 manufacturing establishments (2,000 of them in Northern Ireland). Primary variables in the data set cover each establishment's employment, its sector, its formation date and ownership details (Hart, 1989).

In both Northern Ireland and the Republic, support for small indigenous firms — and particularly for the establishment of new firms — has been an increasingly significant feature of industrial policy. In terms of results, new firms created some 17,000 jobs in Northern Ireland over the period 1973–86, the equivalent figure for the Republic being 30,000 (Table 5.4). By 1986, firms that had started up in the previous 13 years were responsible for 15–16 per cent of all manufacturing employment in both regions. This progress, however, must be placed in the perspective of

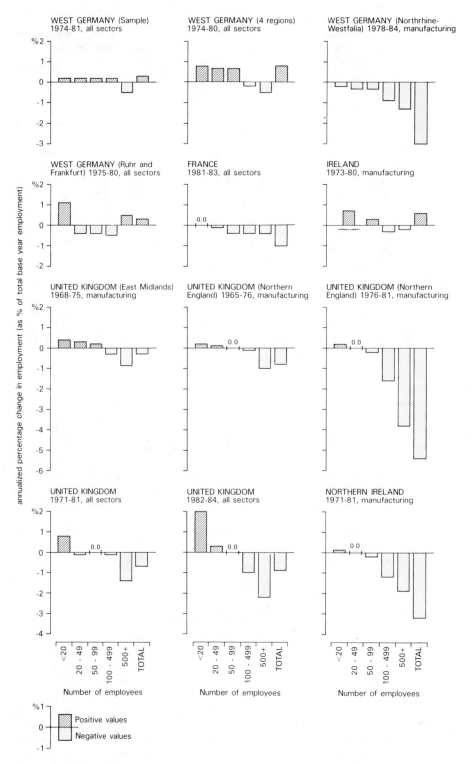

Figure 5.5 *The contribution of small firms to job generation in Western Europe*
Source: Storey and Johnson (1987b, Table 4.18).

Table 5.4 *New-firm formation in Northern Ireland and the Republic of Ireland, 1973–86*

	Northern Ireland	Republic of Ireland
Employment by establishment size, 1973		
1–24 (%)	6.2	13.7
25–99 (%)	15.1	25.9
100–199 (%)	13.3	16.4
200+ (%)	65.4	43.9
New firm formation rate*		
(a)	9.0	11.1
(b)	11.6	16.3
Average employment per firm, 1986	8.4	8.1
Total employment in new firms, 1986 ('000s)	17.0	30.0
Total 1986 employment in new firms		
(a) as % of 1973 employment	9.8	13.3
(b) as % of 1986 employment	15.8	14.6

*Formation rates are net numbers of new firms surviving to 1986 per 1000 employees; (a) in manufacturing plus 12% of non-manufacturing; (b) in manufacturing alone.
Source: Hart (1989).

Table 5.5 *Components of change in manufacturing employment in Northern Ireland and the Republic of Ireland, 1973–86*

Components	Northern Ireland		Republic of Ireland	
	'000s	(%)*	'000s	(%)*
1973 Employment	174.3		225.0	
Companies established by 1973				
Externally owned	−55.8	(−60.6)	−28.7	(−41.9)
Indigenously owned	−35.0	(−42.8)	−64.2	(−41.4)
Net change	−90.8	(−52.1)	−92.9	(−41.6)
New inward investment	+6.9	(+3.9)	+43.1	(+19.1)
Indigenous new firms	+17.0	(+9.8)	+30.0	(+13.3)
Total employment change	−66.9	(−38.4)	−19.8	(−8.8)
1986 employment	107.4		205.2	

*Percentages are of base year employment in manufacturing except for firms established by 1973. For these firms, percentages are of base year employment in each of the two categories.
Source: Hart (1989).

employment trends in the manufacturing sector as a whole (Table 5.5). In the Republic, the 30,000 jobs associated with new firms offset less than a third of all manufacturing job losses over the same period, and a larger contribution to job creation was made by new inward investment. In Northern Ireland the outcome was different in that employment growth attributable to new firms easily exceeded that caused by inward investment. Yet the 17,000 jobs linked to new local firms were still few relative to the 91,000 lost as a result of industrial closures and restructuring. Thus, both north and south of the border, the impact of new firms must be considered modest.

This case study emphasises that the role of small firms in job generation must not be exaggerated, and other indicators also point to this conclusion. Many small businesses close within a few years of start-up; in the UK, for example, VAT data show that half of all new firms have closed by their fifth year (Ganguly, 1985). This partly reflects the fact that new businesses are highly vulnerable to cataclysmic events beyond the control of the owner-manager, such as illness, deception, payment default, or the loss of major customers. Yet, not surprisingly, failure is also related to management ability. Wever's (1986) study of new businesses in the Netherlands shows that significant differences were found in the amount of preparation undertaken by successful and failed founders; the successful ones prepared the start-up more systematically and gave considerably more attention to marketing than did those who failed. Furthermore, even when firms survive, employment generation is unlikely to be substantial. This has been demonstrated by the Irish case studies, in which the majority of firms remained small (Table 5.5), and a broadly similar picture has been found in the UK (Storey, 1985) and West Germany (Hull, 1987). Limited growth in part reflects the characteristics of small-firm founders. Most are strong on technical and production aspects of the business, but lack the managerial skills necessary to plan investment, pricing, marketing and the delegation of responsibilities. Many founders also have little ambition for long-term growth beyond that necessary to achieve a comfortable standard of living, which may reflect the high priority given to the desire for independence as a motive for going into business (Cross, 1981; O'Farrell, 1986).

Employment quality

Relatively little attention has been given to the quality of employment provided by small businesses. Yet the types of jobs created, and the characteristics of the employees recruited to fill the jobs, are of considerable significance for local and regional economic development (Sengenberger and Loveman, 1988).

Locality-based case studies provide one of the main sources of information in this field. Clearly, such evidence may be atypical because of the particular industrial composition or the geographical context of individual studies. Nevertheless, the available information is remarkably consistent in indicating that employment in small firms is fundamentally different in several aspects from that in larger organisations (Curran, 1986; Storey and Johnson, 1987b). Evidence from Britain, Northern Ireland, West Germany, and the Netherlands indicates that, although small manufacturing firms create employment across the occupational spectrum, their main contribution is skilled and semi-skilled blue-collar jobs. The chief exception concerns high-technology firms, in which employment is even more concentrated in the managerial and technical grades. Small firms, in both the manufacturing and service sectors, also employ higher proportions of female workers, and particularly female part-time workers, than do larger companies. In addition, most small firms are unable to match the wages, fringe benefits and opportunities for internal promotion that are offered by large firms. As a result, they tend to recruit younger and less-qualified workers. However, this also means that small businesses provide a significant port of entry into employment for school leavers, the unemployed and people re-entering the labour market.

There is also evidence that employment in small firms is more unstable than in large firms. This may partly reflect personality clashes caused by close management–worker contact. The fact that small-firm employees have been the complainants in almost half the unfair dismissal cases brought to industrial tribunals in the UK tends to support this view (Dickens *et al.*, 1984). But instability is also increased because small businesses are disproportionately concentrated in sectors characterised by seasonal employment and high failure rates (such as construction and tourism).

Finally, there is considerable evidence that the expansion of the small-firm sector has serious implications for employment and working conditions. One recent nine-country study, which included both European Community and other Western European states in addition to the United States and Japan, concluded that, relative to large companies,

in small firms, on average, the levels of wages and [fringe benefits] are lower (even after accounting for workforce characteristics), working time is longer, the standards of job security and of occupational health and safety inferior. The quality of industrial relations is usually poorer, indicated by . . . much lower unionisation and works council representation [and by less extensive] coverage by collective agreement and legal protection . . . This implies that the observed shift towards smaller units of employment could in fact lead to a deterioration in employment standards, social welfare and industrial relations (Sengenberger and Loveman, 1988, pp. 51–2).

These less-satisfactory conditions of employment arise from the poorer average economic performance of the small-firm sector. This is indicated by lower profitability, productivity and innovative capacity, relative to large concerns. However, the attention now being given to the flexible specialisation model of small-firm development, based on high rates of product, process and organisational innovation (Brusco, 1982), suggests that there is an alternative to the low-cost, low-productivity, low-employment-quality cycle which characterises much of the small-firm sector (Piore and Sabel, 1984).

Innovation

If employment impact has been the most widely examined topic in the small-firm debate, then the importance of small businesses in the process of technological change has come a close second. Rothwell and Zegveld (1982) suggest that small firms have certain advantages over large companies in innovation. Entrepreneurial owner-managers can react swiftly to new market and technological opportunities, and are more willing to undertake high-risk innovative projects than managers in large companies. Small firms may also have efficient and informal internal problem-solving strategies which facilitate adaptation to changes in the external environment. However,

Table 5.6 *Share of innovations by firm size, UK, 1945–80 (per cent)*

Number of employees	Share of innovations							
	1945–49	1950–54	1955–59	1960–64	1965–69	1970–74	1975–80	Total
1–199	16	12	11	11	13	15	17	14
200–499	9	6	8	6	7	9	7	7
500–999	3	2	7	5	5	4	3	4
1,000–9,999	36	36	25	27	23	17	14	23
10,000 +	36	44	50	51	52	55	59	52
Number of innovations	94	191	274	405	467	401	461	2293

Source: Rothwell (1984, Table 2).

small firms do have certain inherent disadvantages in innovation, notably their lack of financial and qualified-manpower resources; an inability to obtain economies of scale in production and marketing; and constraints on the identification and use of external sources of scientific and technical information.

The most comprehensive evidence on firm size and innovation comes from the UK and is provided by a database developed by the Science Policy Research Unit (SPRU) at Sussex University. In the early 1980s this contained details on some 2,300 important innovations produced by British companies between 1945 and 1980. On the positive side, the share of innovations accounted for by small firms (those with less than 200 employees) averaged about 12 per cent between 1945 and 1969 but increased to 17 per cent during the 1970s (Table 5.6). Also, the contribution of small firms to innovation has varied quite markedly between industries, and they have consistently made a major contribution in industries such as instruments and machinery, in which capital intensity, development costs and start-up costs are generally low. However, small firms' contribution to innovation must be placed in perspective. For example, firms with 10,000 or more employees increased their share of innovations from 36 per cent in the 1945–9 period to 59 per cent between 1975 and 1980. Similarly, in terms of innovations per unit of output or employment (although not research and development expenditure) the performance of large firms has been consistently better than that of their smaller counterparts (Rothwell, 1984).

Although large firms dominate the innovation process, small innovative firms can nevertheless exhibit above-average rates of growth. Rothwell

and Zegveld (1982) refer to a study in Ireland which found that employment loss was associated with a lack of innovativeness and that high employment growth was strongly correlated with innovation. Similarly, Harrison and Hart (1987) have found that, for the population of small manufacturing firms as a whole, there is a significant link between product and process innovation and business performance in terms of turnover growth and market development. A study of 320 new firms started after 1974 in the UK computer industry provides further evidence of the high rate of employment creation achieved by innovative small businesses (Keeble and Kelly, 1986). By 1984 these firms had on average 28 employees, well above the levels found in other samples of new manufacturing firms. Moreover, over 4 per cent employed more than 100 workers, compared with around 1 per cent in most studies of other industries.

Innovative small firms may therefore have a significant effect on economic development and job creation. However, such firms tend to be spatially concentrated: prominent examples include Cambridge in the UK; the Côte d'Azur, Grenoble and Toulouse in France; and Bavaria and Baden-Württemberg in southern Germany (Keeble, 1989a). Hence, in most regions and sub-regions of every country, the number of innovative new businesses is extremely limited. Moreover, follow-up surveys of innovative firms reveal that, in many cases, their subsequent performance has been characterised by difficulty or failure. Oakey (1984) suggests that this reflects a tendency for profits from an established product to decline at the very time when development of a replacement product demands maximum R & D expenditure. This results in financial stress as

R & D expenditure exceeds profits. Vulnerability may be further exacerbated by rapid changes in demand, by competition, by the high cost of marketing a new product, and by a slow-down in the rate of product innovation as the key founders become more involved in management at the expense of R & D.

It therefore appears that technology-based small firms have been too few in number and have lacked sufficient growth momentum to achieve the economic transformation of urban and regional economies. Even in Cambridge, which many observers claim has the largest concentration of indigenous high-technology firms in Western Europe, such firms remain very small and short-lived. In one study the median employment figure for surviving high-technology firms up to ten years old was 13, and for firms up to six years old the figure was six employees, not significantly different from figures for conventional small firms (Storey and Johnson, 1987b). Moreover, the most successful of the new high-technology firms in the Cambridge area have failed to remain independent; typically they have sold out (sometimes, when difficulties have arisen, as an alternative to liquidation) to foreign firms (Keeble, 1989b).

The wider economic impact

A final issue concerns the wider contribution of small businesses to local and regional economic development. Three aspects of this issue will be considered. Do small businesses make a significant contribution towards diversifying the industrial mix of regions and sub-regions? To what extent are small firms dependent upon large enterprises? And do small firms represent a genuine net economic gain? Are they increasing the export-earning capacity of their regional economies, or are their sales largely locally oriented and therefore potentially displacing other local firms?

Small businesses are often thought to contribute to the diversification, and even the regeneration, of local and regional economies by their ability to provide new goods and services. We noted earlier that the majority of new businesses are established in retailing and other services, but a lack of research means it is unclear whether they perpetuate the existing service structure or broaden its base. In manufacturing, various studies have confirmed that most founders

establish their business in the industry in which they had experience as an employee, thereby limiting the extent to which diversification occurs. Moreover, some studies have indicated that new businesses tend to be concentrated in a narrow range of industries, notably mechanical engineering, metal goods, timber, furniture and, in some regions, paper, printing and publishing. This concentration partly reflects differences between industries in profitability and growth rates (Hudson, 1987). More significantly, these sectoral patterns are positively related to the proportion of small plants in the industries concerned, a measure of the ease of entry (Cross, 1981; O'Farrell, 1986). In addition, in some industries the birth rate of firms may be raised by the fact that, during recession, industry-specific premises and second-hand equipment may be readily available from firms no longer trading (Binks and Jennings, 1986). The combined result of these factors is that most new firms simply perpetuate the industrial structure in their area, rather than contributing to its diversification.

The evidence concerning dependency relationships between small and large firms is again restricted to the manufacturing sector, but its message is clear. A high proportion of the small-firm sector is engaged in what is best described as jobbing work, producing one-offs and small batches of goods to customers' requirements, usually on a subcontract basis. This is particularly common in the metal goods, engineering, electronics and plastics industries; indeed, Fredriksson and Lindmark (1979) noted that in Sweden two-thirds of small firms in the metals and engineering industries operated as subcontractors. No less than 5 per cent of them were linked with Volvo, the biggest manufacturing firm in Sweden. This form of production is also a characteristic of the clothing industry, in which many small firms are subcontractors to major retail chains (Rainnie, 1984; Murray, 1985). Many businesses — especially those engaged in subcontracting — are also dependent on just a small number of dominant customers. For example, O'Farrell (1986) noted that a quarter of all new manufacturing firms in Ireland sold over 60 per cent of their output to four or fewer customers. This dominant-customer relationship has the advantage of offering regular long-run orders and reducing administrative and marketing costs. Clearly, however, it places such firms in a very vulnerable position if dominant customers encounter economic problems, do not pay promptly, or change their input requirements.

Finally, most studies of the market orientation of small businesses relate to the UK, but it is unlikely that they greatly distort the general Western European picture. Their general conclusion is that most small firms do not have the advantage of being export earners for their regional economies. In peripheral regions, dependence on local markets is particularly marked: Lloyd and Dicken (1982) noted that over half the new manufacturing firms surveyed in Greater Manchester and Merseyside made the majority of their sales within their respective conurbations. In Ireland, new manufacturing firms derived just 2 per cent of sales, on average, from exports (O'Farrell, 1986). When the service sector is taken into account, the level of local market dependence is even greater. Four-fifths of the new firms in Storey's (1982) Cleveland study were in non-manufacturing activities (excluding retailing); these derived 78 per cent of their sales from within Cleveland, a further 9 per cent from the remainder of the Northern region, and only one per cent from abroad. This high degree of local market orientation may reflect the production of specialist products and services, which often require personal attention and after-delivery service. However, it primarily stems from the limited marketing efforts made by most owner-managers, who tend to rely on word of mouth, personal recommendations and repeat orders to secure work. Clearly the dependence of both new and established small firms on local markets seriously limits their contribution to regional and national economic development. Moreover, it raises the likelihood that, without a significant increase in demand, newly established small firms may simply be displacing similar businesses which also serve local or regional markets.

Conclusion

There has been a substantial increase in both the absolute and the relative size of the small-firm sector in most Western European countries since the mid-1970s, an increase which largely reflects a rise in the rate of new-firm formation. This trend is the product of a number of factors, the relative significance of which remains a matter of debate.

Arguably, rising unemployment, the rationalisation of large firms, the emergence of new technologies, increased demand for services and rising consumer affluence have all contributed to the trend. A shift in social attitudes towards entrepreneurship has also been helpful, as has the reduction of many constraints on the formation and growth of small businesses, often assisted by government action. But whether the revival of the small-firm sector is a cyclical, and hence temporary, phenomenon or a permanent feature also remains a matter for debate. Both the recession-related and the 'technology-push' explanations imply that the current resurgence of the sector may be cyclical. But if the increase in small businesses is primarily a function of fundamental changes in the economic and social structures of Western European countries, then small firms are likely to continue to increase in significance for the foreseeable future. Longitudinal evidence from the UK confirms that considerable fluctuations in new-firm formation rates have occurred around an underlying secular upward trend (Hudson, 1987).

Despite this industrial restructuring in favour of small-scale units, however, the contribution made to local and regional economic development in Western Europe is relatively modest. Small firms do make a positive contribution to job creation, but they pay lower wages, offer fewer fringe benefits and often provide less-stable jobs than do large firms. Furthermore, while it is true that small firms offer an important port of entry for workers entering or re-entering the labour market, many of the jobs created are only part-time. In terms of working conditions, the sweatshop model is certainly not entirely appropriate, yet it is equally clear that the sector is not a haven of tension-free industrial relations, with open and easy communications, low labour turnover and employers who are concerned for the welfare of their employees (Curran, 1986).

In other respects the contribution of small firms to local and regional economic development is particularly limited. They are not concentrated in profitable or growing sectors, they tend to replicate an area's industrial structure rather than acting as a diversifying force, and their dependence on local markets greatly limits their contribution to net economic growth. Any significant contribution to local and regional economic development is therefore largely confined to the small minority of rapidly growing, often innovative, enterprises. These are the ones that are most likely to contribute to industrial diversification, to operate in growing market niches and to be exploiting markets outside the region or the country. Such firms have little or no local displacement effect and may even create indirect employment locally through their own input and subcontracting requirements. For example, in the Cambridge area just four 'high-flying' companies based on new technologies employed 800 people in 1985, the same number created by all new manufacturing firms in East Anglia between 1971 and 1981 (Keeble and Wever, 1986). However, firms of this calibre are far from common and, as with high rates of new-firm formation in general, they are typically concentrated in certain regions and sub-regions which already constitute the prosperous cores of national space economies in Western Europe (Mason, 1985). Taking a long-term perspective, therefore, it is these regions that are most likely to reap the benefits of the phoenix-like resurgence of small firms. In such areas, the accumulation of investment by these businesses may well make a substantial contribution to economic growth and modernisation. But the small-firm phenomenon cannot be seen as a panacea for the regional economic ills of Western Europe as a whole.

References

Amin, A. (1989) Flexible specialisation and small firms in Italy: myths and realities, *Antipode*, 21, 13–34.

Aydalot, P. (1986) The location of new firm creation. In D. Keeble and E. Wever (eds), *New Firms and Regional Development in Europe*, Croom Helm, Beckenham, Kent, pp. 105–23.

Bannock, G. (1981) *The Economics of Small Firms: return from the wilderness*, Basil Blackwell, Oxford.

Bannock, G. (1987) *Britain in the 1980s: enterprise reborn?* Investors in Industry, London.

Binks, M. and Jennings, A. (1986) Small firms as a source of economic rejuvenation. In J. Curran, J. Stanworth and D. Watkins (eds), *The Survival of the Small Firm. Volume 1: The Economics of Survival and Entrepreneurship*, Gower, Aldershot, pp. 19–38.

Birch, D.L. (1979) *The Job Generation Process*, MIT Program on Neighbourhood and Regional Change, Cambridge, MA.

Bolton, J.E. (1971) *Report of the Commission of Inquiry on Small Firms*, Cmnd 4811, HMSO, London.

British Business (1988) Vat registrations and deregistrations: 1980–87. 12 August, 32–4.

Brusco, S. (1982) The Emilian model: productive decentralisation and social integration, *Cambridge Journal of Economics*, 6, 167–84.

Burns, P. and Dewhurst, J. (eds) (1986) *Small Business in Europe*, Macmillan, Basingstoke.

Carlsson, B. (1989) The evolution of manufacturing technology and its impact on industrial structure: an international study, *Small Business Economics*, 1, 21–37.

Creigh, S., Roberts, C., Gorman, A. and Sawyer, P. (1986) Self-employment in Britain: results from the Labour Force Surveys 1981–1984, *Employment Gazette*, June, 183–94.

Cross, M. (1981) *New Firm Formation and Regional Development*, Gower, Farnborough.

Cuadrado Roura, J.R. (1988) Small and medium sized enterprises and the regional distribution of industry in Spain: a new stage. In M. Giaoutzi, P. Nijkamp and D.J. Storey (eds), *Small and Medium-size Enterprises and Regional Development*, Routledge, London, pp. 247–63.

Curran, J. (1986) *Bolton Fifteen Years On: a review and analysis of small business research in Britain, 1971–1986*, Small Business Research Trust, London.

Curran, J. and Stanworth, J. (1986) Trends in small firm industrial relations and their implications for the role of the small firm in economic restructuring. In A. Amin and J.B. Goddard (eds), *Technological Change, Industrial Restructuring and Regional Development*, Allen & Unwin, London, pp. 233–57.

Curran, J. and Stanworth, J. (1989) Education and training for enterprise: problems of classification, evaluation, policy and research, *International Small Business Journal*, 7 (2), 11–22.

Daniels, P. (1985) *Service Industries: a geographical appraisal*, Methuen, London.

Dickens, L., Hart, M., Jones, B. and Weekes, B. (1984) The British experience under a statute prohibiting unfair dismissal, *Industrial and Labour Relations Journal*, 37, 497–514.

Donckels, R. and Bert, C. (1986) New firms in the local economy: the case of Belgium. In D. Keeble and E. Wever (eds), *New Firms and Regional Development in Europe*, Croom Helm, Beckenham, Kent, pp. 124–40.

Employment Gazette (1987) Numbers of businesses: data on VAT registrations. April, 176–83.

Feldman, M.M.A. (1985) Patterns of biotechnology development. In J. Brotchie, P. Newton, P. Hall and P. Nijkamp (eds), *The Future of Urban Form*, Croom Helm, Beckenham, Kent, pp. 94–107.

Fortune (1987) Europe's new entrepreneurs. 11 May, 20–5.

Fredriksson, C.G. and Lindmark, L.G. (1979) From firms to systems of firms: a study of interregional dependence in a dynamic society. In F.E.I. Hamilton and G.J.R. Linge (eds), *Spatial Analysis, Industry and the Industrial Environment. Volume 1: Industrial Systems*, Wiley, Chichester, pp. 155–86.

Freeman, C. (1986) The role of technical change in national economic development. In A. Amin and J. Goddard (eds), *Technological Change, Industrial Restructuring and Regional Development*, Allen & Unwin, London, pp. 100–14.

Ganguly, P. (1985) *UK Small Business Statistics and International Comparisons*, Harper and Row, London.

Giaoutzi, M., Nijkamp, P. and Storey, D.J. (eds) (1988) *Small and Medium-size Enterprises and Regional Development*, Routledge, London.

Gudgin, G., Hart, M., Fagg, J., D'Arcy, E. and Keegan, R. (1989) *Job Generation in Manufacturing Industry, 1973–86*, Publication no. 1, TNIERC, Belfast.

Hamilton, R.T. (1986) The influence of unemployment on the level and rate of company formation in Scotland, 1950–1974, *Environment and Planning A*, 18, 1401–4.

Harrison, R.T. and Hart, M. (1983) Factors influencing new business formation: a case study of Northern Ireland, *Environment and Planning A*, 15, 1395–1412.

Harrison, R.T. and Hart, M. (1987) Innovation and market development: the experience of small firms in a market economy, *Omega*, 15, 445–54.

Harrison, R.T. and Mason, C.M. (1986) The regional impact of the Small Firms Loan Guarantee Scheme in the United Kingdom, *Regional Studies*, 20, 535–50.

Hart, M. (1989) Entrepreneurship in Ireland: comparative evidence from north and south, *Entrepreneurship and Regional Development*, 1, 129–40.

Hudson, J. (1987) Company births in Great Britain and the institutional environment, *International Small Business Journal*, 6(1), 57–69.

Hull, C.J. (1987) Job creation in the Federal Republic of Germany — a review. In D.J. Storey and S.G. Johnson (eds), *Job Creation in Small and Medium Sized Enterprises*, Commission of the European Communities: Programme of Research and Action on the Development of Labour Markets, Brussels, pp. 216–319.

Illeris, S. (1986) New firm creation in Denmark: the importance of the cultural background. In D. Keeble and E. Wever (eds), *New Firms and Regional Development in Europe*, Croom Helm, Beckenham, Kent, pp. 141–50.

Kayser, G. and Ibielski, D. (1986) The Federal Republic of Germany. In P. Burns and J. Dewhurst (eds), *Small Business in Europe*, Macmillan, Basingstoke, pp. 175–92.

Keeble, D. (1989a) Core–periphery disparities, recession and new regional dynamisms in the European Community, *Geography*, 74, 1–11.

Keeble, D. (1989b) High technology industry and regional development in Britain: the case of the Cambridge phenomenon, *Environment and Planning C, Government and Policy*, 7, 153–72.

Keeble, D. and Kelly, T. (1986) New firms and high technology industry in the United Kingdom: the case of computer electronics. In D. Keeble and E. Wever (eds), *New Firms and Regional Development in Europe*, Croom Helm, Beckenham, Kent, pp. 75–104.

Keeble, D. and Wever, E. (1986) Introduction. In D. Keeble and E. Wever (eds), *New Firms and Regional Development in Europe*, Croom Helm, Beckenham, Kent, pp. 1–34.

Lewis, J.R. and Williams, A.M. (1987) Productive decentralization or indigenous growth? Small manufacturing enterprises and regional development in central Portugal, *Regional Studies*, 21, 343–61.

Lindmark, L. (1983) Sweden. In D.J. Storey (ed.), *Small Firms: an international survey*, Croom Helm, Beckenham, Kent, pp. 179–212.

Lloyd, P.E. and Dicken, P. (1982) *Industrial Change: local manufacturing firms in Manchester and Merseyside*, Inner Cities Research Programme, Department of the Environment, London.

Lowe, M.S. (1989) Never mind the planning . . . feel the initiative: local government policy and worker co-operatives. In D. Gibbs (ed.), *Government Policy and Industrial Change*, Routledge, London, pp. 286–308.

Madsen, O.O. (1986) Denmark. In P. Burns and J. Dewhurst (eds), *Small Business in Europe*, Macmillan, Basingstoke, pp. 1–18.

Martin, R. (1989) The growth and geographical anatomy of venture capitalism in the United Kingdom, *Regional Studies*, 23, 389–403.

Mason, C.M. (1985) The geography of 'successful' small firms in the UK, *Environment and Planning A*, 17, 1499–1513.

Mason, C.M. (1989) Explaining recent trends in UK new firm formation rates: evidence from two surveys in South Hampshire, *Regional Studies*, 23, 331–46.

Mason, C.M. and Harrison, R.T. (1985) The geography of small firms in the United Kingdom: towards a research agenda, *Progress in Human Geography*, 9, 1–37.

Mason, C.M. and Harrison, R.T. (1989) The north–south divide and small firms policy in the UK: the case of the Business Expansion Scheme, *Transactions of the Institute of British Geographers* ns 14, 37–58.

Mazzonis, D. (1989) Small firm networking, co-operation and innovation in Italy, *Entrepreneurship and Regional Development*, 1, 61–74.

Monck, C.S.P., Porter, R.B., Quintas, P.R., Storey, D.J. and Wynarczyk, P. (1988) *Science Parks and the Growth of High Technology Firms*, Croom Helm, Beckenham, Kent.

Murray, R. (1985) Benetton Britain: the new economic

order, *Marxism Today*, November, 28–32.

NEDO (1986) *Lending to Small Firms: a study of appraisal and monitoring methods*, National Economic Development Office, London.

O'Farrell, P. (1986) *Entrepreneurs and Industrial Change*, Irish Management Institute, Dublin.

Oakey, R.P. (1984) Innovation and regional growth in small high technology firms: evidence from Britain and the USA, *Regional Studies*, 18, 237–51.

Pinder, D.A. (1986) Small firms, regional development and the European Investment Bank, *Journal of Common Market Studies*, 24, 171–86.

Piore, M. and Sabel, C. (1984) *The Second Industrial Divide*, Basic Books, New York.

Rainnie, A. (1984) Combined and uneven development in the clothing industry: the effects of competition on accumulation, *Capital and Class*, 22, 141–56.

Rothwell, R. (1983) Innovation and firm size: a case for dynamic complementarity: or is small really so beautiful? *Journal of General Management*, 8 (3), 5–25.

Rothwell, R. (1984) The role of small firms in the emergence of new technologies, *Omega*, 12, 19–29.

Rothwell, R. and Zegveld, W. (1982) *Innovation and the Small and Medium-sized Firm*, Frances Pinter, London.

Segal, N.S. (1979) The limits and means of 'self-reliant' regional economic growth. In D. Maclennan and J.B. Parr (eds), *Regional Policy: past experience and new directions*, Martin Robertson, Oxford, pp. 210–24.

Sengenberger, W. and Loveman, G. (1988) *Smaller Units of Employment: a synthesis report on industrialised reorganisation in industrial countries*, second revised edition, International Institute for Labour Studies, Geneva.

Shutt, J. and Whittington, R. (1987) Fragmentation strategies and the rise of small units: cases from the North West, *Regional Studies*, 21, 13–21.

Stern, P. and Stanworth, J. (1988) The development of franchising in Britain, *National Westminster Bank Quarterly Review*, May, 38–48.

Storey, D.J. (1982) *Entrepreneurship and the New Firm*, Croom Helm, Beckenham, Kent.

Storey, D.J. (1985) Manufacturing employment change in Northern England 1965–78: the role of small businesses. In D.J. Storey (ed.), *Small Firms in Regional Economic Development*, Cambridge University Press, Cambridge, pp. 6–42.

Storey, D.J. and Johnson, S.G. (1987a) *Job Creation in Small and Medium Sized Enterprises*, Commission of the European Communities: Programme of Research and Action on the Development of Labour Markets, Brussels, 3 vols.

Storey, D.J. and Johnson, S.G. (1987b) *Job Generation and Labour Market Change*, Macmillan, Basingstoke.

Sweeney, G.P. (1987) *Innovation, Entrepreneurs and Regional Development*, Frances Pinter, London.

Ward, R. (1987) Ethnic entrepreneurs in Britain and Europe. In R. Goffee and R. Scase (eds), *Entrepreneurship in Europe*, Croom Helm, Beckenham, Kent, pp. 83–104.

Wever, E. (1986) New firm formation in The Netherlands. In D. Keeble and E. Wever (eds), *New Firms and Regional Development in Europe*, Croom Helm, Beckenham, Kent, pp. 54–74.

6

Technological change and regional economic advance

Alfred T. Thwaites and Neil Alderman

Introduction

The economic development of Western Europe is currently seen in terms of increasing trade in goods and services in both home and overseas markets (Commission, 1987a). The capability to do this in competitive markets is increasingly regarded as depending upon the expansion of indigenous technological capacity in order to stave off the growing challenge from the major economic powers of Japan and the United States (Commission, 1986). The perceived need to resist this competition has resulted in a plethora of policies and programmes aimed at stimulating the generation of new technologies and encouraging their adoption by firms in individual member nations and more widely throughout the European Community.

The ability of different regions within Europe to resist these challenges and to take advantage of such initiatives will, to some extent, reflect spatial variability in technological capability and potential. In the absence of remedial action, those regions currently disadvantaged (the less-favoured regions or LFRs) may find themselves subject to the effects of technological change taking place elsewhere. This could occur, for example, through import penetration and substitution in their own markets. At the same time, these regions may be unable to participate in the generation and adoption stages of the technological process, which are assumed to bring positive benefits to the local economy through output growth, income generation and employment opportunities.

There is thus a need to re-create local comparative advantage by designing policies around the specific needs of individual regions. Their unique characteristics may demand alternative approaches to technological development from those appropriate and prioritised at the national or supranational levels. Technological change, of course, is only one of many factors that may be relevant to economic regeneration in any particular instance and needs to be incorporated with others in such strategies.

This chapter opens with a discussion of what we mean by technological change in advanced industrial economies and its effect on local economic development. The debate focuses on industrial research, product and process innovation, and technology diffusion in manufacturing industry. It is argued that technological advance, thus defined, will be spatially variable, and evidence is presented to demonstrate the nature of this variability in terms of the distribution of inputs to and outputs from the innovation and diffusion process. This is followed by an outline of current public and private sector policy responses to the problem of a technologically lagging Europe. The chapter concludes with a section on the need for a regionally oriented technology policy to address local economic development issues and to avoid the creation of a 'two-speed' Europe, with all its attendant economic, social and political problems. In this respect it is complementary to Chapter 11. While the term 'Western Europe' can be taken to apply to all those countries between the Eastern bloc and the Atlantic, including Scandinavia, the bulk of the discussion and evidence presented below relates to the twelve-member European Community.

Defining technological change

Before examining the evidence for spatial varia-tion in patterns of technological change, it is necessary to define the terms that are applied to this phenomenon. Following Schmookler (1972) it may be suggested that the technological capacity of an economy can be defined as the accumulated body of scientific and technical knowledge, weighted by the number of persons who have access to that knowledge and the ability to manipulate, interpret and use it. Technological change, broadly defined in this way, can therefore come about through an increase in the stock of knowledge, through an increase in the number of people with access to it and through an increase in the use of this knowledge to create particular outcomes, for example new products or processes. It may also occur as a result of these factors acting in combination.

The process by which new technological deve-lopments come about is often visualised as a spec-trum of activity that starts with increases in basic scientific knowledge. This is applied to the development of specific products or processes, which are then diffused or spread throughout the economy until they finally reach a state of obsolescence and are rejected from the economic system. Whilst this schema provides a useful set of pegs on which to hang various facets of techno-logical change, it is in reality a recursive system of development rather than the linear progression implied. Much basic scientific knowledge finds no useful commercial application, and many commer-cial innovations either fail economically or undergo further development and transformation before they reach the stage of obsolescence.

The generation of new knowledge, and the development of new products and processes, forms the range of activities termed research and development (R & D) (OECD, 1981). Different forms of R & D are undertaken by different types of organisation and in different locations. Basic research, the quest for knowledge without specific economic goals in mind, is most usually under-taken in institutions of higher education (HEIs) and other publicly sponsored research establish-ments. Commercial firms tend to concentrate upon applied research, giving rise to inventions that may be exploited in the marketplace. This exploitation often requires considerable sums to be expended on what is termed 'development work' to translate ideas and inventions into marketable propositions, including prototypes and pilot developments.

While R & D activities contribute to the economy through effective demand for goods and services, the greatest impact of technological advance comes from the subsequent stages. This is the point at which innovation takes place and it is generally defined as the first commercial applica-tion of some new technique (Freeman, 1974). It is at this stage that technology begins to have a considerable impact on the structure of economic activity, changing the demands for capital, materials, labour and skills, and introducing the possibility of substitution as consumers make choices between competing goods within the limits of scarce resources. It is the diffusion of innova-tions that drives this process — the wider the diffusion, the greater the impact on economies and society — and during this process many incremental changes and modifications to the original technique take place, making the original innovation more widely applicable and acceptable to more consumers (Gold, 1981).

Not all innovations represent the same degree of change or impact upon the economy. Fundamen-tal changes, such as the development of the steam engine or the microprocessor, diffuse widely and influence every aspect of industrial activity (Freeman, 1985). These changes are generally accompanied by further substantial changes that serve to broaden the sphere of influence of the initial development, such as the railway engine or the personal computer. Such changes help to take the fundamental innovation into previously unrelated areas of economic activity and therefore have an appreciable economic impact. Subse-quently there is likely to be further incremental change, increasing the acceptability of the original innovation, for example in the development of microcomputers dedicated to arcade games or word processing. Lastly, but by no means insignificantly, there are changes brought about to meet the demand of individual customers (customisation), such as tailored software developments for the personal computer in the previous example.

This chapter focuses on substantial and incremental technological change in the form of new product or new process developments. There are, however, as Schumpeter (1934) suggested, other forms of innovation, such as organisational, material or market innovations that may be just as important from an economic development point of view, but which are not considered here.

Technological change and local economic development

Technological change is not a new phenomenon; it has long been regarded as important to economic growth, overcoming the limitations imposed by the simple expansion of factors such as capital and labour and the exhaustion of easily extractable natural resources (Bacon, 1625; Defoe, 1797; Mill, 1862; Arndt, 1975). It is seen by some as the key factor underlying the 'long waves' of economic growth that have been identified for most of the advanced industrial nations (Schumpeter, 1934; Mensch, 1979; Freeman, 1985).

In a pioneering study, Solow (1957) estimated that only one-eighth of the growth of the US economy between 1909 and 1949 could be attributed to the increased inputs of factors such as capital and labour, and concluded, somewhat boldly, that the remainder was the result of technological progress. Denison and Poullier (1968), adopting a more sophisticated approach to the analysis, revised Solow's estimates downwards, but still came to the conclusion that advances in knowledge had made a significant contribution to the economic growth of the United States, the United Kingdom and West Germany in the post-war period. Further evidence assembled by Rothwell and Zegveld (1981) confirms this association between technological and economic advance at the national level. (For a more cautionary note see Gold, 1987.)

Technological change is also associated with economic growth at the industrial sector level. In his work for the Bolton Committee which reported in 1971, Freeman showed that those industries which expended a large proportion of their revenues on R & D produced the largest numbers of significant innovations in the UK economy in the post-war period (Bolton, 1971). New products or services generated in this way tend to differentiate the market in favour of the innovator. New process innovations can improve the quality and reliability of goods and services and lower the costs of inputs for each unit of output. Both types of advance can improve the

sales potential of establishments and enterprises (Porter, 1985; Nolan *et al.*, 1980; Odagiri, 1985). Moreover, studies of the UK mechanical engineering industry have shown that technical change is an important ingredient in export success (Rothwell, 1979; 1980). As early as the late 1960s, Mansfield (1968) showed that growth in productivity of the firm was positively related to the cumulative technology-related expenditures made by the firm, findings that have been confirmed at the industry level by the OECD (1986). Advances in technology have also been shown to be associated with the source and scale of profits (Baily, 1972; Schwartzman, 1976; Nolan *et al.*, 1980). Thus, investment in the generation or adoption of new techniques appears on average to bring about increased prosperity in the future, with resultant profits providing firms with an opportunity to retain a technological lead, or at least a leading market position.

These are all important attributes from the point of view of individual regions or nations, particularly those dependent on trade for at least part of their economic expansion. Productivity gains in one enterprise, industry or region can have spillover effects into other areas of the economy as the benefits are passed on through inter-firm, inter-industry and inter-regional trade in goods and services (Schott, 1981). The achievement of technology-led productivity growth in individual regions can therefore be seen as beneficial to both the national economy and the wider European economy.

The assumption that all regions are equally capable of participating in this process is not tenable. The early work of Mansfield, Freeman and others clearly demonstrated that the level and rate of technological change varied not only between industries, but also according to the size of the firm. Small and medium-size enterprises (SMEs) made a major contribution in only a few industries (Townsend *et al.*, 1981). More recent evidence suggests that small enterprises may be particularly important in the subsequent diffusion and adaptation of new techniques to the wider needs of the economy (Rothwell and Zegveld, 1982; Sweeney, 1987). It has long been argued that, because of regional variations in industrial and enterprise characteristics, the type and pace of technological change will vary between regions (Thwaites, 1978), with potentially serious consequences for the development of the local economy in currently technologically disadvantaged regions.

Of major concern in this respect is the nature of the relationship between technological change and employment, this being a key indicator of economic well-being, from the perspective of both the individual and the region. This relationship remains largely unquantified owing to its contingent nature. Employment impacts can be positive or negative, short- or long-term, and direct or indirect (the latter relating to changes in the nature as well as the number of jobs). Identification of these impacts is also hampered by changing macro-economic conditions, and by the nature of the change taking place (for example, product or process innovation). It is also dependent upon whether measurements refer to the individual establishment or the enterprise as a whole, because increased employment within a multi-locational enterprise may reflect decline in some locations, offset by increases in others.

In general, the available evidence points to the conclusion that innovative industries appear to perform much better in employment terms than those in which technology advances more slowly. Evidence from the OECD (1986) suggests that employment generation occurs within industries advancing rapidly in terms of technology. For example, in the production of computer software products and of computer systems, employment grew by 53 per cent and 26 per cent, respectively, between 1983 and 1984 in a group of computer service companies in the UK (European Computing Services Association, 1985). A positive relationship has also been found between product development and employment change within manufacturing industry in the UK, while process change appears more likely to lead to employment loss (Thwaites, 1983; Vickery, 1986).

If advancement in the generation and implementation of technical knowledge, and in the subsequent development and adoption of new techniques, has such positive effects upon enterprises and, in aggregate, local and national economies, it is perhaps not surprising that there should be such a high level of interest in the technological capacity of regions. Technology appears to provide the opportunity for the reconstruction of regional comparative advantage. The evidence suggests, however, that not all regions will be starting from the same technological base, nor will they benefit equally from

current national and supranational technology-based programmes as a result.

The evidence for spatial variation

Input measures

Given our definition of technological change, it is clear that any useful measures of such activity can only be indicators of certain aspects of the process. As already identified, a key input to technological change comes in the form of R & D, and for this there are at least aggregate secondary statistics available, albeit of varying quality. Figures published by the OECD (Tables 6.1 and 6.2) reveal that at the international level the United States dominates gross world R & D expenditure in absolute terms. These data also indicate the position of Japan as a major force in world research.

In Europe research is dominated by West Germany, France and the United Kingdom, who between them account for about 75 per cent of R & D manpower employed within the EEC (Commission, 1987b). Although there is some evidence that smaller countries are starting to devote more resources to R & D, they are building on a small base and may find themselves unable to overcome barriers to entry in research fields dominated by one or more of the larger, higher-spending, countries (in advanced telecommunications equipment, for example).

A further difference between nations lies with the relative shares of government and business funding of R & D (OECD, 1986). Business expenditure on R & D appears to be particularly low in countries such as Greece, Ireland and Portugal. The importance of this type of R & D is its link between technological developments and commercial applications, with attendant economic impacts. The implication is that countries lacking this sort of activity will be less well placed to take advantage of the economic opportunities offered by technological change.

Clearly, levels of R & D vary between nations, but there are also significant regional variations within nations. The most evenly distributed type of research is probably that basic research undertaken by HEIs, which, in the UK at least, tend to be based in most regional centres (Buswell *et al.*, 1985). It must be recognised, however, that there

Table 6.1 *Gross Domestic Expenditure on R & D: national trends*

	1981		Compound real growth rates (%)		1983
	($m)	%[b]	1975–81	1981–83	($m)
United States	73,678.0	46.4	4.2	3.8	88,329.0
Japan[c]	25,574.5	16.1	7.9	8.2	22,493.7
West Germany	15,644.8	9.9	4.7	1.9	18,130.2
France	10,700.8	6.7	4.2	4.7	13,134.4
United Kingdom[a]	11,369.8	7.2	3.1	− 0.7	12,552.8
Italy	4,546.6	2.9	4.6	4.9	5,568.0
Spain	908.0	0.6	3.4		
Netherlands	2,508.4	1.6	0.9	3.5	2,992.4
Sweden	2,166.7	1.4	5.6	7.1	2,776.8
Belgium[d]	1,065.7	0.9	4.1		
Switzerland	1,785.4	1.1	0.8	− 0.5	1,979.8
Austria	765.2	0.5	6.9	4.5	932.9
Denmark	539.8	0.3	2.9		
Norway	593.0	0.4	3.3	6.4	757.6
Greece	102.1	0.1			
Finland	499.3	0.3	7.7	8.5	655.2
Portugal[e]	154.5	0.1	6.9		
Ireland	155.3	0.1	2.7		
Total OECD[a]	158,720	100	4.5	4	
of which EEC[a]	46,940	29.5	4	2.5	

[a] OECD estimates.
[b] Secretariat estimates are provided where 1981 figures are not available.
[c] Data adjusted by OECD would be $23,408 million in 1981 or 15 per cent of the OECD total and, in 1983, $30,750 million.
[d] 1979.
[e] 1982.
Source: OECD/STIIU data bank, December 1985. Derived from OECD (1986).

are inevitably quantitative and qualitative differences in research output between institutions and therefore between regions. Moreover, the development of specific research specialities within particular institutions means that not all regions may have the same basic research capabilities or opportunities for commercial exploitation. Nevertheless, HEIs in the less-favoured areas are frequently regarded as basic building blocks around which to construct future regional technological and economic capabilities. Regional research specialities could form one component of any new regional comparative advantage.

Other government-supported R & D has been shown to be highly spatially concentrated, and this is particularly well illustrated in terms of the UK defence sector (Breheny *et al.*, 1987; Buswell *et al.*, 1985). The importance of this type of research unit from the point of view of local economic development stems from their attraction of 'high-technology' industries, which benefit from close association with such units through government defence-related contracts and their potential for creating spin-offs and conditions for further technological advance. The establishment of a local high-skill labour market for researchers reinforces these tendencies (Breheny *et al.*, 1987).

The identification of the regional patterns of research financed by industry is fraught with problems owing to the paucity of suitable data. It is therefore necessary to rely on more limited survey data. Surveys of manufacturing establishments in a number of metalworking industries suggest that a higher proportion of southern establishments in Great Britain are undertaking R & D of some kind than their northern counterparts

Table 6.2 *Business enterprise R & D expenditure: national trends*

	1981 Million $	%[b]	1983 Million $
United States	51,810.0	50.2	62,816.0
Japan	15,517.3	15.0	21,270.0
West Germany	10,686.3	10.4	12,649.0
France	6,304.4	6.1	7,461.1
United Kingdom	7,029.7	6.8	7,662.9
Italy	2,563.1	2.5	3,179.0
Spain	442.7	0.4	
Netherlands	1,336.0	1.3	1,607.2
Sweden	1,442.1	1.4	1,874.6
Belgium	950.1	0.9	1,148.7
Switzerland	1,324.8	1.3	1,471.0
Austria	427.4	0.4	
Denmark	273.8	0.3	372.3
Norway	309.1	0.3	414.2
Greece	22.9	0.0	
Finland	272.9	0.3	372.3
Portugal[c]	48.3	0.0	
Ireland	67.7	0.1	
Total OECD[a]	103,120	100	
of which EC[a]	29,230	28.5	

[a] OECD estimates.
[b] Secretariat estimates are provided where 1981 figures are not available.
[c] 1982.
Source: OECD/STIIU data bank, December 1985. Derived from OECD (1986).

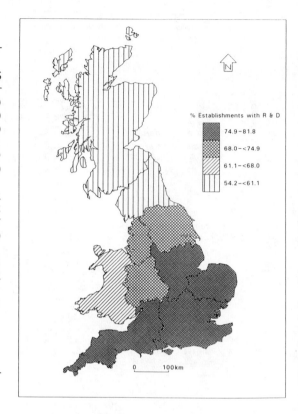

Figure 6.1 *Regional levels of on-site R & D, Great Britain*
Source: Thwaites *et al.* (1981, 1982).

(Figure 6.1). A recent study for the European Commission (Commission, 1987a; 1987b) has, however, attempted to collect data on industrial R & D expenditures and employment for level II regions of the European Community.[1] Notwithstanding problems of data comparability and missing data, this showed, against a twelve-member EC average, the dominance of such regions as Cologne, Darmstadt, Upper Bavaria and Mittelfranken in West Germany; the Ile de France and the Auvergne in France; Utrecht in the Netherlands; and Hovedstads in Denmark. At the other end of the scale, within the same countries, regions such as Weser-Ems and Trier in West Germany, Nord-Pas de Calais in France, and Ost For Storebælt came well below both national and Community averages. Even more problematic, all regions of Spain, Portugal, Greece and southern Italy scored less than a third of the Community average. This evidence suggests that there are wide variations in the extent to which R & D is undertaken by industrial enterprises in different regions of Europe. This can be expected, *ceteris paribus*, to translate into different abilities to generate and use new technologies, with the greatest problems appearing to lie in the southern half of Europe.

Output measures

While employment and expenditure on R & D provide crude measures of the potential to bring about technological change in a region, measures of actual output from this process are even more difficult to find with any degree of reliability. Despite many recognised shortcomings (for a discussion, see Pavitt, 1982) patents are a frequently used indicator of research output, as they represent some degree of novelty. At a

Table 6.3 *Trends in UK regional shares of innovations*

	1945–59 (%)	1960–69 (%)	1970–80 (%)	Number of Innovations
East Anglia	3.8	6.4	3.8 (3.1)	110
South-west England	3.8	4.0	4.5 (5.1)	95
South-east England	36.7	31.2	34.0 (33.1)	770
North England	5.9	5.3	3.7 (4.5)	111
East Midlands	6.8	9.5	8.2 (8.6)	192
Yorkshire and Humberside	6.4	11.5	11.7 (12.7)	237
West Midlands	13.1	8.5	8.4 (9.2)	219
North-west England	12.5	13.3	18.6 (15.8)	346
Scotland	9.3	7.0	5.2 (5.0)	158
Wales	1.3	2.5	1.7 (2.4)	44
Northern Ireland	0.5	0.8	0.1 (0.1)	11
Total %	100.0	100.0	100.0	
no.	559	872	862	2293

Note: Numbers in parentheses for the period 1970–80 are the weighted percentage contributions, assuming the same sectoral mix as in the period 1945–69.
Source: Townsend *et al.*, 1981.

regional level, the worst problems with patents arise, firstly, because not all enterprises and industries place the same level of value on them as a means of protecting short-term monopoly profits, which creates a structural effect; and, secondly, because the registration of patents by head offices may fail to identify the location of the actual research work that gave rise to the patent.

Notwithstanding these difficulties, patents have been used in a recent study in the Italian context (Boitani and Ciciotti, 1988). This reveals that patenting in Italy is dominated by the regions of Lombardy (41.4 per cent of patents granted between 1970 and 1986), Piedmonte (16.1 per cent) and Emilia Romagna (12.4 per cent), all of which lie in the north of the country. Over the same period Calabria and Basilicata in the south both accounted for less than 0.05 per cent. If

patents can be regarded as an indication of research output, it would appear that the north of Italy is advancing rather more in technological terms than the south.

Similar regional variations are apparent when one measures the appearance of new goods or services in the market. The Innovation Data Bank compiled at the Science Policy Research Unit at the University of Sussex, UK, lists innovations which have been important to UK industry in the post-war period (Townsend *et al.*, 1981). These innovations were selected by a number of independent industry experts as having made a significant contribution to the development of their sector, and 2,293 were listed at the time of Townsend's analysis. These data permit a crude regional breakdown of innovative performance over time (Table 6.3). This analysis shows that those UK regions that historically have faced the greatest economic difficulties have an incidence of substantial innovation that has been declining over time and is currently below their relative shares of manufacturing employment and manufacturing establishments. In other words, there is a decline in the share of important technological breakthroughs occurring in the more peripheral parts of the country.

Innovation of a more incremental nature has also been shown to follow a similar pattern, particularly in terms of product change (Thwaites *et al.*, 1981). Such regional differences have been shown to be only partly attributable to such factors as industrial or corporate structure (Alderman *et al.*, 1982; Alderman, 1986). Moreover, product innovation in economically lagging areas often relies heavily on the transfer of new products within corporations or the use of licences or franchises, rather than being based on in-house development. Studies undertaken in a number of European countries confirm the conclusion that there are interregional differences in the generation or modification of products (for example, Northcott and Rogers, 1984; Kleine, 1982). Brugger and Stuckey (1987, p. 246) conclude that:

The collective results . . . demonstrate that significant territorial disparities in innovative behaviour exist in Switzerland, as in other countries. Firms and plants in centrally located, highly populated regions are, as a rule, more innovative than firms in rural and peripheral regions.

The adoption of new processes

Complementing the introduction of new products or services to the marketplace is the development and introduction of new equipment and processes. In this way, the products of one industry often become the processes of a number of others. The importance of new production techniques lies with the possibilities for improving the quality and reliability of products, together with increases in efficiency in terms of the output achievable for each unit of input. In certain circumstances the development of new processes is a prerequisite for new product production at a competitive price (for example, in semiconductors). Such process innovation is essential if enterprises and regions are to remain competitive in price and quality

terms and not be left behind on the production learning curve.

Clearly, any regional comparison of the uptake of new process technology can only be made in a meaningful sense if the technology is reasonably widely applicable, otherwise differences in industrial structure will tend to obscure any patterns. For the purposes of this discussion we concentrate on two aspects of process technology that have important implications for a wide range of sectors: computer-numerically controlled (CNC) machine tools and programmable robots. Their importance arises partly as a result of fundamental changes in market conditions (which have resulted in saturation and overcapacity in many sectors; see, for example, Piore and Sabel, 1984) and partly because of the need to respond to both market fragmentation and rapidly changing consumer demands (Mensch, 1979). This has meant that many manufacturers who previously organised their production on a flow-line or massproduction basis have had to adjust to handle small or medium-sized batches of components or products. Programmable equipment allows manufacturers to shift from one product to another in response to different customer orders in a much more flexible, and hence efficient, way. The rate of uptake of such technologies is therefore, *ceteris paribus*, likely to be a good indicator of the ability of different regions to adjust successfully to such changes in global market conditions.

CNC machine-tools, principally applied to metalworking operations, offer important improvements in speed, quality, flexibility (reprogrammability) and scope (the ability to handle complex shapes) (Ray, 1984). There has been a phenomenal growth in the stock of CNC machine-tools throughout the 1970s and 1980s. In West Germany, for example, the number of these tools (including NC machine-tools[2]) is estimated to have risen from around 25,000 in the early 1980s to over 60,000 in 1985. As a proportion of the total machine-tool stock (5.7 per cent), this rate of uptake exceeded that of Japan, with the UK rate lagging behind that of Japan (Metalworking Production, 1988). The same sources suggest that the proportion of Italian machine tools that are NC or CNC is much higher (15.2 per cent in 1985), but this is on a much smaller total machine-tool population. Although the surveys on which such figures are usually based

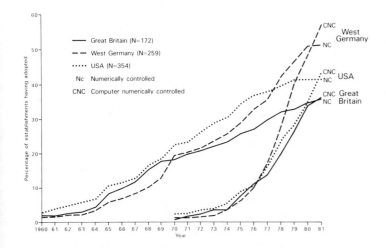

Figure 6.2 *The adoption of NC and CNC machine tools, Great Britain, the USA and West Germany*
Source: Gibbs and Thwaites (1985).

have to make quite bold assumptions (see Prais, 1987), nevertheless the growing importance of this technology is undeniable.

Comparison of adoption rates is complicated by differences in industrial structures among nations and regions, but results of a survey undertaken in Great Britain, West Germany and the United States in 1981 have been standardised to deal with this variation. Based on this survey Figure 6.2 shows the comparative rates of adoption by establishments within the agricultural machinery, metalworking machine-tools and construction equipment sectors. The apparently superior performance of West Germany reinforces the picture provided by other sources.

These international differences are influenced by numerous factors, such as the respective establishment and enterprise size distributions, organisational structures and technical considerations. The fact that new technologies may or may not be introduced into particular establishments for a whole range of reasons means that regional variations should be expected. Figure 6.3 shows how the proportion of surveyed establishments adopting CNC machine-tools varied regionally within Great Britain in the international study referred to above. Data provided by Ciciotti (1982) illustrate the case of the Italian mechanical engineering and vehicles sectors (Table 6.4). These data all provide evidence to suggest that, in the early to middle stages of the diffusion process, adoption levels are likely to be higher in the industrial heartlands of Western European nations than in

the peripheries. More recent work demonstrates how, as the technology matures over time, regional convergence can be expected as the number of non-adopting establishments decreases (Alderman *et al.*, 1988).

This is not to imply that regional variations are no longer important. Rather, it reinforces the fact that lagging regions are likely to be catching up in terms of the adoption of mature technologies, after the initial economic benefits of leadership have been realised elsewhere. Furthermore, it is likely that regional disparities will be reappearing in the adoption of newer technologies. Regional convergence also reflects the fact that the way technological change is commonly measured may be increasingly inappropriate for identifying differences. In the case of technologies such as CNC, what matters is not so much the numbers of establishments that have adopted in different regions, but the ways in which they are using the technology. This latter aspect of technological change, needless to say, presents a much more challenging measurement problem.

The term 'programmable robot' is widely used, but is often loosely defined. Following Cox (1984, p. 75) a useful definition is: 'a reprogrammable multi-functional manipulator designed to move material parts, tools, or specialised devices through variable programmed motions for the performance of a variety of tasks'. Programmable robots can increase speed and flexibility in the handling of materials or in the performance of operations such as welding or painting, and they

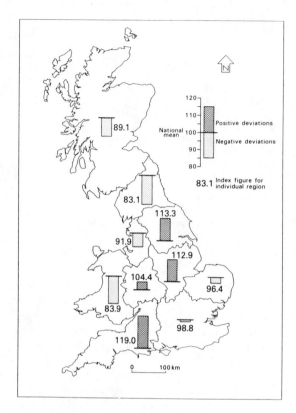

Figure 6.3 *Regional index of CNC machine-tool adoption, Great Britain, 1981*
Source: Thwaites *et al.* (1982).

Table 6.4 *Regional levels of NC/CNC adoption, Italy*

	% of establishments
Lombardy	59.1
Emilia Romagna	41.1
South	25.0

Source: Ciciotti (1982).

have progressed the most in recent years and it is estimated that, if simple pick-and-place robots were to be included in these figures, Japan had some 80,000 robots by 1984 (Lewis *et al.*, 1985). The extent to which this type of equipment has penetrated the economy in these countries should not be overstated, however, since the majority of programmable robots will have been introduced by large manufacturers, notably the vehicle producers. Nevertheless, it is anticipated that by 1990 the market for such equipment will be of the order of $3 billion, a twelvefold increase over the 1980 level (Lewis *et al.*, 1985).

Evidence of robot adoption at the regional scale is extremely fragmentary. In Italy, Camagni (1984) has estimated that the north dominates the robot scene, especially in and around the major industrial centres of Milan and Turin (home of the Fiat car company). The south of Italy, or Mezzogiorno, is almost completely lacking in this form of advanced manufacturing technology. In the UK, Northcott *et al.* (1986), on the basis of a survey of 1,200 establishments, discovered that the North-West region had the highest level of robot penetration, with 3 per cent of factories using them, followed by Yorkshire and Humberside and the East and West Midlands (Figure 6.4).

An interesting discovery of this study was the expectation that the highest increases in robot numbers would be in the peripheral areas of the north and not in the more prosperous core regions of southern Britain. The introduction of robots could prove to be a double-edged sword from the point of view of these peripheral economies. On the one hand, investment in new technology is seen to be a good thing and something to be encouraged. On the other hand, the introduction of robots may be a new reflection of the 'branch plant economy' syndrome (Watts, 1981), whereby investment is made by remote parent companies

are particularly useful in hostile environments. When used in conjunction with CNC machines and automatic material-transport systems, they form part of what are known as 'flexible manufacturing systems', in which all the hardware is controlled by a central computer, permitting flexible scheduling of jobs and optimal routing of component parts. In these situations the robot is more than simply a cost-cutting technology; it has the potential to add value to production processes.

Although robots of some description (for example, pick and place) have been available for decades, their diffusion and use has really only taken off with the advent of the microprocessor and the development of the programmable robot. The data in Table 6.5 suggest that, despite inevitable problems of comparability between national statistics, ten years ago the United States and Japan had a commanding lead in the application of these technologies. It is the Japanese that

Table 6.5 *Robot population worldwide**

	1974	1978	1980	1981	1982	Average Annual Growth Rate (%)	
						1981–82	1974–82
Japan	1,500	3,000	6,000	9,500	13,000	37	31
USA	1,200	2,500	3,500	4,500	6,250	39	23
West Germany	130	450	1,200	2,300	3,500	52	51
Sweden	85	800	1,133	1,700	1,300[†]	–[†]	(41)
Great Britain	50	125	371	713	1,152	62	48
France	30		580	790	950	20	54
Italy	90		400	450	790	76	31
Netherlands	3	4	49	62			

* Based on a narrow definition of robots.
† Change in the definition used for statistical purposes.
Source: Fletcher (1984).

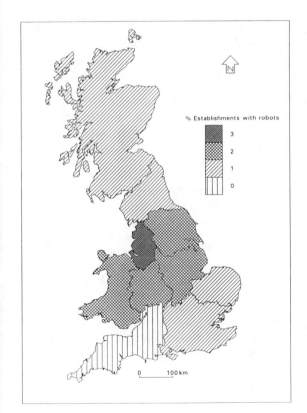

Figure 6.4 *Robot adoption rates, Great Britain, 1985*
Source: Northcott *et al.* (1986, Table 6).

with the objective of replacing labour in order to increase productivity in the manufacture of mature products. Achieving this objective would allow companies to siphon off the economic benefits to the headquarters region. The evidence, as suggested earlier, is conflicting on this account. In some instances robots are clearly intended as labour-saving (and hence job-displacing) devices, but other evidence suggests investment in robots can lead to future profitability and job growth (*Computer Weekly*, 1987).

Summary

The evidence presented here indicates that the rate of technological advance, and the potential for future change, vary between countries and regions within Europe. As a whole, Europe appears to be at something of a disadvantage in technological terms with respect to its major competitors, the United States and Japan. Most notably, despite the difficulty of comparison, some regions within Europe, and particularly those in southern Europe, appear to be especially vulnerable to the technological challenges from other regions in Europe and elsewhere.

Observation would suggest that, in the absence of strong ameliorative policy, LFRs in Europe will benefit from new processes of technological change mainly as recipients of their outputs, in the form of imported products and services, rather than as generators of new technology and its wealth-creating potential (Williams *et al.*, 1986). Similarly, the ability to introduce advanced

101

technologies into the production process, and advanced components into existing products, to achieve growth in output, market share, or improvements in other appropriate performance indicators, seems limited. Such observations underline the challenges facing those aiming to rebuild the comparative regional advantages of the LFRs.

Public and private sector policy responses

An obvious response from the public sector, at supranational, national or local levels, has been the identification of specific areas of technology for special encouragement and support, for example through IT Year (British Business, 1982; Commission, 1986). These technologies include electronics, biotechnology and new materials, such as ceramics. They have been widely promoted by many nations. This kind of targeting has tended to be extended to cover anything that may be remotely regarded as 'high technology' and (largely by definition) research-intensive. By implication, therefore, such technological support will only benefit LFRs to a limited extent, as this type of activity has been shown to be generally lacking in such areas. These technology-based programmes therefore appear largely to militate against participation by the LFRs.

In order to try to combat the technological challenges from the United States and Japan, the public and private sectors in Europe have come together in R & D ventures, such as the ESPRIT programme, to encourage the collaboration of research institutes and large and small firms from research-intensive and research-constrained regions of the European Community. What remains largely untested is whether or not such activities divert scarce resources in LFRs away from other research projects of more fundamental value to the local economy. Moreover, it is not known whether there may be a loss of sovereignty over any resultant research outputs, because of the lack of a local capability to exploit them. Anecdotal evidence suggests that this may indeed be the case in certain parts of the UK (Commission, 1987a; 1987b).

In many of these supported research areas, and perhaps in part because of public sector interest, competition is intense. The costs of participating in some high-technology areas are rising rapidly to the extent that most enterprises, and even some nations, cannot support the expenditure necessary to keep pace with developments (Charles, 1988). Enterprises based in LFRs may therefore be effectively barred from entry to many parts of the new industries. One outcome of this has been the establishment of joint ventures and research agreements linking enterprises around the world (Williams *et al.*, 1986; Charles, 1988). Yet the majority of enterprises party to these agreements appear not to be located in LFRs. Further research is needed to investigate the forms of these agreements between the large firms and their implications for smaller firms and local-area development (Charles, 1988), particularly in the light of the Single European Act (Commission, 1987c).

There is also increasing pressure in Europe to improve the links between HEIs and commercial firms in the belief that this will help improve the competitive position of industry, both nationally and internationally, by reorienting HEI research towards more practical outcomes. (An additional motivation for this trend is to provide HEIs with much-needed revenue.) Once again, however, the ability of individual regions to respond to such opportunities as may arise through this process is likely to vary.

Although considerable research has been carried out to demonstrate the extent of (or limits to) the relevance of HEI links to industry in the innovation process, (see, for example, Oakey, 1984; Howells, 1986) there is a need to establish who benefits where from such collaboration. Similarly, the implications of vigorously pursuing such a policy require clarification. Flood (1984), for example, suggests that when such links are forced on HEIs for economic reasons this may jeopardise their long-term future in research terms, because they become involved in unstable short-term research projects, rather than stable long-term programmes of work. The criteria that are used in determining selectivity in HEI research programmes are critical, and there is always a danger that recommendations will be inappropriate for particular regions and regional institutions. High-technology, high-risk research areas are unlikely to be pursued successfully by all regions.

At a general level, it is argued that policies designed at the macro scale may not realistically meet the technological or economic needs of individual LFRs delineated at the subnational scale (Commission, 1987; 1987b). However, there have been moves in recent years to address more closely the technological problems of local areas through the regionalisation of technology policy and the introduction of initiatives at the regional level. This development has perhaps gone furthest in France, where the government offers incentives for research decentralisation outside the Paris region, but recent research for the European Commission (1987a, p. 145) concluded:

Overall, however, national regional policies which have a specific RTD [technology] component appear not to have spread rapidly within the Community. Instead regionalisation of RTD programmes and more fundamental shifts towards providing greater regional autonomy, as in Spain, appear to have been more common.

Such approaches tend to promote new technologies but, because what is best practice at the national level may not be suitable at the local level, there is a danger that in the LFRs they may increase confusion as to what are appropriate technology strategies. Partly in response to this problem, in recent years attempts have been made — particularly by local authorities and other agencies — to satisfy local technological needs. There are difficulties, however. In countries such as Greece or Portugal, limits to the national technological capacity transcend any problems that may be identified at the local level. On the other hand, in countries such as the UK, constraints on the nature and direction of local authority expenditure have meant that the amount devoted to industrial and technological support has necessarily been small. This often fails to overcome thresholds of scale and scope in technological advance. In some regions a further problem is that the expertise needed to support local technological development is absent. With the exception of West Germany, which operates a local support system for technological advance, and similar developments in local autonomy for technology policy in Spain and the Netherlands, there is currently only a limited policy response throughout the rest of Europe to the particular technology needs of LFRs.

Towards an understanding of the role of technology in the rebuilding of regional comparative advantage

The previous sections have attempted to demonstrate, firstly, the importance of technological change to local economic development and, secondly, how this can be seen to vary between nations and regions within Western Europe. It has also been suggested that current policy responses to these spatial differences are likely to reinforce the present disparities, because their perceived role is primarily to offset competition from the United States and Japan. It should be apparent by now that attempting to stimulate developments in so-called 'high-technology' industry will not be a universal panacea for all disadvantaged regions, for the reasons alluded to above.

Furthermore, it may be argued that part of the problem lies with the fact that, so far, research has failed to provide a complete explanation of the differences in technological capability which have been found to exist between different regions within Europe. In part this is a consequence of the way in which technological change has been conceptualised and measured. Typically, it has been treated as a series of discrete events — the introduction of some new innovation (product or process) — rather than as the continuous process that in reality it is. Much research has also stopped short of true explanations, relying on statistical associations for its understanding of the factors that influence technological change. While there can be no doubt of the relevance of many of these factors, the strength of their causality, and their role at different stages of the process of technological change, remain assumed rather than proven.[3]

Some authors (for example, Thwaites and Alderman, 1988; Molle, 1988) are beginning to argue that what is necessary is not simply a description of changing patterns of technological activity at the regional level. Instead, a fuller understanding is required of the precise process that produces these patterns, in order that the role of technological change in the rebuilding of regional comparative advantage may be better understood and effective policy leverage points identified. Blanket policy prescriptions based upon perceived common goals fail to recognise that specific regional conditions for effecting technological change are likely to vary and that, in consequence, policy needs to be tailored to these regionally specific requirements.

Technological change, as we refer to it here, occurs within individual workplaces or establishments. The specific form that it takes will reflect conditions internal to the establishment, particularly with respect to the processes of decision-making, bargaining and conflict resolution that take place within and between different interest groups. But decisions or choices about what technology is developed or introduced will also be influenced by conditions external to the establishment. These conditions will reflect the aims of the corporation and its structure, within which the establishment is embedded, or, in the case of the independent establishment, the nature of its relationships with other establishments, especially large corporate bodies. Through the influence of corporate decision-making, the impact of global economic changes will be felt at the establishment level. External influences will also be manifested through the local (as opposed to the global) economic environment, such as the effect of the local labour market, institutional frameworks and public policy at the regional level.

Only through a comprehensive understanding of the interrelationships of these various conditions, and their influence at different points in time during the process of technological change, will it be possible to develop policies that are sensitive to the needs and aspirations of individual regions. This framework begins to convey the complexity of the process involved. This very complexity, however, has the advantage of explicitly recognising the individuality of regions; individuality which not only brings varying responses to technological advances, but also varying capabilities and needs for such advances in the first place.

Some indication of the individuality of regions is provided in the Third Periodic Report, in which the European Commission (1987d) classifies the lagging regions of the EC into six types on the basis of common characteristics and problems:

1. less-developed or (backward) regions;
2. declining industrial regions;
3. agricultural regions;
4. urban problem regions;
5. peripheral regions (including islands);
6. frontier regions.

The less-developed, agricultural and peripheral regions are characterised by low population density, below average industrialisation, high dependence on agriculture, low labour productivity and below average incomes. The declining industrial and urban problem regions are characterised by high population density, high unemployment and 'a certain weakness of incomes'. The frontier regions have specific problems arising from their proximity to regions with different legal, social, economic and planning systems (Commission, 1987d).

The Third Periodic Report argues that such typologies provide a useful mechanism for providing an indication of the types of policy measures that a particular region will require. Examples cited (Commission, 1987d, p. 51) include

the development of industries for the further processing of agricultural products in agricultural regions, the establishment and extension of time-saving, efficient and cheap transport and communications systems in peripheral regions, the removal of obstacles to urban renewal and development in urban problem areas, and the reclamation and conversion of derelict industrial sites in declining industrial regions.

This serves to illustrate the great diversity of regional characteristics and problems in Europe, from which it should be clear that blanket programmes for selected technologies will not provide solutions in all cases. Rather they will benefit those regions most able to respond.

Clearly, the nature and role of technological change in these different types of problem region will be different, and the use and encouragement of technological change as a policy instrument needs to recognise these differences. The individuality of regions, and their capabilities and potential with respect to local, national and international markets, need to be identified so that they are not automatically put in direct competition with every other region for technological supremacy. The challenge to Europe is not only to stave off competition from the United States and Japan, but also to devise schemes for individual regions that generate and apply technology in accordance with their needs, without leaving them totally open to technological competition from the more advanced regions within Europe. If this is not achieved, the emergence of a 'two-speed Europe' will be an inevitable outcome.

At the supranational level the European

Community has already commissioned research designed to provide details of the needs of individual regions in technological terms, and the mechanisms necessary to support technological activities at the local level (Commission, 1987a; 1987b). Any emergent programme will be designed to support national and local efforts within member states. It will be interesting to see how quickly this approach to local technological change diffuses and is taken up within national and regional economies.

Notes

1. A full definition of level II regions is to be found in Commission (1987d, Annex 1, p. 11). In essence, they are 'basic administrative regions': in France, for example, they are the 22 *régions*; in Belgium, the nine provinces; and, in Italy, the 20 *regioni*.
2. NC (numerically controlled) machine-tools are operated by means of programs prepared off-machine and coded into punched paper tape. They were developed and went into commercial use in the 1950s. Nowadays they have been largely superseded by CNC machine-tools, which possess a microprocessor control unit that enables programs to be prepared, edited and stored on the machine itself.
3. The implications of this work are that, if all regions successfully developed to some acceptable level the factors believed to be causally related to technological progress, then they would all be engaged in the production of the same goods and services in high-technology sectors, such as microelectronics or biotechnology. This is predicated on the notion of some idealised regional industrial structure, technological trajectory and output, regardless of whether or not the market could absorb this output and whether consumer choice might, as a result, become more limited. This idealistic vision of technologically-led regional regeneration ignores the fact that, given a scarcity of resources, factors other than technological change may produce greater multiplier effects in the short or medium term in some regions.

Acknowledgements

The authors would like to thank all those in CURDS who have contributed to the programme of research on technological change upon which this chapter draws. Secretarial assistance was provided by Denise Weites, and the financial support of the ESRC, under CURDS' Designated Research Centre funding, is also gratefully acknowledged. Digitised map outlines were provided by courtesy of Drs H. Mounsey and R. Baxter.

References

Alderman, N. (1986) A case study of the application of log-linear and logit models in industrial geography. Unpublished Ph.D. thesis, Department of Geography, University of Newcastle upon Tyne.

Alderman, N., Goddard, J.B., Thwaites, A.T. and Nash, P. (1982) *Regional and Urban Perspectives on Industrial Innovation: applications of logit and cluster analysis to industrial survey data*. Discussion Paper no. 42, Centre for Urban and Regional Studies, University of Newcastle upon Tyne.

Alderman, N., Davies, S. and Thwaites, A.T. (1988) *Patterns of Innovation Diffusion: technical report*, Centre for Urban and Regional Studies, University of Newcastle upon Tyne.

Arndt, H.W. (1975) Limits to Development, *Australian Quarterly*, 47, 79–89.

Bacon, F. (1625) *The Essays on Counsils Civil and Morale*, London.

Baily, M. (1972) Research and development costs and returns, *Journal of Political Economy*, 80, 70–85.

Boitani, A. and Ciciotti, E. (1988) Patents as indicators of innovative performances at the local level. Paper presented at the European Regional Science Association Summer Institute, Arco, Italy, July.

Bolton, J.E. (1971) *Report of the Committee of Inquiry on Small Firms*, Cmnd 4811, HMSO, London.

Breheny, M. and McQuaid, R.W. (eds) (1987) *The Development of High Technology Industries: an international survey*, Croom Helm, Beckenham.

British Business (1982) Government projects and initiatives in support of IT, *British Business Information Technology Review*, 1, 12 March, 10–11.

Brugger, E.A. and Stuckey, B. (1987) Regional economic structure and innovative behaviour in Switzerland, *Regional Studies*, 21, 241–54.

Buswell, R.J., Easterbrook, R.P. and Morphet, C.S. (1985) Geography, regions and research and development activity. In A.T. Thwaites and R.P. Oakey, (eds), *The Regional Impact of Technological Change*, Pinter, London, pp. 36–66.

Camagni, R. (1984) *Il Robot Italiano*, Il Sole 24 Ore, Milan.

Charles, D. (1988) The globalisation of R & D: technology and the formation of international strategic alliances. Paper presented to the ESRC, Urban and Regional Economic Seminar Group, Newcastle upon Tyne. (Available from CURDS, University of Newcastle upon Tyne.)

Ciciotti, E. (1982) La diffusione regionale delle innovazioni, *Politica ed Economia*, 4, April.

Commission of the European Communities (1986) *Proposal for a Council Regulation Concerning the Framework Programme of Community Activities in the Field of Research and Technological Development (1987 to 1991)*, COM (86) 430 final, Commission of the European Communities, Brussels.

Commission of the European Communities (1987a) *Research and Technological Development in the Less Favoured Regions of the Community (STRIDE)*, Commission of the European Communities, Brussels.

Commission of the European Communities (1987b) *Science and Technology for Regional Innovation and Development in Europe*, Commission of the European Communities, Brussels.

Commission of the European Communities (1987c) *Single European Act*, Commission of the European Communities, Brussels.

Commission of the European Communities (1987d) *Third Periodic Report from the Commission on the Social and Economic Situation and Development of the Regions of the Community*, COM (87) 230 final, Commission of the European Communities, Brussels.

Computer Weekly (1987) Revving up for takeover?, 23 April, 26.

Cox, A.R. (1984) Technology and manufacturing industry: United States of America. In Department of Trade and Industry, *Technology and Manufacturing Industry: an overseas review*, DTI, Overseas Technical Information Unit, London.

Defoe, D. (1797) *An Essay on Projects*, London.

Denison, E.F. and Poullier, J. (1968) *Why Growth Rates Differ*, Allen & Unwin, London.

European Computing Services Association (1985) *The Ninth Annual Survey of Computing Services Industry in Europe*, International Data Corporation, London.

Fletcher, A.M. (1984) Technology and manufacturing industry: the Federal Republic of Germany. In Department of Trade and Industry, *Technology and Manufacturing Industry: an overseas review*, DTI, Overseas Technical Information Unit, London.

Flood, J. (1984) The advent of strategic management in CSIRO: a history of change, *Prometheus*, 2, 38–72.

Freeman, C. (1974) *The Economics of Industrial Innovation*, Penguin, Harmondsworth.

Freeman, C. (1985) The role of technological change in national economic development. In A. Amin and J.B. Goddard (eds), *Technological Change, Industrial Restructuring and Regional Development*, Allen & Unwin, London, pp. 100–14.

Gibbs, D.C. and Thwaites, A.T. (1985) The international diffusion of new technology in manufacturing industry: a comparative study of Great Britain, the USA and West Germany. Paper presented at the IBG/CAG Symposium on 'Technical Change in Industry — Spatial Policy and Research Implications', Swansea. (Available from CURDS, University of Newcastle upon Tyne.)

Gold, B. (1981) Technological diffusion in industry: research needs and shortcomings, *Journal of Industrial Economics*, 29, 247–69.

Gold, B. (1987) Technological innovation and economic performance, *Omega*, 15, 361–70.

Howells, J. (1986) Industry–academic links in research and development: some observations and evidence from Britain, *Regional Studies*, 20, 472–76.

Kleine, J. (1982) Location, firm size and innovations. In D. Maillat (ed.), *Technology: a key factor for regional development*, Georgi, Saint-Saphorin, pp. 147–74.

Lewis, A., Nagpal, B.K. and Watts, P.L. (1985) Robotics: market growth, application trends and investment analysis, *Proceedings of the Institution of Mechanical Engineers*, 199, 35–40.

Mansfield, E. (1968) *The Economics of Technical Change*, Longman, London.

Mensch, G. (1979) *Stalemate in Technology: innovations overcome depression*, Ballinger, New York.

Metalworking Production (1988) *The Sixth Survey of Machine Tools and Production Equipment in Britain*, Morgan Grampian, London.

Mill, J.S. (1862) *Principles of Political Economy*, Vol. II, London.

Molle, W. (1988) Technology, trade and differential growth in the European Community. Paper presented at the European Regional Science Association Summer Institute, Arco, Italy, July.

Nolan, M.P., Oppenheim, C. and Withers, K.A. (1980) Patenting, profitability and marketing characteristics of the pharmaceutical industry, *World Patent Information*, 2, 169–76.

Northcott, J. and Rogers, P. (1984) *Microelectronics in British Industry: the pattern of change*, Policy Studies Institute, London.

Northcott, J. with Brown, C., Christie, I., Sweeney, M. and Walling, A. (1986) *Robots in British Industry: expectations and experience*, PSI Research Report no. 660, Policy Studies Institute, London.

Odagiri, H. (1985) Research activity, output growth and productivity increase in Japanese manufacturing industries, *Research Policy*, 14, 117–30.

Oakey, R.P. (1984) *High Technology Small Firms*, Pinter, London.

OECD (1981) *The Measurement of Scientific and Technical Activities, Frascati Manual 1980*, OECD, Paris.

OECD (1986) R & D, invention and competitiveness, *Science and Technology Indicators no. 2*, OECD, Paris.

Pavitt, K. (1982) Patent statistics as indicators of innovation activities: possibilities and problems, *Scientometrics*, 7, 77–99.

Piore, M.J. and Sabel, C.F. (1984) *The Second Industrial Divide: possibilities for prosperity*, Basic Books, New York.

Porter, M.E. (1985) *Competitive Advantage: creating and sustaining superior performance*, Free Press, New York.

Prais, S.J. (1987) Some international comparisons of the age of the machine-stock, *Journal of Industrial Economics*, 34, 261–77.

Ray, G.F. (1984) The diffusion of mature technologies, *NIESR Occasional Paper no. 36*, Cambridge University Press, Cambridge.

Rothwell, R. (1979) The relationship between technical change and economic performance in mechanical engineering. In M.J. Baker (ed.), *Industrial Innovation: technology, policy, diffusion*, Methuen, London, pp. 36–59.

Rothwell, R. (1980) Innovation in textile machinery. In K. Pavitt (ed.), *Technical Innovation and British Performance*, Methuen, London, pp. 126–41.

Rothwell, R. and Zegveld, W. (1981) *Industrial Innovation and Public Policy: preparing for the 1980s and 1990s*, Pinter, London.

Rothwell, R. and Zegveld, W. (1982) *Innovation and the Small and Medium Sized Firm*, Pinter, London.

Schmookler, J. (1972) *Patents, Invention and Economic Change: data and selected essays*, posthumously edited by Z. Griliches and L. Hurwicz, Harvard University Press, Cambridge, MA.

Schott, K. (1981) *Industrial Innovation in the UK, Canada and the United States*, British–North American Committee, London.

Schumpeter, J.A. (1934) *The Theory of Economic Development*, translated by R. Opie, Oxford University Press, London.

Schwartzman, D. (1976) *Innovation in the Pharmaceutical Industry*, John Hopkins University Press, Baltimore, MD.

Solow, R. (1957) Technical change and the aggregate production function, *Review of Economics and Statistics*, 39, 312–20.

Sweeney, G. (1987) *Innovation, Entrepreneurs and Regional Development*, Pinter, London.

Thwaites, A.T. (1978) Technical change, mobile plants and regional development, *Regional Studies*, 12, 445–61.

Thwaites, A.T. (1983) The employment implications of technological change in a regional context. In A.E. Gillespie (ed.), *Technological Change and Regional Development*, London Papers in Regional Science, Vol. 12, Pion, London, pp. 36–53.

Thwaites, A.T. and Alderman, N. (1988) The location of industrial R & D: retrospect and prospect. Paper presented at the European Regional Science Association Summer Institute, Arco, Italy.

Thwaites, A.T., Edwards, A. and Gibbs, D.C. (1982) *Interregional Diffusion of Production Innovations in Great Britain*, final report to the Department of Trade and Industry and the Commission of the European Communities, Centre for Urban and Regional Development Studies, University of Newcastle upon Tyne.

Thwaites, A.T., Oakey, R.P. and Nash, P.A. (1981) *Industrial Innovation and Regional Development*, final report to the Department of the Environment, Centre for Urban and Regional Development Studies, University of Newcastle upon Tyne.

Townsend, J., Henwood, F., Thomas, G., Pavitt, K. and Wyatt, S. (1981) Science and technological indicators for the UK: innovations in Britain since 1945, *SPRU Occasional Paper Series no. 16*, Science Policy Research Unit, University of Sussex.

Vickery, G. (1986) Technology transfer revisited: recent trends and developments, *Prometheus*, 4, 25–49.

Watts, H.D. (1981) *The Branch Plant Economy: a study of external control*, Longman, London.

Williams, H., Gillespie, A.E., Howells, J., Pywell, C. and Thwaites, A.T. (1986) *New Information Technology Production in the Regions of Europe*, final report to the Commission of the European Communities, Centre for Urban and Regional Development Studies, University of Newcastle upon Tyne.

7

Producer services and economic development

P.W. Daniels

Introduction

The producer services are now regarded as 'one of the keys to the dynamic process of transformation and internationalisation of the economy in recent decades' (Ochel and Wegner, 1987, p. 26). They are at the leading edge of the widely recognised structural shift from manufacturing to service employment in advanced economies (Fuchs, 1968; Stanback, 1979; Gershuny and Miles, 1983; Giarini, 1987). The manifestations of this change are not only evident in the overall behaviour of national economies but also have measurable impacts on spatial patterns of economic activity. Some of the established patterns of economic activity are being reinforced; others are being rewritten.

Producer services generate output which is used by other firms in the production of a good or service. Some examples include market research, accountancy, advertising, management consultancy, corporate legal services, banking and corporate insurance. They make an important contribution to the export income of individual cities and regions, as well as to the balance of payments. To these services can be added the headquarters of manufacturing companies, together with R & D functions, because although the latter may be classified as manufacturing industry their locational preferences, recent growth and contribution to economic development all parallel, and indeed interact with, those of 'pure' producer services. This 'basic' role of producer services in economic development has only recently been measured and recognised (see, for example, Beyers *et al.*, 1985; Pederson, 1986; Ley and Hutton, 1987). Their locational

behaviour is therefore of vital importance in relation both to the contribution of export revenues to local incomes and economic welfare and to the spin-offs from any multiplier effects.

The growing prominence of services is perceived as a symbol of economic progress. There is some kind of economic development continuum but there is no agreement about its theoretical basis. A number of theories are part of the common currency. Until relatively recently the most common theory was the *three-sector model* (Clark, 1940), which posits that economic development occurs sequentially, commencing with primary activities, then secondary activities and finally tertiary (or service) activities. Thus, service activities will account for well over half of total employment as per capita income improves and creates an increasing demand for them. This effect is enhanced by the tendency for the productivity of services to grow more slowly than for primary or secondary activities. This is a rather mechanistic explanation for the shift to services (see, for example, Gershuny and Miles, 1983) and is now viewed less favourably as a number of competing theories have been formulated. *Deindustrialisation* theorists (Singh, 1977; Blackaby, 1978) see the emergence of service industries as occurring by default; manufacturing industry's share of total employment has been contracting in both absolute and proportional terms, leaving a residual of less-efficient, less-productive and less-attractive service industries. Such a change certainly does not equate very well with the notion of progress in economic development. Consequently, there has been a growing chorus of 'manufacturing matters' (see, for example, Cohen and Zysman, 1987). Whereas deindustrialisation is predominantly an economic theory of structural change, *post-industrialism* is more akin to a theory of social and occupational change consequent upon the emergence of service industries as the principle source of employment. Bell's (1974) vision of a society in which information is the main technology used in the production of a good or service (rather than energy in industrial society) is accompanied by an expansion of white-collar, professional and technical occupations and the structural changes accompanying economic development (see, for example, Howells, 1988).

On their own, none of these theories provides an adequate basis for explaining the rise of services in the economies of Western Europe. Indeed, in some respects they are more useful as loose frameworks for analysis. They do not really advance our understanding of processes of service industry growth. In relation to the latter it is necessary to consider the effects of internationalisation of manufacturing and service industry production, trade and organisation (Dunning and Pearce, 1985; Markusen, 1988; Petit, 1986) as well as deregulation of financial markets, the growing specialisation of service outputs, corporate efforts to achieve greater flexibility of production, and advances in global telecommunications (to name but a few examples). All have encouraged greater division of labour, especially into white-collar, knowledge-intensive, information-processing occupations (Noyelle, 1983). Manufacturing and service enterprises have also found it expedient to externalise the consumption of certain services which had previously been produced using their own internal resources. Producer services, in particular, have benefited from, as well as contributed to, all of these processes. The demand for the production of these services has been stimulated in parallel with new opportunities for suppliers to be innovative in the refinement and development of new services.

Recent growth of producer services

The service sector as a whole, on the basis of numbers employed, is now the most important in almost all the countries of Western Europe. Although great care must be taken when comparing statistics on service employment in different countries (see, for example, Elfring, 1987; Bailly and Maillat, 1988) the majority had more than 50

Table 7.1 *Service employment as a percentage of total economically active population, 1970–82*

	1970	1975	1982	Difference (1970–82)
Austria	43.2	47.9	52.8	9.6
Belgium	52.1	56.5	64.7	12.6
Denmark	50.7	58.8	64.0	13.3
France	46.4	51.1	57.2	10.8
West Germany	42.9	47.6	51.8	8.9
Greece	34.2	36.8	41.9	7.7
Ireland	43.1	45.8	51.6	8.5
Italy	40.3	44.2	50.6	10.3
Luxembourg	46.4	50.3	58.9	12.5
The Netherlands	54.9	59.4	66.3	11.4
Portugal	37.1	32.2	37.0	– 0.1
Spain	37.4	39.7	47.8	10.4
Switzerland	45.3	50.9	54.5	9.2
United Kingdom	52.0	56.7	62.6	10.6

Source: Derived from Bailly and Maillat (1988, Figure 1, p. 55).

per cent of total employment in services by 1982 (Table 7.1). Only Greece and Portugal remained below this figure. By 1984 almost three out of every four jobs in Denmark, the UK, the Netherlands and Belgium were in service industries (Howells, 1988) but the highest relative growth rates between 1977 and 1983 occurred in Portugal, Ireland, Luxembourg and Italy. A form of 'catching up', this accounted for over half (56 per cent) of the employment expansion in service industries in the European Community during the period.

A significant feature of the high average annual compound growth of service employment (relative to manufacturing) has been the even faster expansion of producer services (Ochel and Wegner, 1987; Elfring, 1987). Estimates of the share of producer services in total employment vary considerably according the definition used (see, for example, Marshall *et al.*, 1988) and the quality of statistics available. Using a definition confined to business and professional services, financial services, insurance services, and real estate services, Elfring (1987) shows that although producer services comprised less than 10 per cent of total employment in France, West Germany, the Netherlands, and the UK in 1984, they were responsible for some 25–30 per cent of new service jobs since 1973 (Table 7.2). During the last decade professional and business services have

Table 7.2 *Producer services' percentage share of total employment (1984) and of service employment growth (1973–84)*

Producer Services	France		West Germany		The Netherlands		UK	
	a*	b†	a	b	a	b	a	b
Business and Professional	4.5	13.6	2.8	13.9	5.7	19.0	4.9	20.7
Financial	2.1	4.8	2.4	8.6	2.4	6.1	2.3	9.3
Insurance	1.0	1.5	1.1	2.1	1.2	1.0	1.3	3.1
Real Estate	0.3	0.9	0.4	0.0	0.7	2.1	0.6	2.9
Total	7.9	20.8	6.7	24.6	9.9	28.2	9.2	36.0

Notes: *Share of total employment, 1984.
  † Share of service employment growth, 1973–84.
Source: Elfring (1987).

been growing more than twice as fast as the average for services as a whole. This has occurred even though their use of information technology (and therefore scope for substituting capital for labour) has expanded rapidly.

Whatever recent statistics may show, however, there remains a knowledge gap. The processes underlying the locational behaviour of producer services are not as well understood when compared with consumer services or manufacturing activities. This means that there are probably more questions than answers at present. What of the existing role and future potential of producer services in the economic development process? What is their contribution to the differentials in the performance of individual cities or regions within countries or between the states of Western Europe? What kinds of policy instrument are required to harness their contribution more effectively to regional development? (See, for example, Commission of the European Communities, 1987.) The location patterns that can be equated with the emergence of producer services will be outlined in the next section of this chapter, and the discussion then progresses to an assessment of what these patterns mean for economic development at the urban and regional scales.

The location of producer services

The locational behaviour and office-space needs of individual producer service firms depend upon a number of factors. These include employment change (expansion and contraction); reorganisa-tion or changes in the division of labour; firm size; the location decisions made by competitors; and the extent to which information technology is used for production. The effects of changes in operating costs, such as office rents and related overheads, or the needs created by changes in organisational form and structure (following merger or takeover, for example) are also relevant. Firms may require to adjust the standards of their accommodation in order to remain competitive or, as a response to employee needs, to maintain their ability to attract employees of the required quality.

This list is not comprehensive, but it distinguishes between requirements created by factors internal to the firm (such as employment change, size, or the impact of changes in organisational structure) and those arising from external factors (such as the behaviour of competitors, changes in office accommodation standards or changes in rents and related overheads). Some of these factors can be accommodated through *in situ* adjustments such as employment change, adoption of information technology or changes in accommodation standards. Others, such as the behaviour of competitors, will most likely generate a change in location, the establishment of new branches, or mergers or acquisitions in order to gain access to markets not formerly served.

Centrality, in relation to markets, clients, labour and organisation is a ubiquitous requirement. Agglomeration economies are therefore highly significant. Proximity to other specialist services, to highly-qualified labour, to competitors or to national institutions and government departments

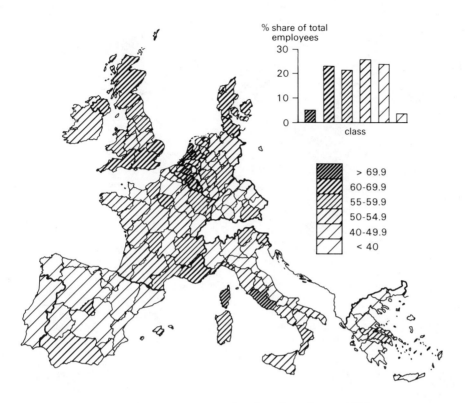

Figure 7.1 *Service employees as a proportion of total employees, 1983*
Source: Howells (1988).

ensures that individual firms remain competitive. Effective orientation to the business environment is the key to attracting and keeping clients. Face-to-face contacts, which have yet to be rendered redundant by telecommunications technology, remain extremely influential in the location of decision-makers and of 'front-office' staff, thus further reinforcing the demand for centrality. Accessibility underpins the pull of centrality, with good nationwide and international communications being particularly important. Locational imperatives result in a considerable degree of functional ordering as the headquarters/control functions of producer service firms seek positions in major agglomerations, leaving regional or divisional control functions in smaller centres.

Services in general are therefore unevenly distributed within Western Europe (Figure 7.1; see also Howells, 1988). This is particularly true of producer services, but comparable employment or occupational data are difficult to assemble and many studies rely on surrogate evidence to demonstrate these concentration tendencies. Thus, Howells (1988) shows that almost 30 per cent (149) of the top 500 banks (by assets) in the world are controlled from the 12 member states of the European Community, but 70 per cent of them are located in just four countries (France, Italy, West Germany and the UK). The majority of these banks have headquarters in Paris, Milan, Munich/Frankfurt and London respectively (see, for example, Olbrich, 1984). In proportional terms service employment has actually been growing most rapidly in peripheral regions (Figure 7.2; see also Keeble *et al.*, 1982) but in real terms certain core regions have been retaining a disproportionate share of services. This is conveyed most clearly by Figure 7.3, which shows service structures in 1983 and has strong core-periphery overtones. (The index used is the ratio of producer to consumer service employment in each sub-region, and a relatively high producer-consumer ratio is an indication of a favourable long-term service industry structure.) It is

111

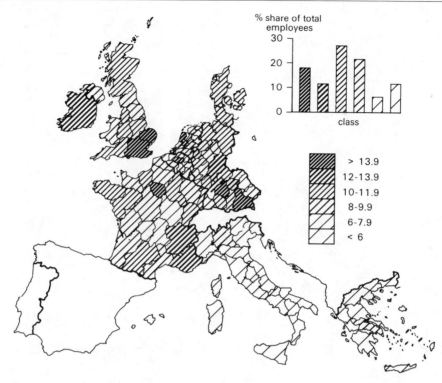

Figure 7.2 *Employment in financial and business services as a percentage of total service employment, 1981*
Source: Howells (1988).

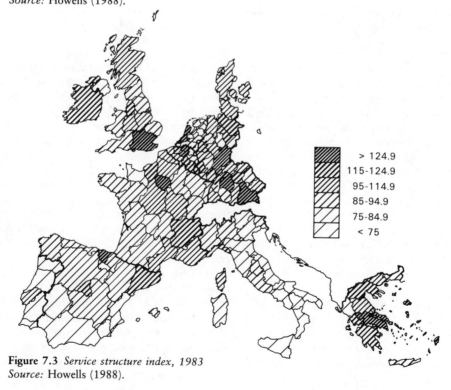

Figure 7.3 *Service structure index, 1983*
Source: Howells (1988).

generally the more dynamic, growth-oriented, regions that have the largest bias towards producer services. Examples include Ile de France, the Randstad, South-east England and the Rhein-Main region. Bailly and Maillat (1988) use a coefficient of specialisation to demonstrate the concentration of the branches of banks, business services and insurance offices in Ile de France. Values of 1.31, 1.78 and 2.29, respectively, compare with 1.26 and 1.38 for transport and retail services.

At the inter-urban level, producer service location incorporates a strong hierarchical component which, if anything, is becoming stronger. Although the figures are for the service sector as a whole, Philippe (1984) has used coefficients of specialisation to show that if the value for towns of between 5,000 and 10,000 inhabitants in France is 1.0, it rises to 1.2 for towns of between 10,000 and 100,000, 1.25 (over 100,000) and 1.4 for the Paris agglomeration. Similar statistics for Dutch cities are provided by van Dinteren (1987) and by Buursink (1985), and for the major urban agglomerations in West Germany by Bade (1985). We can safely infer from these and similar studies that larger cities will have even higher location quotients (or coefficients of specialisation). This is particularly true for producer service enterprises, which are increasingly operating international networks, as well as expanding within their own national markets. Factors such as connectivity between international business centres (telecommunications, air travel) have therefore become very significant. Large cities (for the reasons outlined earlier) have always attracted a disproportionate share of producer services. While, in certain circumstances, there may be some downward filtering to smaller towns and cities (see below), the aggregate effect seems unlikely seriously to challenge the well-established dominance of the major European banking, finance and commercial centres such as London, Paris, Amsterdam, Milan and Frankfurt.

Agglomeration of producer services is prominent but we should not loose sight of the dispersal or decentralisation of certain functions, especially of 'back-office' routine activities. This is a response to the rising costs of centrality (rents, salaries, congestion) and the readier prospects in non-central locations for attracting part-time labour, which has characterised job structures in some producer services (such as insurance) during recent years. The influence of decentralisation is largely confined to intra-metropolitan or intra-regional location decisions. Although in absolute terms city centres may still have the largest share of producer service employment, the most dynamic areas are the suburbs and smaller towns in metropolitan regions. In the UK there has been a substantial spillover of producer services into 'subdominant' towns and cities around London and the provincial conurbations, and into smaller free-standing towns in southern England (Gillespie and Green, 1987; Howells and Green, 1988). Professional and business services in the Netherlands grew at a substantially higher rate between 1973 and 1985 in the 'intermediate zone' around the Randstad, a zone which comprises several small and medium-sized cities (van Dinteren, 1989). Taking 1973 as 100, the values for 1985 were 166 for the Randstad and 284 for the intermediate provinces. Bade (1985) has also identified similar processes for 'production-oriented' services in metropolitan rings, as well as in the intermediate and peripheral areas of West Germany.

Dispersal takes different forms. It may create, for example, office clusters in suburban business centres in locations adjacent to major motorway intersections or in carefully landscaped, low-density office parks. These, and others, are the antithesis of congested and costly city-centre producer service complexes. Very little is known about the factors determining the success of business or office parks, since they do not apparently offer the kinds of agglomeration economies associated with suburban business centres or city centres. It seems reasonable to assume that they are occupied either by producer services serving primarily local or regional markets, or by the 'back offices' of major service and manufacturing companies. The latter are engaged primarily in routine, computerised functions that can be adequately performed away from the contact-intensive, high-cost environment of major city centres. Some of the major relocations from central Paris, central London or the Randstad cities have been 'partial' moves of 'back office' functions by insurance companies, banks, or the corporate headquarters of manufacturing firms to suburban centres or to smaller towns, such as Reading or Basingstoke in the London Metropolitan Region. Telecommunications have enhanced the scope for locating 'back offices' out of centre. However, it should be noted that most of the expansion of producer services in 'intermediate' regions and towns does not result from relocation but from the growth of indigenous firms or the location of new establishments by national, or even international, service firms expanding into new markets.

Dispersal is undoubtedly important but there has also been a resurgence in the importance attached to the location of the 'front office' activities of producer service firms. In order to survive, many of these services must provide state-of-the-art and immediate advice to sophisticated business clients in increasingly dynamic and innovative markets. Deployment of the best information technology and the most specialised labour inputs is essential; the latter are most readily accessible from city-centre office complexes, while the former is also most readily assembled and utilised by high-technology office buildings. These are invariably constructed first in the information-rich, 'traditional' city-centre locations (see below). Recent years have therefore witnessed substantial demand for 'intelligent' or 'wired' city-centre offices occupied by corporate and merchant banks, insurance companies and other financial, business and professional services. But this has been, and is likely to continue to be, a highly selective process; only major financial centres such as London, Paris, Milan, Amsterdam and Frankfurt will be major participants.

In the peripheral regions of Europe the central areas of major cities (Manchester, Hamburg, Barcelona) continue to exert a disproportionate pull on the location of producer services. Establishments that are part of multi-locational operations, rather than activities that are indigenous to the local or regional economy, attach particular importance to such locations. This does not, however, reflect the renaissance of 'front offices', but the attraction of city centres as the only areas with sufficient critical mass to generate agglomeration economies attractive to office-based producer services. In many cases the diseconomies of such city-centre locations are not enough to encourage substantial redistribution of even routine functions to suburban centres. Indeed, there appears to be a threshold city size (perhaps 1.5 million) below which centralisation (i.e. city-centre location) is preferred by producer services. Above this threshold it becomes necessary, or at least feasible, to contemplate suburbanisation of 'back-office' functions.

On average the suburbs offer lower rents and wages, easier access to labour (especially the part-time labour which comprises an ever-increasing proportion of the office labour force), generally superior journey-to-work conditions, lower staff turnover, and improved staff reliability. Together with the peripheral regions, they are undoubtedly an attractive location for certain kinds of service function. Yet prestige, the greater prospect of good-quality and specialised inputs, maximum access to the range of qualified human resources, and good external transport connections, still make city centres or core regions highly attractive locations for producer services.

Information technology and producer services

Information and communications technology has been a vital catalyst in the development of producer services and the spatial transformation of the European economy (Ochel and Wegner, 1987; Howells, 1988). It has brought about four kinds of transformation in the production of services (Ochel and Wegner, 1987). The range of products has been diversified (product-mix effect), the market for products has been extended, the place of production has been influenced (locational effects), and the way in which production is performed has been changed (production processes). Technology has permitted greater flexibility in the production and distribution of services that are largely knowledge- or information-intensive. This, in turn, has stimulated demand for better-educated labour and a related shift away from blue-collar to white-collar occupations. It is unlikely, however, that information technology will, on its own, cause changes in the spatial pattern of producer services. It certainly advances the potential for decentralisation or counterurbanisation, but this may well be displaced by the effects of other factors (see, for example, Funck and Kowalski, 1987). In economic environments where deregulation and privatisation of services is increasingly the norm, for example, there may well be important locational implications. Perhaps counter-intuitively, deregulation of telecommunications and related information-handling services will be likely to encourage further centralisation of producer services. This is because many are not only major users of information technology but are also leading innovators in its application and development. They therefore need to locate in those parts of Western Europe attracting investment in state-of-the-art technology such as fibre-optics networks, teleport facilities (for satellite transmission of data etc.), or local-area networks. Inevitably, such locations are found in or around the metropolitan or core regions that are already endowed with high (above average) levels of diversified producer service employment. Thus 'the major . . . technological effects on the EC services will have a centralising effect in inter-regional terms, although on a sub-regional basis there appear to be few constraints on [producer] service activity dispersal' (Howells, 1988, pp. 96–7). While peripheral cities or regions will, of course, engage in service transactions using information and telecommunications technology, they will be net exporters rather than importers. This will reduce the potential for the growth of producer services in those areas.

The way in which information technology is used is crucial for future producer service location. The potential for job saving has, thus far, been counterbalanced by an increase in the demand for these services, as well as by the use of information technology in new product development. However, it has been demonstrated that the propensity to utilise technology in service innovation is positively related to the level of concentration of service activities (see, for example, Foord and Gillespie, 1985 (UK); Cappellin and Grillenzoni, 1983 (Italy)). The number of routine workers has tended to remain static or to grow more slowly, while substantial growth has taken place in the more specialist service-producing occupations. Provided that demand stays ahead of productivity, producer service employment will therefore continue to grow. But information technology is accentuating the distinction between highly-skilled managerial and professional workers and those with more routine skills. The change in the ratio between these two groups could be an important determinant of future location trends; the former generating demand for more centralised locations in the major European cities or the core regions, and the latter able to function more effectively from dispersed locations. Nevertheless, a cautious view of the impact of technology on producer service location seems justified. We should not overestimate the ability

or the willingness of the labour force to adapt to information technology at the same rate as the appearance of new hardware and applications. Because there are so may variables involved, location requirements are certainly changing; but this is occurring at a much slower rate than the innovations that are taking place in the use of, and access to, the services rendered by information technology (Hepworth, 1987).

Role of structural and organisational changes in producer service firms

During the last decade a notable trend in the development of European service businesses has been the emergence of multinational and transnational corporations. These are the result of merger and takeover activity, as well as direct expansion by some firms into other national territories (Clairmonte and Cavanagh, 1984; Dunning, 1985; Enderwick, 1989). Global telecommunications networks, the relative ease of international business travel and deregulation of telecommunications and financial and professional services have encouraged this trend. As firms become larger and more diversified, economies of scale, high thresholds for specialised products and services and the need to 'protect' specialised services formerly distributed via agencies or representative offices become significant. These circumstances have helped to redefine the national and European horizons of many companies. These will be even further extended by the planned introduction of a single or unified European market after 1992 (see below).

The location of administrative and managerial or 'control' functions, together with the development of free-standing branch offices, is a crucial feature of the producer service response to these new circumstances. This is linked to the combined effects of the expansion of markets, the growth of firms, increases in personal and corporate income, product differentiation, and the changing role of technology. The transformation of businesses into larger organisations, increasingly operating outside their own national boundaries, has created a demand for large, specialised staffs with innovation, planning, marketing and control responsibilities. With the nature of consumption changing the character of demand, production is itself changing and creating an increase in the level of dependence on producer services. The extent of externalisation of demand for specialised services is partially governed by firm size (Marshall, 1982; Pederson, 1986). The specialists required can be recruited for in-house use, but external sources are becoming more prominent because they can utilise economies of scale in the production of a highly specialised service made available to a wide range of client organisations.

An examination of the way in which producer service production and organisation have evolved helps us to understand the locational outcomes (Daniels *et al.*, 1988; 1989). During the decade 1976–86 the evolution of producer services has been dominated by concentration of ownership and control in large firms as a result of mergers and takeovers involving primarily medium-sized competitors; it has entailed internationalisation (which is partially linked to concentration); and it has brought about diversification. Large producer service firms are increasingly prepared to supply an ever-widening range of specialised services, provided that they will lead to fee income or meet the requirements of clients who are becoming far better informed about the kind and quality of services they require. Services that were once the domain of specific activities such as insurance, property consultants or management consultants are being increasingly included in the portfolios of non-specialists. UK property consultants are diversifying, for example, into adjacent skills in financial services, and this may involve recruiting accountants or successful industrialists into key management roles. Similarly, financial services are now providing property investment advice, accountants and solicitors are providing property management and valuation services and architects and engineers are offering construction management expertise (Leyshon *et al.*, 1987a).

Convergence in the organisation, competitiveness and range of services offered by large producer service firms provides the context for considering, briefly, their impact on spatial patterns. Certain localities within Europe — mainly in and around the large cities — are clearly preferred for new office openings by accountancy conglomerates, property consultants and investment banks (Figure 7.4). There are at least four reasons. Each of the major firms is trying to protect, or even to enhance, its share of a growing market. As competitors develop the geographical coverage of their office networks, those with less-

comprehensive coverage tend to follow; prior to 1960 some firms were entirely based in key cities such as London, Paris, Brussels or Frankfurt. They are therefore now trying to catch up. Related to this is an effort to protect market share by identifying and, hopefully, exploiting locations with latent business development opportunities. The large producer service firms have become market-makers, positively seeking new clients and creating markets for new services. There is also some pull exerted by small firms, which are seen as a valuable source of new and expanding fee income since most are still not aware of the scope for using advanced services. The threshold for establishing new offices has therefore fallen and makes it possible for the large firms to begin operations in places where they do not already have any major clients.

Locational preferences and/or differential growth by producer services are therefore crucial for the economic health of cities and regions. Differential growth of office-based producer services is exaggerating the dichotomy between central and peripheral regions or cities. Some of the processes responsible have been illustrated in studies of the inter-regional flow of business service transactions in Switzerland, Spain, Denmark and the Netherlands (Cuadrado, 1986; Jeanneret *et al.*, 1984; Pederson, 1986; Perret-Gentil, 1984; van Dinteren, 1987). One of the most important findings from these studies is the role of head offices in determining the location of externally purchased goods and services. The probability is that these purchases will focus on the same region or city, but there will also be spillover of demand for specialised producer services to higher-order urban areas if these services are not available locally. The performance of indigenous producer services will therefore be dependent to some degree upon the demand generated by head offices (representing both manufacturing and service firms) and on that arising from other locally controlled establishments. This traditional perception of the link between office-based services and other components of the local economy is now changing as it has become apparent that business and professional services, for example, export a growing proportion of their output to other regions in Europe and elsewhere.

116

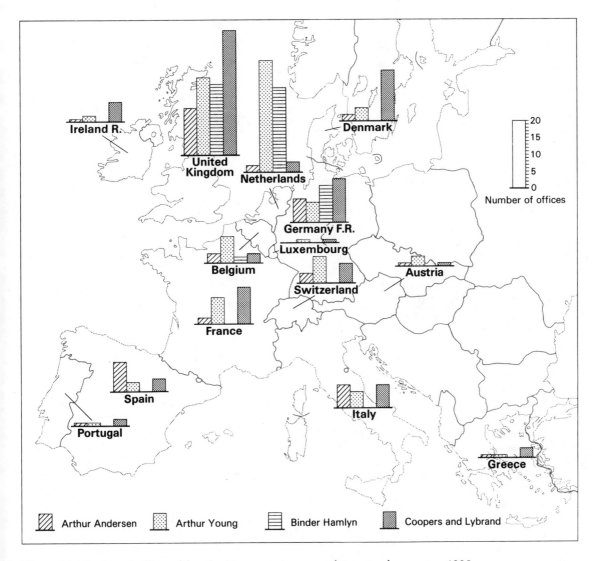

Figure 7.4 *Number of offices of four leading accountancy conglomerates, by country, 1985*

The example of financial services

The growth and location of financial services (insurance, banking, finance and leasing) has been governed by a number of factors. Developments in information technology have greatly improved both the internal and external efficiency of financial markets. The latter has been enhanced by the higher speed and handling capacity of financial information systems. Data production and distribution are increasingly subject to computer applications and are supplied 'on line' by a growing number of electronic information companies such as Reuters and Telerate. Internal efficiency has been improved through better links between markets and between geographically separate parts of the same firm, and between institutions and individuals. Dealing and settlement systems for financial services are increasingly being automated; remote trading is practicable. Consequently, turnover in financial markets has increased dramatically, aided by the development of automatic execution systems for trading small lots, by progressive introduction of expert

117

Table 7.3 *Trade in invisibles (services), 1970 and 1984*

	Balance of trade in invisibles ($bn)		Specialisation index*	
	1970	1984	1970	1984
West Germany	− 2.21	− 5.2	20.1	22.3
France	+ 0.68	+ 6.97	27.2	37.2
Italy	+ 1.01	+ 2.03	30.7	26.4
United Kingdom	+ 2.48	+ 9.69	37.5	34.1

*Value of service exports as a percentage of total exports of goods and services.
Source: Commission of European Communities (1987).

Table 7.4 *The top six banking centres in Western Europe, 1986*

	No. of firms	Income* ($m)	Rank	Assets* ($bn)
Paris	6	1,712	2	659
London	5	2,934	1	390
Frankfurt	3	1,003	3	307
Zurich	2	826	4	–
Amsterdam	3	739	5	193
Basel	1	415	6	–
Total	20	7,629	–	1,198
Total top 12 banking centres†	65	22,119	–	5,378

Notes: *Cumulated net income and cumulated assets of the top 50 commercial banks and top 25 securities firms in the world represented in each city.
†The other cities appearing in the rankings (on the basis of income, assets, or both) are: Tokyo, New York, Osaka, Hong Kong, Los Angeles, Montreal, Munich, Nagoya, Kobe, San Francisco.
Source: Derived from Noyelle (1988, Table 5.5, p. 101).

systems, and by a reduction in time and costs for the execution of deals. All of these technology-related developments encourage institutional investors to be more 'active' and have pushed up the volume of business that can be transacted during trading hours.

Although technological advances have, in theory, increased the locational flexibility of financial services, in practice they display a clear trend towards spatial concentration. This is true in relation both to the broadest international patterns of growth and location, and to the situation within Western Europe. Tokyo, New York and London, along with a larger number of 'second tier' cities, are attracting the lion's share of international financial services; in the UK there is well-documented evidence detailing the concentration of financial services in and around the City of London (Leyshon *et al.*, 1987b). Globalisation of markets and the advent of 24-hour trading have further encouraged centralisation. Thus, just six European cities are included in a list of the top 12 global banking centres (Table 7.4) in 1986. Although they top the European ranks, London, Paris and Frankfurt have far fewer firms than Tokyo (22) or New York (16) and less than half the income. On the basis of assets, Paris heads the European list but is only just greater than one-third the size of Tokyo. With one or two exceptions, such as the headquarters of insurance offices, most major provincial cities are heavily dependent on financial services provided by branch establishments. Foreign banks, for example, tend to establish branches outside national financial centres in locations where foreign direct investment is taking place. Thus Japanese banks are well represented in Manchester and Dusseldorf.

The demand from large national and multi-national firms for participation in highly dynamic financial markets using the best methods available to ensure that capital, securities and loan raising can take place in a way that maximises returns, has also driven the expansion of financial services. Pension funds and insurance companies have amassed large capital funds that need to be invested to cover future disbursements and have developed into highly professional institutional investment vehicles. Deregulation of financial markets (such as allowing outsiders to take 100 per cent ownership in Stock Exchange firms, the removal of fixed commission rates and the ending, in the London Stock Exchange, of the traditional demarcation between jobbers and brokers) has created new opportunities for foreign and domestic firms that had formerly been excluded.

Many of these developments have triggered new alliances between financial service firms. Some have adopted 'global' strategies, i.e. providing a full suite of financial services. Others (often the smaller firms) have opted for a niche strategy, i.e. identifying a particular gap in the market or refining an existing service to a level far superior to that of competing suppliers. Little wonder, in either case, that the growth of financial services has been so place-specific. The increasingly specialised labour demand, specialist information and knowledge, or simply a heightened awareness of trends and innovations, are most readily realised in, or very near to, large-scale agglomerations such as Amsterdam, Paris, Zurich and London. The buoyancy of office markets has been closely linked to the demands generated by financial services during the last five years. However, the impact on provincial office markets has been much more limited, despite recent claims by cities, such as Bristol, that they are now significant financial service centres.

The crisis on 'Black Monday' (19 October 1987) has introduced much uncertainty into the future prospects for financial services. Overcapacity (too many firms and banks chasing business) was identified as a problem before the crash occurred; it has brought forward the labour-shedding and rationalisation that observers considered inevitable as intense competition pushed commissions so low that it would be impossible for some firms to survive. Both foreign and UK investment banks have already begun to shed labour; 50,000 jobs (out of over 200,000 in financial services in the City of London) are expected by some analysts to disappear by the early 1990s. It is difficult to know how to translate these losses to provincial cities, perhaps much will depend on the importance attached by the financial services already there to a physical presence in regional markets. The fall in financial services jobs will, in turn, reduce the demand for office space (it has recently been estimated that job losses following the crash on Wall Street have been accompanied by a decline in office rents and rental values of 20–25 per cent). Repeated in the European context, a slow-down in office rent growth will reduce the pressure to search for less costly locations for back-office functions. It is notable, for example, that a record 0.48 million square metres of office space was completed in central London in 1987. With more than 50 per cent of occupier demand emanating from banking and finance, it is not difficult to anticipate some slack in the market which will give the survivors rather more room for manoeuvre.

Producer services in Western Europe post-1992

A new factor likely to affect the location of

European producer services is the creation of an integrated market after 1992. The barriers that currently exist in relation, for example, to recognition of one member state's professional qualifications in another member state, to the movement of labour between member states and to the sale of some services across national boundaries will be removed.

In June 1988 there was agreement in principal by the Council of Trade Ministers in Luxembourg on the mutual recognition of diplomas; this means that more than 10 million professional employees such as actuaries, accountants, corporate lawyers, bankers and surveyors will be able to practice in member states without having to requalify. Many of these are employed by producer service organisations. There remains uncertainty about how precisely these new freedoms will operate for specific types of producer service. Corporate legal firms in the Netherlands and United Kingdom for example, would like to be free to set up branch offices in member states. This would enable them to provide a more direct service to clients and to undertake 'unreserved' work (work not reserved to the legal profession of the host country by law). Both countries are major exporters of these services and can see further opportunities to expand. Such growth will probably be restricted to major European cities; Paris already has some 17 branches of large City of London law firms, Brussels has eight. Since these firms already provide more diverse advice on business, financial and legal matters than German or French firms, they are better placed to extend their office networks throughout member states and to compete more effectively with the accountancy conglomerates that are providing similar services.

It can also be expected that the single European market will lead to further merger and takeover activity. Non-life insurance firms, accountants, and financial services will seek access to markets in countries where they have not operated before. France's second largest insurance firm has recently completed a cross-border link-up with an Italian insurer which is one of many in progress or anticipated. Mergers to create larger national firms are also occurring. Since small firms will be the target for large, well-capitalised groups, control functions will become further centralised in Europe's principal financial centres. On the other hand, London's status as the premier European insurance centre could be threatened if firms do not remain competitive with other centres.

In the case of financial services the market will give securities firms and banks freedom to establish branches and sell financial services across national borders. Taxes and regulations that discriminate against residents of the member countries will be eliminated, controls on the transfer of capital will be removed, and a common regulatory framework for financial firms will be adopted. As a result, competition will be greatly increased and prices for financial services will fall, especially in countries like Spain, where they have operated in a very protected national market. This, combined with deregulation of capital markets (in the Netherlands, Italy, France and Spain); with the superior 'local knowledge' of banks and securities firms; and with high costs (office rents, staff salaries, housing) in some locations (most notably London), could well lead to the redistribution of wholesale financial markets within Europe. Should this happen it may also cause foreign banks (which have flocked to London during the 1980s in preference to other European centres because the City deregulated first and had a concentration of financial expertise) to engage in second-phase establishment of branches elsewhere in Europe, where they can be better attuned to client needs.

Environmental/quality-of-life factors and producer service location

Quality-of-life improvements arising from better housing, education facilities, shorter and less-congested journeys to work, and access to superior outdoor recreation opportunities were a key selling point in the promotion of service (office) relocation during the 1970s in France, the Netherlands and the UK. They will continue to be an influential factor in producer service locational behaviour, especially in view of the shifting balance between higher-paid information-producing employees and other staff. Employers must be cognisant of the value that such employees attach to rural or semi-rural lifestyles in small towns and villages, and to an ability to commute comfortably to workplaces with high-standard internal and external environments. There is little doubt that these considerations have been important for the expansion of high-technology industries (many of which include a substantial (attached) office component) in areas

such as New England, Massachusetts or 'Silicon Valley' (California) in the United States, or along the M4 Corridor and in south Hampshire in the UK. It is possible for large cities in Europe to compete for producer services that have a genuine choice of location on the basis of the quality of office accommodation available; but it is perhaps more difficult for them to provide such space in environments that rival smaller towns in the outer areas of the Ile de France, in the rim of Randstad or in outer south-east England. On the other hand, they are able to fulfil many quality-of-life requirements from a residential, rather than workplace, perspective.

Conclusion

In so far as producer services have come to comprise the knowledge capital of a region or city, they are now key actors in economic development. Producer services are reservoirs of knowledge, information and expertise that comprise an ever-larger proportion of value added during the creation of a marketable good or service for final consumption. Innovation in production, product refinement, organisational efficiency, effective use of factor inputs, distribution for intermediate and final consumption, and the forms of finance capital sustaining the market economy all require the consumption of producer services. Since decisions to utilise such services are made in international, regional or local headquarters of manufacturing and service enterprises (many of them large, multinational, multi-establishment and multi-product organisations) that are invariably located in the major cities of Western Europe, it is also the case that producer services are gravitating towards similar locations. The need to service the particular (and often unique) requirements of clients for financial advice, management consultancy or advertising strategy means that proximity, face-to-face contact and customer confidence are important location factors. The resulting corporate complexes are further reinforced by the tendency during recent years for producer services to engage in merger and takeover activity. This has concentrated control and production in small numbers of large firms which are invariably controlled from, or have the largest numbers of partners in locations with the largest markets, i.e. core regions and metropolitan areas. The resulting uneven development of producer services in Western European space economy *may* be disadvantageous for the development of less-favoured regions, but research does show that these services do contribute to the basic component of local economies. This can, of course, benefit the peripheral locations provided that producer services, probably indigenous enterprises, can be encouraged to grow and remain there. Herein lies one of the great challenges for regional policies in Western Europe. Investment in human capital, educational resources and information networks will be crucial for meeting the needs of areas that have so far been excluded from the growth impetus provided by a dynamic producer service sector. They must become part of the capacity of local firms to exploit new technologies and to modernise production (see also Chapters 6 and 11). The EC is certainly aware of these needs, and a number of initiatives such as ESPRIT, COMETT and IRIS are already in progress or planned. These relate, respectively, to the availability of research and development, to individual skill training and to the interaction of local partners — such as producer service suppliers, their clients, local authorities and educational and research establishments — in innovation (Commission of the European Communities, 1987).

At the other end of the spectrum, the spatial concentration of producer services also creates a set of problems that require just as much attention as those found in marginal regions and cities. Apart from the well-documented problems such as congestion, high wage levels and escalating office rents, the interaction between information technology and the functioning of many producer services has created demands for changes in the built environment, most notably in offices. Buildings that allow large numbers of workers to be accommodated on a small piece of valuable real estate are being made redundant by a demand for buildings accommodating large numbers on large pieces of city-centre land. The combination of planning procedures, building conservation in historic cores and lack of space has made developers and occupiers look for alternative, but still central, locations. Will these become successful alternatives for producer services unable to find space or to expand in 'traditional' locations? If not, will demand be diverted to less-congested second-tier cities? Or will the necessary infrastructure be put in place to ensure that the overspill locations are,

as far as possible, an integral part of the 'core' producer service complex?

There is, of course, no single solution suitable for all locations. It will be necessary to match need with the most appropriate strategy in the context of the wider international factors that govern the apparently inexorable progress of producer services. Since they are now acknowledged as a vital input to economic development, solutions will surely be found to some of the problems outlined in this chapter. Hopefully these will emerge sooner rather than later.

References

Bade, F.J. (1985) *Changes in the structure of the economy and its spatial implications*, Institut für Wirtschaftsforschung, Berlin.

Bailly, A.S. and Maillat, D. (1988) *Le Secteur Tertiare en Question*, Economica, Paris.

Bell, D. (1974) *The Coming of the Post-industrial Society*, Heinemann, London.

Beyers, W.B., Alvine, M.J. and Johnson, E.K. (1985) *The Service Economy: export of services in the central Puget Sound region*, Central Puget Sound Development District, Seattle, WA.

Blackaby, F. (ed.) (1978) *Deindustrialisation*, Heinemann, London.

Buursink, J. (1985) *De Diensten Sector in Nederland*, van Gorcum, Assen/Maastricht.

Cappellin, R. and Grillenzoni, C. (1983) Diffusion and specialization in the location of service activities in Italy, *Sistemi Urbani*, 2, 249–82.

Clairmonte, E. and Cavanagh, J. (1984) Transnational corporations and services: the final frontier, *Trade and Development*, 5, 215–73.

Clark, C. (1940) *The Conditions of Economic Progress*, Macmillan, London.

Cohen, S. and Zysman, W. (1987) *Manufacturing Matters: the myth of the post-industrial economy*, Basic Books, New York.

Commission of the European Communities (1987) *Services, Europe: the essential change*, Directorate — General Science, Research and Development, Brussels. (English translation of synthesis of the results of FAST research into Changes in Services and New Technologies.)

Cuadrado, J.R. (1986) *Supply and Demand of Services and Regional Development: the case of Ciudad Valenciana (Spain)*, EEC FAST Programme, Occasional Papers 93, Brussels.

Daniels, P.W., Leyshon, A. and Thrift, N.J. (1988) Large accountancy firms in the UK: operational adaptation and spatial development, *The Service Industries Journal*, 8, 317–47.

Daniels, P.W., Thrift, N.J. and Leyshon, A. (1989) Internationalization of professional producer services: accountancy conglomerates. In P. Enderwick (ed.), *Multinational Service Firms*, Routledge, London, pp. 79–106.

Dunning, J.H. (1985) *Multinational Enterprises, Economic Structure and International Competitiveness*, Wiley, Chichester.

Dunning, J.H. and Pearce, R.D. (1985) *The World's Largest Industrial Enterprises, 1962–83*, Gower, Aldershot.

Elfring, T. (1987) *New Evidence on the Expansion of Services in Seven OECD Countries*, School of Management, Erasmus University, Rotterdam.

Enderwick, P. (ed.) (1989) *Multinational Service Firms*, Routledge, London.

Foord, J. and Gillespie, A.E. (1985) *Reorganization, New Technology and Office Jobs*, CURDS Discussion Paper no. 75, University of Newcastle upon Tyne, Newcastle upon Tyne.

Fuchs, V. (1968) *The Service Economy*, Columbia University Press, New York.

Funck, R. and Kowalski, J. (1987) Innovation and urban change. In J. Brotchie, P. Hall and P.W. Newton (eds), *The Spatial Impact of Technological Change*, Croom Helm, Beckenham, pp. 229–39.

Gershuny, J. and Miles, I. (1983) *The New Service Economy*, Pinter, London.

Giarini, O. (1987) *The Emerging Service Economy*, Pergamon, Oxford.

Gillespie, A.E. and Green, A. (1987) The changing geography of 'producer services' in Britain, *Regional Studies*, 21, 397–411.

Hepworth, M.E. (1987) Information technology as spatial systems, *Progress in Human Geography*, 11, 157–80.

Howells, J. (1988) *Economic, Technological and Locational Trends in European Services*, Gower, Aldershot.

Howells, J. and Green, A. (1988) *Technological Innovation, Structural Change and Location in UK Services*, Avebury, Aldershot.

Jeanneret, P., Hussy, J., Bailly, A., Maillat, D. and Rey, M. (1984) *Le Tertiaire moteur dans les petites et moyennes Villes en Suisse: le cas d'Aigle et de Delemont*, Communauté d'Etudes pour l'Amenagement du Territoire, Lausanne.

Keeble, D.E., Owens, P.L. and Thompson, C. (1982) *Centrality, Peripherality and EEC Regional Development*, HMSO, London.

Ley, D.F. and Hutton, T.A. (1987) Vancouver's corporate complex and producer service sector: linkages and divergence within a provincial staples economy, *Regional Studies*, 21, 413–24.

Leyshon, A., Thrift, N.J. and Daniels, P.W. (1987a) Large commercial firms in the UK: the operational development and spatial expansion of general practice firms of chartered surveyors, *Working Papers on Producer Services 5*, University of Bristol and University of Liverpool, Liverpool and Bristol.

Leyshon, A., Thrift, N.J. and Daniels, P.W. (1987b) The urban and regional consequences of the restructuring of world financial markets: the case of the City of London, *Working Papers on Producer Services 4*, University of Bristol and University of Liverpool, Liverpool and Bristol.

Markusen, J.R. (1988) Service trade by the multinational enterprise. In P. Enderwick (ed.), *Multinational Service Firms*, Routledge, London, pp. 35–60.

Marshall, J.N. (1982) Linkages between manufacturing industry and business services, *Environment and Planning A*, 16, 437–50.

Marshall, J.N., et al., (1988) *Services and Uneven Development*, Oxford University Press, Oxford.

Noyelle, T.J. (1983) The implications of industry restructuring for spatial organisation in the United States. In F. Moulaert and P. Salinas (eds), *Regional Analysis and the New International Division of Labour*, Kluwer Nijhoff, Boston, pp. 113–33.

Noyelle, T.J. (1988) New York's competitiveness. In T.J. Noyelle (ed.), *New York's Financial Markets: the challenge of globalization*, Westview Press, Boulder, CO, pp. 91–114.

Ochel, W. and Wegner, M. (1987) *Service Economies in Europe: opportunities for growth*, Pinter, London.

Olbrich, J. (1984) Regional policy and management jobs: the locational behaviour of corporate headquarters in West Germany, *Environment and Planning C*, 2, 219–38.

Pederson, P.O. (1986) *Business Service Strategies: the case of the provincial centre of Esbjerg*, EEC FAST Programme, Working Paper 19, Brussels.

Perret-Gentil, J.C. (1984) Les acquisitions de services par les enterprises et leur implication spatiale, Programme Nationale de Recherche 'Problèmes Regionaux en Suisse', *Working Paper 43*, Berne.

Petit, P. (1986) *Slow Growth and the Service Economy*, Pinter, London.

Philippe, J. (1984) Les Services aux entreprises et la politique de développement regional. Paper presented at Colloque de l'Association de Science Regionale de Langue Francaise, Lugano.

Singh, A. (1977) UK industry and the world economy: a case of deindustrialisation, *Cambridge Journal of Economics*, 2, 113–36.

Stanback, T. (1979) *Understanding the Service Economy*, John Hopkins University Press, Baltimore, MD.

van Dinteren, J.H.J. (1987) The role of the business-service offices in the economy of medium-sized cities, *Environment and Planning A*, 669–86.

van Dinteren, J.H.J. (1989) The enlargement of the Dutch metropolitan complex: the case of business services, *Tijdschrift voor Economische en Sociale Geografie*.

Part II
Inheritance
and response:
the urban dimension

8

Urban decay and rejuvenation

Louis Shurmer-Smith and David Burtenshaw

'New cities for old' may seem, at first sight, an apt comment when reviewing the relentless pace of recent urban renewal in Western Europe. In the space of one generation, more change has been wrought on the European city than occurred to most in all the preceding centuries of their existence. Nothing, it seems, can remain untouched for very long, as large tracts of the city have undergone comprehensive redevelopment. In Paris, in the 20 years from the mid-1950s to the mid-1970s, more than a quarter of the built-up area was affected by demolition and reconstruction. Yet, over a comparable period during the Second Empire (1852–70), Haussmann created a vast building site covering an even greater area of the French capital, as the medieval city was chiselled apart by a new boulevard system (among other *grands travaux*). In other words, urban renewal, be it slow and piecemeal or rapid and comprehensive, is by no means new.

Change, of course, is endemic in any city, but no less so is continuity. Simply to juxtapose decline and decay, on the one hand, and rebirth or rejuvenation, on the other, is to risk misunderstanding the pattern of constant structuring and restructuring of urban space over time. Seen historically, the imprint of this cumulative legacy in terms of highly varied and distinctive townscapes has on balance proved enriching. The boulevards of Haussmann's Paris (whatever transport planners may think) today provide the linear components of the much-visited, modern tourist city. The Piazza San Marco in Venice is equally a townscape of international repute because of the diverse architecture of its imposing façades, dating from several centuries.

In view of the antiquity of many of the surviv-

ing historical townscapes, it is perhaps only to be expected that the conservation ethic remains so peculiarly European. Yet, while many historic buildings commonly outlive the functions for which they were first created, nevertheless the physical decay of the urban fabric is an inevitable consequence of the ravages of time. In the face of changing social and economic demands, the pressures to rebuild anew — rather than to rehabilitate and preserve existing structures — have often proved irresistible in the name of progress, at least until recently. It is not surprising, therefore, that urban renewal, as an integral part of inner-city regeneration or rejuvenation policies, should now feature so strongly on political agendas at both national and local levels. The implications of this renewal often prove contentious as conflicts between conservation and other aspects of planning become increasingly evident.

The search for appropriate responses to the problems of urban growth and decline across Western Europe has not been helped by forces for change, both regional and global, which have had profound effects on the future of the city. The optimism and certainty of the immediate post-war period — when economic and technological expansion seemed unlimited and the future looked 'controllable' — was within just two decades to give way to gloom and uncertainty, against a background of economic crisis. By the mid-1970s, it appeared that many European cities were in trouble, and the ability of some to change function and to adjust was also in doubt. According to Hall and Cheshire (1987) cities, faced with decentralisation and deindustrialisation, were reverting to their old roles as centres of commerce, administration and culture — hence the link made by some observers to the emergence of the so-called 'post-industrial society' (Bell, 1973). Daniel Bell's vision of the shape of this new phase in the social, economic and political structure of Western society was clear in at least one respect. Several recognisable processes were simultaneously at work in determining both the momentum and the type of urban restructuring and change; these included the decline in industrial production and employment in the city and compensating rapid growth of the service economy, an important aspect of which was information-oriented employment (see also Chapter 7). As successful cities made this transition, so the rationale behind the allocation of urban land to different uses also changed. It was now the turn of the consumption economy to dictate priorities. That the consumption choices being made were often conspicuous ones — extending from the traditional symbolic investment in housing (i.e. gentrification) through to new-found tastes in fashion and design, entertainment and restaurants — only served to emphasise the emerging distinctiveness of the new, inner-city lifestyle. Here indeed, historically, was a new phase in urban development and the primacy of consumption over production (Smith and Williams, 1986).

Realignment of the urban system

The challenge presented by the changing momentum of growth and decline in Western Europe's city regions is already well documented. A succession of major studies adopting similar spatial frameworks of analysis (Functional Urban Regions), details the various stages in a cyclical model of urban change (Hall and Hay, 1980; van den Berg *et al.*, 1982; Cheshire and Hay, 1989). Each phase is described in terms of the relationship between change in the urban core and in the suburban ring (Figure 8.1), thereby providing 'a convincing account of the progress of an urban region from innovatory youth to industrial senescence' (Hall, 1988a, p. 116). The dominant post-war trend has been one of *decentralisation* of population and jobs away from the city to suburbs and, most recently, into surrounding small towns and rural areas. It was the latter trend, that of *counter-urbanisation* discussed in Chapter 14, which appeared to augur a progressive abandonment of

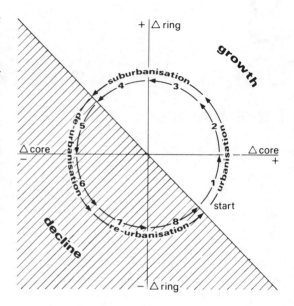

Figure 8.1 *The urban life-cycle model*
Source: Court (1987).

cities, or *deurbanisation*. The former self-sustained growth of many large agglomerations was now replaced by one of self-sustained contraction. With manufacturing jobs in decline, major parts of the inner city appeared in jeopardy.

Not unexpectedly, in view of different histories of social and economic development, the precise nature and timing of various phases within this cyclical process varied regionally across Europe. In southern Europe, often uncontrolled suburbanisation has remained a major challenge to planners well into the 1980s (Wynn, 1986) while counterurbanisation as a dominant trend is already firmly established in much of north-west Europe (Burtenshaw and Court, 1986). In short, Western European countries appear to be arranged on a continuum from north to south, with the UK and West Germany leading the movement through the various stages to eventual decline (Hall, 1988a).

Reasons for this differential, but by no means regular, progression through the urban cycle across Europe relate in large part to the wider regional social and economic contexts. In attempting to classify general types of problem city, Cheshire and Hay (1989) conclude that urban decline is primarily concentrated in the old cities of the industrial revolution, and also in the port cities that handled the output of these industries. Even so, they readily acknowledge that in such broadly based, urban area classifications, inner-city decline is often masked by the general prosperity of the region as a whole. Accordingly, the accepted image of, for example, Bonn as a 'growth' city tends to ignore the problems of Tannenbusch at the intra-urban scale. Equally, at the interregional scale, the problem status of some cities is further compounded by their increasing peripherality relative to dynamic core regions, despite much-improved trans-European communication links. Liverpool, as both a declining manufacturing city and a port located on the deindustrialising edge of the European Community, provides the archetypal example.

In complete contrast, the rapidly growing 'poor' cities of southern Europe present a very different type of problem. Agricultural restructuring still remains a dominant force and major factor in urban development. Although the pace of rural–urban migration is slowing down, the enormity of housing, service and infrastructural deficits in many urban areas remains. Cities themselves have become nodes of disadvantage, in regions of Europe where the formal planning system and its surrounding bureaucracy lack not just resources but also credibility and public confidence (Wynn, 1986).

Clearly, with constituent countries at different stages of population shift within the urban system, no useful generalisation is possible for Western Europe as a whole. Indeed, it is far from certain that all countries will see the majority of their cities, like those in the UK and to a lesser extent West Germany, run full course through to the stage of deurbanisation and decline. What does now appear certain, however, is that this 'final' stage of urban decline, currently represented by a narrow band of nineteenth and early twentieth-century industrial cities stretching in an arc from northern Italy to northern Britain, does not mark an end to the cycle of urban change. On the contrary, the evidence for *reurbanisation* is growing, giving a 'second break' in the long-run process of urbanisation, a return to the city (Breheny, 1987). This is directly related to rejuvenation of the physical and social fabric of the European centres over the past decade. As cities change roles and reassert both their political and economic dominance, traditional measures of urban growth must be reinterpreted. With the contraction of

urban-based manufacturing, the increasingly strong emphasis on the service base of urban economies appears irrevocable. Here is the real engine now driving economic growth in advanced industrial countries (Hall, 1988a), a theme examined in depth by Daniels (Chapter 7).

The adjustment costs of making the successful transition to having a prosperous service-based economy may be high. Seen nationally, or even internationally, the ability of all cities to attract new activities, as well as residents, is not equal. What has now emerged as critical to cities' success is not least a matter of creating the right image, a culturally brighter urban scene. In short, urban living must be fun, as well as profitable for those who invest in it. Not unexpectedly, then, the process of urban regeneration across Europe is of variable success. While places like Manchester and Liverpool continue to face particularly difficult adjustment problems, other strong provincial cities like Munich and Grenoble, or smaller national capitals like Copenhagen, appear to be making the transition successfully (Hall and Cheshire, 1987).

Post-war shifts in state intervention and political ideology

At any time, urban policy-making will be simultaneously shaped by the contingencies of necessity, on the one hand, and the demands of ideology, on the other. Over 40 years of economic and political change in the post-war period have allowed sufficient time for both processes to make their mark. For example, few would deny the impact of the international economy on urban issues as growth turned to decline. No less apparent has been the gradual disengagement of the state from urban restructuring. The general trend, which provides an essential backcloth for approaching the problems of urban rejuvenation, has been for the ambitious thinking of the early post-war growth period to give way to a much more pragmatic, retrenched view of what planning could achieve (Batty, 1984).

Pressures to rebuild and start life again dominated the immediate post-war period. But the first task of providing shelter and, not least, the means of economic survival, meant that the chance to redesign urban morphology was often sacrificed to speed and economy. This usually amounted to utilitarian building along existing street lines, as was largely the case in Essen and Caen; only in some cities were ambitious schemes introduced to build for a new age. Bochum attempted to create broader streets in the centre, Le Havre was rebuilt on a new grid plan layout, and the main shopping areas of Rotterdam and Plymouth were redesigned. The emphasis was on rehousing those who needed homes, and planning for a resurgence of migration into the cities which the war had halted.

It was in this post-war climate of rebirth that the infant discipline of planning grew to prominence. By the 1960s, positive powers of public sector reconstruction (as well as negative powers of control) were in place in most European countries. Moreover, the modernisation of the physical fabric of the city was viewed as part of a wider, social and economic mission. This was the era of 'heroic' planning on the grand scale, with the inherited problems of urban decay and neglect simply excised in relentless slum clearance, city-centre redevelopment, urban motorways and, not least, massive public housing projects. Needless to say, the transformation was not completed overnight, and often the expectation of large-scale urban renewal — whether for housing or roads — inevitably brought in its path long periods of physical degradation through planning blight.

Professionalised town planning, with its roots at the time firmly fixed in the disciplines of architecture and civil engineering, readily adopted new technological and design solutions. Cheap, industrialised construction — inspired by new architectural fashions for high-rise, high-density living influenced by Le Corbusier, Gropius and Mies van de Rohe — combined to create such townscapes of redevelopment as Neue Vahr (Bremen) or Parkhill (Sheffield). There is no shortage of other examples of this pervasive international style, which would bring a first element of uniformity to the development of post-war European cities.

The full onslaught of the world economic crisis in the early 1970s heralded a period of reaction and reappraisal. The period of rapid population growth which had fuelled housing demand was clearly over. Projections were being scaled down, as in the Paris region where eight new towns were cut to five, and their original target populations halved. With this demographic and economic

downturn came financial stringency, and the whole tenor of urban planning changed. Yet it was not simply the end of the boom which brought retrenchment; growing disillusionment with urban-policy making itself was also influential. Looking back, the results of two decades of large-scale, social experimentation were quickly rejected. First to be indicted were the vast and bleak housing estates surrounding most cities. The social problems of modern, mass suburbia had produced their own vocabulary: 'Sarcellitis' and 'new town blues' were already bywords of the caring professions.

Reactions went further as sections of the urban population mobilised locally against specific schemes, challenging the right of new financial and property companies to redevelop, without proper public participation, areas such as Westend (Frankfurt) and Italie (Paris). The resulting civil unrest in cities in both West Germany and France may not have rivalled the larger-scale urban social movements in Spain and Italy but, linked to a growing, grassroots concern for environmental issues, it undoubtedly influenced a real shift in planning philosophy. Large-scale redevelopment was now superseded by selected rehabilitation in the urban renewal process across most of Western Europe.

As 'the custodian of Europe's cultural patrimony' (Hall, 1969), the city now found itself protected as never before. Increasing support, both nationally and internationally, was readily available for urban conservation policies which everywhere quickly became an integral part of the formal planning process (Kain, 1980). Indeed, by the early 1970s evidence from a wide sample of West European cities (Burtenshaw *et al.*, 1981) revealed a remarkable degree of uniformity in the overall response. With the Netherlands' *Monumentenwet* (1961), France's Malraux Act (1962), Britain's Civic Amenities Act (1967) and Italy's Housing Act (1971) the previous system of individually listed buildings worthy of preservation gave way to comprehensive conservation planning of urban areas of historic interest. But given the costs involved, it soon became clear that in historic quarters new functions would increasingly displace old uses, altering the social and economic balance of the central areas in question. *'Renovation oui, expulsion non'* was a vain cry in the early 1970s by residents of the Marais district in central Paris, the first and largest of France's *secteurs-sauvegardés.* The example of Bologna's *centro storico* in avoiding the social effects of conservation planning and retaining its residential population mix would prove somewhat exceptional (see below). Almost everywhere, gentrification became a highly visible component of a fundamental transformation within the inner city, in turn reflecting wider changes in the reshaping of advanced capitalist societies (Smith and Williams, 1986).

Against this background of sharpening social divisions within the city, the continued rationale for urban policies is no longer as unequivocal as it seemed a few decades ago. Gone today is the broad consensus of the past that interventionist planning should be undertaken for the 'public good'. Financial feasibility rather than social desirability provides the new emphasis and, in effect, the real dilemma for urban governments. Faced with these new urban realities, policy-makers have turned increasingly to the private sector to supply much of the new investment stimulus. The sometimes uneasy relationship between planners and developers is hardly unexpected as multinational finance capital begins to dictate the overall terms of urban regeneration.

At first sight, the ventures themselves have often appeared highly speculative, but there is now little doubt that the building boom generated by the renewal of depressed and derelict inner-city areas, not to mention decaying docklands, has become big business. Value is now perceived in many of these locations. Politically, there is a new convergence of interests as the New Right allies itself to other major economic and social developments in the 1980s. Cities are back in fashion as places to live, a recolonisation led by upwardly mobile, professional, two-income, childless households in search of a new lifestyle in urban living. The formula may vary, but not the distinctive quality of the 'designer community' in question. What better, for example, than a water-side setting as redundant wharves, warehouses, even factories, are rehabilitated to bourgeois taste? Inevitably, the whole process remains controversial in political, social and economic terms. Without doubt waterfront redevelopment has become the flavour of the present decade, and in many Western European cities it is a key element in the wider processes of urban renewal (Hoyle *et al.*, 1988).

Present directions of urban change

The structural imperatives underlying the process of urban change are enormously complex and defy recourse to any one theoretical paradigm. In itself, the problem of trying to blend the specificity of different places at the micro-scale with the generality of national/international influences at the macro-scale remains a major challenge in the area of comparative research on urban areas (Bourne, 1986). This, of course, is not to deny that some generalisations are possible in identifying the key parameters influencing the development of modern cities. The previous discussion has already touched on several of the substantive forces involved in the reshaping of urban areas — the changing function of cities, the economic crisis, new patterns of consumption, and the role of state and capital. It is not the intention here to suggest some exclusive list of determining 'factors', but more simply to establish the broader context within which the detailed typology of urban rejuvenation presented later in the chapter may be understood. Neither should the use of conventional umbrella headings — economic, social, political and technological — be seen as underestimating the whole chemistry of inter-action between different agents of change.

Economic shift: late capitalist cities and the new urban hierarchy

Within a more global economy, new international regions are being created. The former insularity of regions inside national economies is no longer decisive as cities become part of 'a more extensive, less-rooted kind of economic life, competing for footloose capital, for manufacturing plants, research laboratories and corporate headquarters' (Mulgan, 1989). Not surprisingly, competition within a newly emerging urban hierarchy is fierce as leading cities negotiate directly with trans-national corporations and institutions. So Paris loses to Barcelona in the bidding to stage the 1992 Olympics, but the positions are reversed as the French capital successfully clinches the prestigious new Euro-Disneyland site. Lesser prizes are available further down the urban hierarchy, but are no less sought after as a potential catalyst for economic regeneration. Thus Sheffield hopes to see itself back on the international map as the venue for the 1991 biennial World Student Games, the biggest sporting event seen in the UK. With vast sums of public and private money absorbed by such speculative ventures, some observers remain sceptical at what they see as simply a political diversion from the real issues in question. Nevertheless, parts of the Lower Don Valley — the former steel and engineering heart of Sheffield — are already benefiting from the commencement of major construction schemes, just as Munich in 1972 exploited the opportunity to build its *U-Bahn* and the Olympic village as a new, social housing area.

Of course, for international and most national cities, it is the concentration of money capital and the gamut of financial, administrative and professional services that defines status. Few cities have emerged so far as truly global cities, as hubs in world financial markets and centres of advanced producer services. The map of Western Europe certainly reveals no shortage of candidates aspiring to join the ranks of New York, London and Tokyo. *Randstad Wereldstad* (Randstad World City), the slogan promoted by the Dutch National Physical Planning Agency in the late 1980s, quickly found its counterpart elsewhere as great provincial cities like Frankfurt, Barcelona and Milan vied for international status. Sensitive to perceived weaknesses in her own country's position within this West European hierarchy, the French government's regional planning agency, DATAR, has recently commissioned a major research programme on what is often simply dubbed 'Manhattanisation', symbolising the transformation of national/international city economies.

Social change: post-modern culture and new urban lifestyles

Today, inner-city rejuvenation is becoming an important stage for promoting the new stylisation of urban life (Jager, 1986). New quarters of galleries, studios, boutiques, theatres and restaurants provide a now all-too-familiar mix in regenerated urban areas right across Western Europe. Generally, this is but part of a broader strategy to use art and culture as the means to reverse urban economic decline, the promotion of a 'soft' infrastructure supporting new activities. In an attempt to offset recent, and often traumatic,

histories of decline, many UK industrial cities now strive to change their image: Liverpool ('Tate of the North'), Bradford (Museum of Film, TV and Photography), Cardiff ('Media City'), Portsmouth ('Flagship of Maritime England'). Arguably the most successful of these has been Glasgow, designated 'European City of Culture' for 1990. The multiplier effects of mobilising the economic potential of such arts activities are not easily quantifiable. What is undeniable, however, is the critical mass of activity required in order to create a thriving 'scene' (Mulgan, 1989).

'Conviviality' and 'liveability' of cities are now the passwords of the developer who seeks to package this new urban lifestyle. The new middle class is evidently a willing partner in the promotion of a consumption ethic which emphasises such aesthetic-cultural themes as a means of self-expression (Jager, 1986). Not all cities, of course, will offer the right combination of environmental possibilities to ensure success but, as many dockland and urban waterfront redevelopment schemes have already shown, even the most unlikely redundant spaces and buildings can be imaginatively recycled — and very profitably, too. Much more than just a form of investment (as well as a status symbol), such urban conservation and rehabilitation serves also to maintain class boundaries as contrasts between upgraded and downgraded neighbourhoods lead to further polarisation of living conditions and social well-being. It is this increasing duality in urban structure which has remained a major challenge to policy-makers and planners.

Political coalition: public and private sector in new pro-growth enterprises

The state, whether locally or nationally represented (with the particular balance of influence between the two often becoming a crucial political determinant) remains the important arbiter of change within the urban arena. While in the past the involvement of the public sector in urban renewal has appeared largely reactive, the growing evidence of a proactive role, whether alone or in partnership with the private sector, is now widespread across Western Europe. However, in line with current New Right orthodoxy, the emphasis is increasingly placed on 'leverage', the use of minimal public resources to prime the private sector pump. But such attempts to create a favourable (which is to say, profitable) investment climate as part of a major urban regeneration scheme risk provoking contradictions as between the dictates of a market-led philosophy, on the one hand, and a socially led policy on the other (Law, 1988). Relative profitability may become the deciding issue, inevitably undermining the social component of any scheme. So, for example, old warehouses in Cardiff's redeveloped Atlantic Wharf have not been converted into small business workshops, as originally demanded by the County Council, but instead into luxury apartments yielding much greater profits to the Urban Development Corporation's private sector partner. Yet, contrary to the much-quoted view that this particular redevelopment was led by the private sector, it is estimated that two-thirds of the total cost was funded by the public sector (Tweedale, 1988).

Nevertheless, the desire to stimulate free enterprise and the profit motive in order to accelerate economic regeneration in Western European cities is now evidenced by the increasing willingness of municipal authorities everywhere to enter into corporatist-type relationships with the private sector. The creation in the UK during the 1980s of Urban Development Corporations, and notably their acquired jurisdiction in the planning process over existing statutory authorities, provides an extreme example of this trend which, without doubt, was strongly politically motivated by a desire to increase direct central government control (Lawless, 1988). But this ideological dimension to the political economy of cities is not just confined to the UK. In France, despite the major decentralisation reform of 1982, designed to strengthen local authorities in their urban development activities, it is clear that only the larger cities — through the political strength of their mayors — are able to act independently of the other levels of territorial authority. Nevertheless, the various mechanisms for associating private capital more closely with public investment are particularly well adapted to all scales of urban development projects. The possibility of establishing joint stock companies (*sociétés d'économie mixte* or SEM) goes back to 1951 legislation, but it was the introduction of urban action zones (*Zones d'Aménagement Concerté* or ZAC) under the 1967 Land Act which allowed a more flexible, mixed-economy approach at the

urban scale. By no means the least achievement of ZACs has been to force a more rapid completion of development projects. However, as with the British UDC, it is difficult to resist the conclusion that the French SEM formula has also caused municipal authorities to lose a significant measure of control to the specialist (albeit semi-public sector) agency in which technocratic values tend to dominate over popular preferences. Accordingly *rentabilité* dictates that communal facilities and low-cost public housing are scaled down in prestigious city-centre redevelopments, whether in provincial (and socialist-controlled) cities like Rennes and Lille or, as might be expected, in Paris.

When viewed against a long-established *dirigiste* tradition, it may perhaps appear startling that a demand-led philosophy should today so dominate the planning ideologies of municipal authorities spread right across the French political spectrum. Yet entrepreneurship is apparent in many rapidly growing, medium-sized cities of the provinces. For example, for more than a decade now, Montpellier has sought to emulate the aggressive success of its much larger 'twin' city, Barcelona, even to the point of using the internationally renowned Catalan-born architect, Bofill, to provide the master-plan for its ambitious expansion to the east and south. Achieving a higher profile within the single European market is without doubt adding a timely boost as newly created urban infrastructures exploit to the full the latest in technology and design.

The impact of technology: new locations and the working landscape

While new technologies continue to permeate the physical fabric of city life, their ultimate spatial impact remains ambivalent. A true prototype of 'technopolis' or 'teletopia' the new information city of tomorrow, has yet to emerge. Some elements are already visible as, in the leading cities, the new information economy establishes itself alongside older ones based on manufacturing and services. Emphasis on 'smart' communications infrastructure has become a decisive selling point for the post-industrial city. Above all it is the teleport, with its cluster of satellite and microwave dishes, which symbolises the latest technology: the London Docklands Teleport, Amsterdam's Telecenter, Cologne's Media Park, Hamburg's

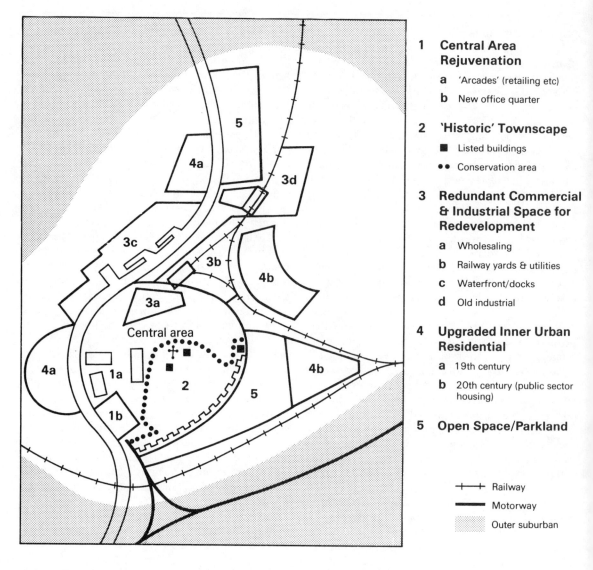

Figure 8.2 *Typology of locations of inner-urban rejuvenation and change*

Telecommunications Centre and Paris's planned *Carrefour de la Communication* at La Défense are all members of this exclusive club. Not unexpectedly, the costs of building and equipping this new environment are high. In the case of ultra-modern office complexes, anything up to one-third of the cost is now accounted for by electronics and control systems for heating, air conditioning, lighting, communications and security.

However, in terms of the architecture and design of this new working landscape, the state-of-the-art changes rapidly. By the 1980s, demand was already partially shifting towards 'ground-scrapers', whether newly constructed on redundant railway land (as at Charing Cross and Gare d'Austerlitz) or accommodated in refurbished industrial buildings (as in Billingsgate and La Villette). And the 1990s may well see further change as classic, historic office buildings, presently regarded as unconvertible, become user-friendly again. Flat wiring, now used for computers in the USA, could accelerate this trend since it has made special floor-voids redundant. New building designs allowing for natural light

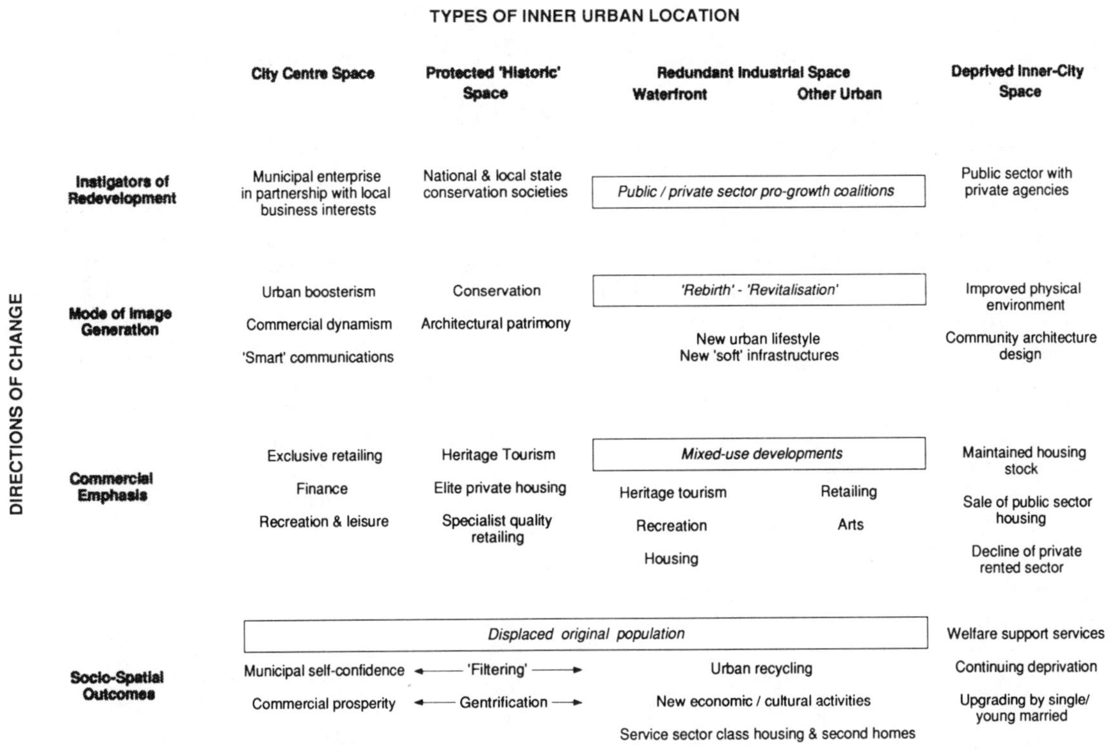

TYPES OF INNER URBAN LOCATION

	City Centre Space	Protected 'Historic' Space	Redundant Industrial Space — Waterfront	Redundant Industrial Space — Other Urban	Deprived Inner-City Space
Instigators of Redevelopment	Municipal enterprise in partnership with local business interests	National & local state conservation societies	*Public / private sector pro-growth coalitions*		Public sector with private agencies
Mode of Image Generation	Urban boosterism / Commercial dynamism / 'Smart' communications	Conservation / Architectural patrimony	*'Rebirth' - 'Revitalisation'* / New urban lifestyle / New 'soft' infrastructures		Improved physical environment / Community architecture design
Commercial Emphasis	Exclusive retailing / Finance / Recreation & leisure	Heritage Tourism / Elite private housing / Specialist quality retailing	*Mixed-use developments* / Heritage tourism / Recreation / Housing	Retailing / Arts	Maintained housing stock / Sale of public sector housing / Decline of private rented sector
Socio-Spatial Outcomes	Municipal self-confidence ← / Commercial prosperity ←	*Displaced original population* / 'Filtering' → / Gentrification →	Urban recycling / New economic / cultural activities / Service sector class housing & second homes		Welfare support services / Continuing deprivation / Upgrading by single/ young married

(DIRECTIONS OF CHANGE)

Figure 8.3 *Processes of inner-urban rejuvenation and change: a suggested framework*

and natural ventilation may well obviate the now habitual provision of air-conditioned, open-plan working spaces, shown by research to be linked to the 'Sick Building Syndrome'. Increasingly, people are demanding to work in spacious, well-planned and well-maintained surroundings with — though this may be difficult in city-centre locations — adequate car-parking provided at the workplace itself.

At first sight, this points to greenfield sites for both factories and offices away from the congested old cities and crowded suburban trains, seemingly adding to the competitive pressures of rejuvenating inner cities. However, city centres represent established concentrations of capital, structures and cultural inertia. In weighing up the relative advantages and disadvantages of such locations, the historic role of cities as centres for the congregation of diverse interests and informal networks is not easily ignored. No wonder that financial and banking services, otherwise so dependent on the very telecommunications that allow businesses to disperse away from cities, cling to the prestige of central locations. Seventy

per sent of Milan's 45,000 employees in this service sector category, for example, are found in offices clustered within the heart of the old city, around the Duomo and La Scala. Far from contradicting such trends, new technologies may be seen as reinforcing existing lines of urban development.

Towards a typology of urban rejuvenation

That Western European cities have entered a significant period of change in the last two decades of the twentieth century is already clear. Equally in evidence, however, are the forces of inertia that affect both the direction and pace of urban restructuring, in places with differing socio-economic bases and situated in varying political and cultural settings across the continent. The overall picture is far from clear, but even half-formed images of the changing internal configuration of cities offer a number of common themes. Most obvious among them is that urban decline itself has engendered profitable opportunities in so

133

far as relatively cheap and strategic central locations have become available, bringing in turn a
renewed focus on the inner city. An attempt to
depict schematically the series of locations whose
redevelopment and rejuvenation prospects have
been shaped by the cycle of post-war urban
change is mapped in Figure 8.2. In addition,
Figure 8.3 seeks to relate the spatial outcomes to
the wider processes that were addressed in the
previous section. The main focus of the following
discussion remains firmly on inner-urban areas of
change.

Boosting the city centre

Prominent in all cities throughout the period has
been the continuing revitalisation of the central
area. The desire to create or maintain a strong,
dynamic and visually attractive core now comes
top in any city's agenda, in order to attract new
commercial and business activities and, where
possible, to exploit to the full the associated
tourist potential. In large and small cities alike,
boosting the image of the central area can become
a matter of urban survival. The potential for
success may be clearly illustrated by reference to
two examples.

In Hamburg, the destabilising effects of a major
shift of commercial and service activity left the
centre to decay both physically and psychologic-
ally. By the mid-1970s, the media had declared
the city centre dead (Kossak, 1988). Purchasing
power had been lost to successive rings of
expanded shopping centres on the periphery of
this urban agglomeration of 2.5 million people.
Corporate head offices had moved out to City
Nord, a new satellite business centre. With a
much-weakened cultural scene, the spiral of
decline appeared complete. That it was not only
halted, but spectacularly reversed, owed much to
a new climate of entrepreneurship spearheaded by
a dynamic *Land* authority in partnership with
local business interests. By the mid-1980s, an
elegantly redesigned city hall square (Rathausmarkt) had provided a link between the western
and eastern halves of the central area, as well as
a thriving new venue for numerous city festivals.
At the same time, the realisation of the Hamburg
Arcade Quarter brought a much-revived cultural
and commercial focus, and not least a considerable boost for national and international tourism.

Similar dramatic change seems likely to be
achieved in Lille, in northern France, where a
strategic, trans-frontier, crossroads location will
be reaffirmed by the opening of the Channel
Tunnel in 1993 and the completion of the
northern European TGV rail link (diverted to pass
through the centre of the city). As a result, the
fortunes of this old industrial capital are being
dramatically revived as the massive Part Dieu
redevelopment project completely transforms the
central city area between the old and new railway
stations. Providing a centrepiece for the first phase
of this 100 hectare scheme are the new European
headquarters of the Maxwell Communications
Corporation, to be housed in the reconverted old
Souham army barracks. Masterminding the entire
project is the Euralille consortium, backed by five
major banks, the SNCF and the local Chamber of
Commerce and Industry.

Defending historical townscapes

Inevitably, within such historic cities the possible
contradiction of reconciling new demands for
space with an established consensus to conserve
became quickly apparent. With the strong backing
of national legislation in the 1960s, urban conservation soon represented a powerful movement, as
the list of cities chosen to serve as architectural
showpieces lengthened. What was particularly
clear was the essentially defensive purpose of the
exercise adopted by Western historical centres in
the first wave of urban conservation in the late
1960s and early 1970s. It is only more recently
that the dilemma posed by conservation policy
within a more general urban renewal policy,
sometimes characterised by strong social goals,
has become a recurrent question (Ashworth,
1984).

In the context of social goals, Bologna provides
an outstanding example of urban rehabilitation
(Appleyard, 1979). Since the late 1960s, inner-
urban policies in this city have aimed both to
create a congenial living environment for local
people and to maintain a vigorous commercial
life. According the the original (1969) Bologna
development plan, emphasis was to be placed on
the quality of urban life, and growth for its own
sake was to be opposed. Achieving these goals
provided a massive challenge, but sensitive
restoration of old buildings has gone hand in hand

with measures to maintain a stable, socially mixed population. These measures have included the control of rents in rehabilitated properties, and official resistance to property speculation. In addition, the creation of 18 neighbourhood councils has facilitated the task of providing for the needs of the historic core's population of over 70,000. Today, although it is less frequented than other great historic cities in Italy, old Bologna — with its 35 kilometres of medieval arcades — is nevertheless the country's best conserved city after Venice.

Yet, while the achievements of planned rehabilitation in this city demonstrate the constructive role which enlightened city administration may play, for two reasons they must be placed in perspective. First, it is striking that Bologna has proved a difficult example for other Italian cities to follow (Venuti, 1986). It is tempting to explain this in terms of political factors: at first sight, Bolognese initiative appears to have flowed from the left-wing viewpoint of the city government. But many other Italian cities, such as Florence, Pisa and Venice, have had similar left-wing coalitions yet have placed much less emphasis on social goals. Given this similarity, it may well be that Bologna's particular policy owed much to the influence of key local individuals. Second, despite the social orientation of its policy, Bologna has been the subject of increasing commercial pressures (Ward, 1988). Although the city council took the lead in financing the restoration of the most run-down streets, in the long run private capital was also necessary. In some quarters, at least, this has initiated the gentrification process, a development that is unlikely to be temporary since the city's communist administration now accepts that private funds will provide the main source of future protection for the historic core. In this new situation, the 1980s saw the emphasis of the city's urban development strategy switch from the core to the periphery of this thriving agglomeration of over half a million people.

Recycling redundant space and disused buildings

Immediately beyond the protective girdle of conservation areas, but usually still within the inner city, the continuing demands for development space have often been accommodated in what at first seem unlikely locations. From the smallest sites (such as disused churches, cinemas and hospitals) to vast areas formerly developed for transport and utilities (railway goods yards and gasworks), these redundant leftovers from the nineteenth and early twentieth centuries provided a timely (indeed, possibly a last) opportunity to bring a new balance of activity back to the central areas of many old industrial cities. Former national wholesale food markets, like Covent Garden in London or Les Halles and La Villette in Paris, clearly also offered areas sufficiently extensive and accessible for large-scale redevelopment projects. But major, multi-functional schemes, often demanding a sustained effort of both political will and financial resources, are not always easily brought to fruition. All three of the above-named projects were the subject of lengthy planning battles through the 1970s over the future use of such strategic central sites, reflecting above all the complexities of local versus national state political power.

In the case of the Forum les Halles, the municipality of Paris (as majority shareholder in the powerful SEMAH[1] consortium) imposed a final project in which the dictates of commercial viability were paramount. This was done despite the opposition of a conservationist-minded state president whose preference was for a large formal garden. Today, the sunken shopping arcades of the Forum, with its three labyrinthine tiers below ground, are surrounded by a zone of high-quality mixed residential and office development as might befit such a prestigious site in the heart of the city (Bateman and Burtenshaw, 1983). Even so, within this massive redevelopment scheme, the local community has not been forgotten: a wide range of cultural, sports and leisure facilities now exists, as well as some subsidised housing.

Equally well sited, the outcome for Covent Garden, at least initially, was to prove totally different. Here an elected residents' forum successfully resisted comprehensive, large-scale redevelopment in favour of the small-scale 'market', with its busy boutiques, restaurants and street theatre. The refurbished old Central Market building opened in 1980 and, set in a piazza, is now the second most popular tourist attraction in London, after the Tower of London. But with increased prosperity came renewed property speculation in a politically charged conflict between the claims of locals and outsiders, as

135

escalating rents threatened Covent Garden's welfare and community structure, and not least its remaining open space. Most recently, the clashes have focused on proposals to demolish the still unused floral hall in order to build offices as well as an extension to the Covent Garden Opera House. Outside the old central market area, the wider programme of urban renewal appears even more fragmented as cutbacks in public spending have allowed indiscriminate private sector developments to undermine agreed community objectives still further, especially in relation to housing (Spence, 1985).

La Villette and Bercy in eastern Paris provide examples of recycling redundant space and buildings on a much larger scale, and effectively illustrate the potential of such renewal in terms of broad urban planning goals. The Villette dock basin, located at the former junction of three canals, housed the old Paris abattoir quarter; the now largely moribund Bercy river-port quays and depots were traditionally associated with the wine trade. Such was the scope for large-scale development offered by these sites, that both now represent the very corner-stones of the East Paris Plan (Figure 8.4). This aims to give much-needed impetus and coherence to the programme of urban regeneration in this long-disadvantaged half of the capital. Although the project is not yet completed, the pace of change around La Villette since the mid-1980s has proved impressive. Within what will be Paris's largest municipal park, a vast culturo-scientific museum and arts complex has taken shape. In addition, the ingenious conversion to 'high-technology' uses of many redundant industrial buildings, of varying dates spanning over a century, has become an immense and immediate success.

Already, the indications are that developments such as these, and not least that of the vast new Ministry of Finance building sited on the river-front at Bercy, will lead to a significant shift in inner-metropolitan structure by restoring some balance of prestige between west and east Paris. Not surprisingly, however, physical renewal on this scale has already brought signs of social transformation in eastern Paris. The changing electoral map of recent municipal elections reveals the crumbling of the last bastions of the inner-city 'red belt', a clear indicator of the displacement of local population and communities.

Transforming urban waterfronts

By the 1980s derelict docklands, not only in large seaport cities but also in smaller coastal and inland ports, constituted a very important category of urban redevelopment projects (Hoyle *et al.*, 1988). Port decline, and related industrial abandonment of waterfront areas, offered quite unprecedented opportunities in many cities for extensive, planned, social and economic regeneration. That is not to say that the areas in question were without their particular development problems and challenges: the physical complexity of sites with large areas of water to manage; the pressures and constraints imposed by residual industrial activity; stocks of handsome buildings requiring conservation; and, not least, the frequent emergence of hostility in local communities in response to proposed redevelopment. Whether hostility has been satisfactorily assuaged or not — and here a clear distinction needs to be drawn between the physical, social and economic achievements of revitalisation — highly visible and prestigious initiatives have proved irresistible to most port city authorities in their attempts to adjust to harsh economic change.

Not unexpectedly, the mixed-use strategies adopted for these 'water cities' vary enormously, both in scale and emphasis. In the UK, Liverpool's Albert Dock provides perhaps the conventional response of attempting to create a high-quality commercial, residential, cultural, tourist and media complex. Similarly Glasgow's Princes Dock combines a major residential development, a business park and an exhibition and conference centre, all within much of the retained landscaping of its initial highly successful Garden Festival site. In contrast, Dunkerque Neptune will see the French city's nineteenth-century dock transformed into a major urban node which, in addition to a business park and tourist resort, will boast a new university campus. Higher education and applied research also feature in a Dutch project for an informatics university to provide the central focus of a future high-technology campus in the obsolete, inland port-industrial area of Laakhaven in The Hague (Jobse, 1987).

Despite the major physical changes which such schemes bring about, their regenerative effects, particularly with regard to their contribution to the solution of inner-city problems, have frequently been seriously questioned. In demand-

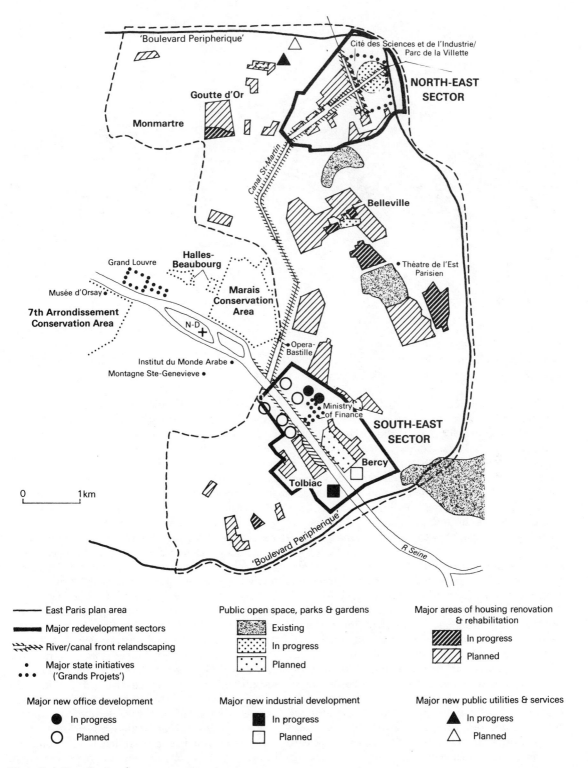

Figure 8.4 *East Paris urban regeneration programme*

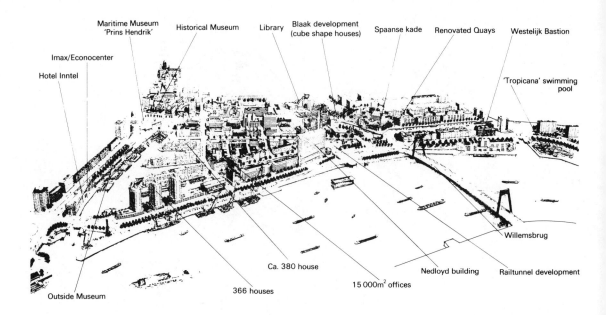

Figure 8.5 *The Waterstad, Rotterdam*
Source: Gemeente Rotterdam.

led planning strategies, the balance of power between public authorities and private investors can all too easily shift, with the original social objectives cast aside as commercial interests take hold (Pinder *et al.*, 1988, pp. 254–6). Without that strongly monitored social and economic dimension, as already demonstrated in the much-debated experience of the London Docklands, the planning strategy in question becomes little more than a large-scale physical development process (Church, 1988).

Rotterdam perhaps offers a near-model example in which social goals and commercial interests have been kept in balance (Pinder and Rosing, 1988). Providing a central focus to the waterfront redevelopment area, the inner-urban zone known as the Waterstad displays a variety of innovative architectural and design features (Figure 8.5). At first sight, this visually spectacular development of mainly pre-nineteenth-century dock basins may appear immediately reminiscent of many such schemes where market forces have been allowed to dictate the direction of change. However, the city authority has defined and implemented a clear planning framework within a policy that has an overt social dimension. Leisure provision, which

assumes a high profile within the scheme as a whole, has been primarily non-commercial in the early development stages, with a focus on museums and a new public library. Again, away from the immediate waterfront apartment complexes, new housing has also retained a significant public sector element. In short, the city of Rotterdam has remained a heavy investor, albeit in partnership with private sector interests which have been allowed to develop many of the prime sites. While redevelopment of this core area provides an interesting example of successful public sector–private capital relationships, however, the strategy adopted must be seen within the broader inner-city planning context. Pinder and Rosing (1988) show that a heavy commitment has been made to socially led urban revitalisation policies in disadvantaged inner-urban communities outside the Waterstad. Here the large majority of refurbished and newly constructed development takes the form of housing for the original inner-city population. In this way, the public authority has been able to accommodate powerful commercial pressures and residential demand from higher-income groups largely within the Waterstad, without provoking serious community friction.

Upgrading inner-urban backwaters

In contrast to the rapid pace of redevelopment achieved in old city centres or around derelict waterfronts, the problems of renewing the outworn fabric of large expanses of the surrounding built environment have at times appeared intractable. Although by no means confined to the UK, it is here in the older industrial cities, with their cumulative backlog of urban deprivation and decline, that the 'inner-city problem' has reached crisis proportions. At various dates over the postwar period, different components of the problem have been given emphasis and specific policies have been developed in response (Edwards, 1984). The earliest strategy focused on demolition in areas of bad housing and environmental decay but, with few exceptions, the vast slum-clearance programmes of the 1950s produced environments that were at best nondescript, at worst disastrous. This was a legacy that would challenge the planners of the next generation. In Liverpool, for example, one-third of council housing (totalling 25,000 dwelling units) from that period was officially defined as hard to let by the 1980s (Hall, 1988b).

With the benefit of hindsight it is evident that, in the UK and in Western Europe, by the 1970s the emphasis had switched from demolition and replacement to rehabilitation. Government-sponsored programmes such as the West German *Sanierungsgesetze*, the French *Habitat et Vie Sociale* and the British 'Housing Action Areas', provided a comprehensive, area-based framework which left municipal authorities free to initiate the required remedial action. This policy of general environmental improvement, often involving the active participation of residents in community design, has been described by Allison (1975) as 'modified Benthamism' because of its aim of spreading well-being. Certainly the concentration of effort everywhere on publicly funded physical renewal was very much reflected in the various contributions to the European Campaign for Urban Renaissance (Eversley, 1981).

But, in the UK, the perception of the problem of multiple urban deprivation was already shifting significantly, away from a 'pathological' to a 'structuralist' interpretation. No longer were environmentally focused solutions in the forefront. Economic regeneration was now the crucial issue, promoting new investment in infrastructure and jobs. The Glasgow Eastern Area Renewal Project (GEAR), a much-publicised showpiece development which ran from 1976 to 1986, typifies the problems associated with the multi-purpose approach which would also come to characterise the strategy of the UK's urban development corporations in combating urban decline. Led by the newly formed Scottish Development Association, in partnership with the local authorities, this co-ordinated programme was designed to regenerate a large area of older housing, declining industries and derelict land, as well as to provide employment specifically for local people. Certainly the resulting environmental transformation was dramatic; but many of the jobs created went to inhabitants of Glasgow as a whole, rather than of the GEAR area, where unemployment when the project ended was 30 per cent. This was 10 per cent higher than when the scheme commenced, partly due to a further wave of industrial closures. In such a setting it becomes difficult to escape the conclusion that, for urban renewal to reverse the downward spiral of decline, a more robust economy than the Strathclyde region currently enjoys will be required.

Alternative futures

Public policy and private capital are reshaping the major cities of Western Europe at an unprecedented rate. The balance of influence may vary as much within individual countries as between them, but nowhere can it be argued that the decline of some cities and the growth of others is merely the irreversible outcome of economic processes. The rhetoric of an emerging post-urban society, which finds such a ready echo in the United States, and which sees 'the old cities of the industrial revolution already consigned to the graveyard of urban history' (Newton, 1986, p. 214) is hardly applicable to this side of the Atlantic. Even in the previously cited example of Glasgow, the continuing malaise of parts of the city's East End contrasts sharply with the new prosperity of its centre. However, it is the long-apparent bipolarisation of the range of urban prospects in American cities that does find an echo in Western Europe as it enters the last decade of the twentieth century.

Many cities find themselves in a half-way position, caught in the 'cruel dialectic of decay and opportunity' (Williams and Smith, 1986, p. 209). Nowhere is this as true as in the older and larger industrial agglomerations as they experience a painful shift in their economic base. The dual scenarios are often found in close proximity. The rejuvenated city centre, having attracted a mixture of public and private investment, flourishes on its influx of workers, shoppers and visitors. But in large parts of the inner-city beyond, the older urban fabric and its associated community will continue to deteriorate in the face of neglect and reduced public spending. Social processes at work in this environment of deprivation are not normally the concern of private capital. Yet such prospects are not immutable: the capacities of the state's urban interests and policies to compensate for the stresses of urban change are greater than it is now often fashionable to suppose. If that potential is not being fully realised, then it is because of the demotion of regional or local interests in the political agenda.

Set firmly against this background of the interplay between state and market forces, Gurr and King (1987) have attempted to generalise about future directions of urban change in Western cities. Four general types of city, each shaped by distinctive combinations of public and private activity, are suggested. *Old* and *new industrial cities* provide the first two types, with the former stabilised by active state intervention at the national level, while in the latter case the public sector role is more passive and indirect. The vitality of the third type, the *administrative and service city* depends on 'a synergistic mix of public and private sector activity' (Gurr and King, 1987, p. 192). Finally, *welfare cities* have no significant growth sectors and are dependent to a large extent on the public sector; these represent the economic backwaters of advanced capitalist societies.

The implications for the future of cities, as revealed by this categorisation, are substantial. Whether directly or indirectly influenced by public policies and spending, functionally the survival of all modern cities remains dependent on a threshold of continued public funding and control. Any significant shift in political ideology, as described earlier, which leads to a disengagement of the state from the urban arena, can only serve to accelerate the retreat that is still under way in many cities.

Note

1. *Societé d'Economie Mixte d'Aménagement des Halles.*

References

Allison, L. (1975) *Environmental Planning: a political and philosophical analysis*, Allen & Unwin, London.

Appleyard, D. (ed.) (1979) *The Conservation of European Cities*, MIT Press, Cambridge, MA.

Ashworth, G.J. (1984) The management of change: conservation policy in Groningen, the Netherlands, *Cities*, 1, 605–14.

Bateman, M. and Burtenshaw, D. (1983) Commercial pressures in central Paris. In R. Davies and T. Champion (eds), *The Future for the City*, Academic Press, Cambridge, pp. 205–19.

Batty, M. (1984) Urban policies in the 1980's: a review of the OECD proposals for managing urban change, *Town Planning Review*, 55, 489–98.

Bell, D. (1973) *The Coming of Post-Industrial Society: a venture in social forecasting*, Basic Books, New York.

Bourne, L.S. (1986) On the future of urban areas: challenges for research and public policy, *Tijdschrift voor Economische en Sociale Geografie*, 77, 408–16.

Breheny, M.J. (1987) Return to the city?, *Built Environment*, 13, 189–92.

Burtenshaw, D., Bateman, M. and Ashworth, G.J. (1981) *The City in West Europe*, Wiley, Chichester.

Burtenshaw, D. and Court, Y. (1986) Suburbanisation and counterurbanisation: the case of Denmark. In G. Heinritz and E. Lichtenberger (eds), *The Take-Off of Suburbia and the Crisis of the Central City*, Steiner, Stuttgart, pp. 52–69.

Cheshire, P.C. and Hay, D.G. (1989) *Urban Problems in Western Europe: an economic analysis*, Unwin Hyman, London.

Church, A. (1988) Demand-led planning, the inner-city crisis and the labour market: London Docklands evaluated. In. B.S. Hoyle, D.A. Pinder and M.S. Husain (eds), *Revitalising the Waterfront: international dimensions of dockland redevelopment*, Belhaven Press, London, pp. 199–221.

Court, Y.K. (1987) Counterurbanisation: the case of Denmark. Unpublished Ph.d. thesis, CNAA, Portsmouth Polytechnic.

Edwards, J. (1984) UK inner cities, *Cities*, 1, 592–604.

Eversley, D. (1981) Retrospect and prospects, *The Planner*, 67, 148–9.

Gurr, T.R. and King, D.S. (1987) *The State and the City*, Macmillan, London.

Hall, P. (1969) *London 2000*, Faber and Faber, London.

Hall, P. (1988a) Urban growth and decline in western Europe. In M. Dogan and J.D. Kassarda (eds), *The Metropolis Era: a world of giant cities*, (Vol. 1), Sage, London, pp. 111–26.

Hall, P. (1988b) *Cities of Tomorrow: an intellectual history of urban planning in the twentieth century*, Blackwell, London.

Hall, P. and Cheshire, P.C. (1987) 'The key to success for cities', *Town and Country Planning*, 56, 50–2.

Hall, P. and Hay, D.G. (1980) *Growth Centres in the European Urban System*, Heinemann, London.

Hoyle, B.S., Pinder, D.A. and Husain M.S. (eds) (1988) *Revitalising the Waterfront: international dimensions of dockland redevelopment*, Belhaven, London.

Jager, M. (1986) Class definition and the esthetics of gentrification: Victoriana in Melbourne. In N. Smith and P. Williams (eds), *Gentrification and the City*, Unwin Hyman, Boston, pp. 78–91.

Jobse, R.B. (1987) The restructuring of Dutch cities, *Tijdschrift voor Economische en Sociale Geografie*, 78, 305–11.

Kain, R. (ed.) (1980) *Planning for Conservation*, Mansell, London.

Kossak, E. (1988) The city is dead — long live the city. In C. Hass-Klau (ed.), *New Life for City Centres: planning, transport and conservation in British and German cities*, George Over, London, pp. 71–6.

Law, C.M. in association with Grime, E.K., Grundy, C.J., Senior, M.L. and Tuppen, J.N. (1988) *The Uncertain Future of the Urban Core*, Routledge, London.

Lawless, P. (1988) Urban development corporations and their alternatives, *Cities*, 5, 277–89.

Mulgan, G. (1989) New times: a tale of new cities, *Marxism Today*, March, 18–25.

Newton, K. (1986) The death of the industrial city and the urban financial crisis, *Cities*, 3, 213–18.

Pinder, D.A. and Rosing, K.E. (1988) Public policy and planning of the Rotterdam waterfront: a tale of two cities. In B.S. Hoyle, D.A. Pinder and M.S. Husain (eds), *Revitalising the Waterfront: international dimensions of dockland redevelopment*, Belhaven, London, pp. 114–28.

Pinder, D.A., Hoyle, B.S. and Husain, M.S. (1988) Retreat, redundancy and revitalisation: forces, trends and a research agenda. In B.S. Hoyle, D.A. Pinder and M.S. Husain (eds), *Revitalising the Waterfront: international dimensions of dockland redevelopment*, Belhaven, London, pp. 247–60.

Smith, N. and Williams, P. (1986) *Gentrification in the City*, Unwin Hyman, Boston.

Spence, A. (1985) Carpet-baggers move into the Garden, *Town and Country Planning*, 54, 96–7.

Tweedale, I. (1988) Waterfront redevelopment, economic restructuring and social impact. In B.S. Hoyle, D.A. Pinder and M.S. Husain (eds), *Revitalising the Waterfront: international dimensions of dockland redevelopment*, Belhaven, London, pp. 185–98.

van den Berg, L., Drewett, R., Klaassen, L.H., Rossi, A. and Vijverberg, C.H.T. (1982) *Urban Europe: a study of growth and decline*, Pergamon Press, Oxford.

van de Cammen, H. (ed.) (1988) *Four Metropolises in Western Europe*, Von Gorchum, Maastricht.

Venuti, G.C. (1986) Bologna: from expansion to transformation, *Built Environment*, 12, 138–44.

Ward, C. (1988) Trying to get it right, *Town and Country Planning*, 57, 293.

Williams, P. and Smith, N. (1986) From 'renaissance' to restructuring: the dynamics of contemporary urban development. In N. Smith and P. Williams (eds), *Gentrification in the City*, Unwin Hyman, Boston.

Wynn, M. (1986) *Planning and Urban Growth in Southern Europe*, Mansell, London.

9

From small shop to hypermarket: the dynamics of retailing

S.L. Burt and J.A. Dawson

The retail industry is a major component of the Western European economy, accounting for around 13–14 per cent of GDP in most countries. Despite this, there are few studies of this activity on a continental scale, the principal exceptions being Jefferys and Knee (1962), Distributive Trades EDC (1973), Dawson (1982) and Burt and Dawson (1988). This chapter examines the broad changes taking place in Western European retailing, and places particular emphasis on factors influencing locational change. The importance and role of location in retail strategy have recently been established by several authors (Ghosh and McLafferty, 1987; Laulajainen, 1987; and Jones and Simmons, 1987). The argument advanced in this chapter is that the spatial components of retail change can only be assessed and understood through consideration of the behaviour of the institutions involved. The strategies and operational policies followed by Western European retailers, and the legislative and environmental contexts within which policies must be implemented, shape the location of retailing. What must always be remembered, however, is that these strategies and policies are inextricably interwoven with trends in consumer preferences. As a service industry, changes in retailing are closely related to changes in the groups that are served. Consequently, the identification of significant shifts in demand and consumer behaviour is essential if recent investment patterns are to be fully understood.

The nature and form of the retail industry have evolved considerably during the last 30 years in response to the changing social, economic and political environments in which it operates. The structure of the industry, the companies and their operations have all been transformed in response to the socio-economic changes which have occurred as Western European society has moved towards the post-industrial phase (Bell, 1974). The position of retailing, as the final stage in the distribution channel, accounts for much of the adaptation that has taken place (Stern and El-Ansary, 1977), but it is important to appreciate that the character of the response has changed steadily with the passage of time. In the immediate post-war era, retailers sold the products which they were able to obtain from manufacturers. Today, many of them are far more consumer-driven and respond closely to consumer demand through the implementation of the marketing concept (Walters and White, 1987). The companies that have identified consumer needs, and have then obtained from manufacturers the right goods to satisfy those needs, have become the large successful retailers in Europe. Those that have continued to sell what the manufacturer thinks the consumer should have, have declined or even closed. As retailers have attempted to become more responsive to the consumer, they have been assisted by advances in management information and control systems. These have provided them with the ability to respond to, and even pre-empt, consumer change and to adjust their retail operations to particular regional and local factors.

Changes in the consumer environment

Understanding changes in consumer activity is central to understanding both why and how Western European retailing has evolved.

Table 9.1 *Rates of population change (per cent)*

	1950–59	1960–69	1970–79	1980–89*	1990–1999*
Belgium	+ 5.9	+ 5.6	+ 1.7	+ 0.5	+ 0.8
Denmark	+ 7.3	+ 8.1	+ 4.2	− 1.2	− 2.4
France	+ 9.4	+ 11.1	+ 6.8	+ 3.4	+ 0.4
Greece	+ 10.3	+ 5.5	+ 9.6	+ 2.5	+ 5.7
Ireland	− 4.7	+ 3.9	+ 14.6	+ 10.7	+ 9.5
Italy	+ 6.6	+ 6.9	+ 5.8	+ 0.8	+ 1.1
Luxembourg	+ 6.9	+ 9.7	+ 4.7	+ 3.9	+ 0.8
Netherlands	+ 13.6	+ 13.5	+ 8.5	+ 5.9	+ 4.5
Portugal	+ 5.1	+ 2.4	+ 9.8	+ 6.5	+ 4.4
Spain	+ 8.8	+ 10.9	+ 10.8	+ 5.9	+ 3.7
UK	+ 3.5	+ 5.8	+ 1.0	+ 0.8	+ 1.9
West Germany	+ 10.9	+ 9.5	+ 2.5	− 1.5	n/a

n/a: not available
*Estimate
Source: Calculated from Eurostat data and UN *Demographic Yearbook*
(various dates)

Consumption, consumer behaviour and shopping behaviour have all evolved as part of broad shifts in European society. At one level, there has been a trend towards mass demand that has encouraged the growth of retail strategies based on mass merchandising; the supermarket, superstore, hypermarket and department store have been developed as a response. At another level, and particularly since the early 1980s, the search by the consumer for choice and individuality has created specialist stores. All of these changes have implications for the location of retailing. Yet, while consumer preference has been of prime importance, the retailing and manufacturing industries have not been totally passive, and some changes in demand have been generated by the retailer or manufacturer. The consumer responds to product innovation, and not all new product development has been consumer-led, for example with products in the snack market.

The broad changes in European society which have influenced consumer activity may be categorised as demographic, socio-economic and attitudinal in nature. Demographic changes include the fall in population growth rates (Table 9.1) and a decline in the average size of households, partly resulting from falling birth rates. New family structures, together with increases in longevity and the number of young adults, have expanded the number of households, particularly one-person households. In some regions of Europe over 25 per cent of households now contain only one person. With respect to the changing age structure of the population, Tynan and Drayton (1985) emphasise the importance of the 'Methuselah' market, and although their paper considers only those due to retire within five years, or who have been retired for up to five years, the elderly in general now comprise a major retail target group. The implications of the expansion of this market mean not only more spending on goods such as pharmaceutical products, but also changes in retail methods and locations. These include a need for more in-store services and easily accessible stores. Population projections suggest that, by the year 2000, over 15 per cent of the Western European population will be aged 65 or over and, at the regional scale, spatial concentrations of this age group can be identified (Figure 9.1). The Mediterranean rim of southern France and northern Italy, and the south-west and south coast of the UK have notably older populations than elsewhere in Europe. Retailers are already responding to these new spatial patterns by adjusting product ranges in these areas and by locating experimental stores directed at the elderly in them.

Alongside the spatial concentration of demographic groups there are other important spatial changes in demand to which retailers are responding. The suburbanisation of population remains a strong trend throughout Western Europe, and has been particularly significant because it has been dominated by the movement of affluent consumers,

143

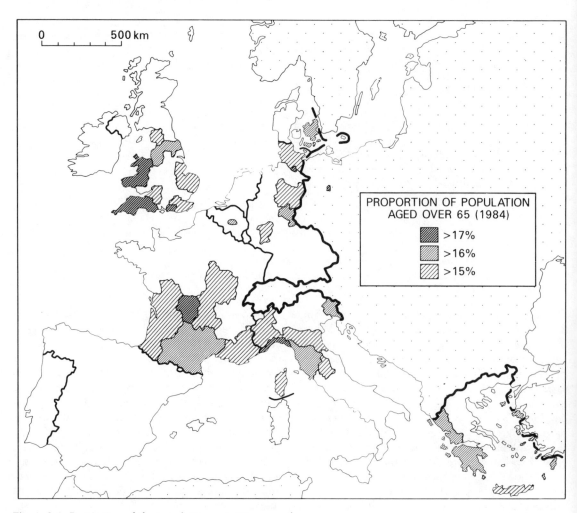

Figure 9.1 *Proportion of the population over 65 years of age*
Source: Based on Eurostat data.

with high disposable incomes, to the suburbs. More recent relevant trends have been counter-urbanisation to selected smaller town environments (Chapter 14; Fielding, 1982) and the gentrification of certain areas by relatively wealthy sections of the community. In each case the spatial change in population has provoked retailer responses, such as movement into smaller towns by large fashion retailers and the redevelopment of old, often blighted, property to create specialist shopping centres in major urban areas as, for example, with Princes Square in Glasgow.

The emergence of post-industrial society has also influenced consumer activity through its effect on occupational structures. The service sector has become the major employer (Chapter 7; Ochel and Wegner, 1987) and there has been a growth in part-time employment, a shortening of the working week and an increase in telecommuting (Kinsman, 1987). As the nature of employment has changed, there has been an increased involvement of women in the workforce and the creation of a pool of unemployed among certain age groups, particularly the young. Incomes have consequently polarised: a general growth in real incomes for those in employment (especially in the growing number of dual-income households) has been paralleled by rising numbers of low-income people who are outside the workforce because of age, lack of skills or simply an absence of job vacancies (Chapter 11). As with the evolution of age structures, these changes have resulted in strong spatial patterns at a variety of scales within Western Europe. For example, at the regional level they are apparent in the differentials between north and south in both England and Italy; at the sub-regional scale, they are to be seen in the disparities between the inner city and outer city in many major centres; and, at the urban level, new suburban housing occupied by middle- and upper-income people have attracted retailers to locate new stores in these suburban districts. Levels of consumer spending, and also its composition, reflect these spatial differences in disposable income.

Changes in individual lifestyles and attitudes deriving from the shift to post-industrial society (Dahrendorf, 1975; McNulty, 1989) have been of particular interest to retailers. Three examples illustrate their significance. The first is the change that is under way in family structure and, in particular, in the family life cycle. The traditional life-cycle sequence associated with raising a family has been substantially modified by the addition of new stages resulting from divorce and non-marriage (Murphy and Staples, 1979; Lawson, 1988). There have also been adjustments to the length and duration of the various stages of the family cycle. These have contributed to the emergence of smaller households, to an increase in the number of households and to a change in spending patterns associated with the various life-cycle stages.

Second, there have been changes in personal values. These have been monitored through a variety of social survey and market research projects (Mitchell, 1978) that have revealed the emergence of a common pattern. There is a growing number of people for whom personal and individual values are more important than group norms based on the mass consumption of conspicuous products. Consumers with alternative values seek individual and personalised products which reflect their sensations and experiences. Various lifestyle typologies have been applied by academics and commercial organisations to characterise these new types of consumer (Dubois, 1987; Europanel and CCA, 1989) and they are often known collectively by descriptive titles such as 'yuppies', 'dinkies', 'emulators' and 'achievers'.

Their implications for retailing are considerable, and have led to the development of specialist stores aimed specifically at these emergent sections of the market.

Third, lifestyle changes have occurred as a result of an increase in real and perceived personal mobility. In addition to the widely seen rise in car ownership, for some groups of people perceptual horizons have broadened markedly. Meyrowitz (1985) has pointed out that, in general, increased holiday travel and exposure to the visual media have heightened consumers' awareness of distant places and cultures, causing a response in their shopping behaviour. This has had interesting implications for the sourcing policies of retailers, who have had to look further afield to find new products, from Greek yoghurt to Scottish knitwear.

As has been indicated, these interrelated demographic, socio-economic and lifestyle trends have far-reaching implications for the retail industry. The size, composition and location of the various consumer groups and markets have changed within nations, regions and urban areas. There has been a fragmentation of consumer demand and behaviour and, because this has occurred throughout Western Europe, retailers have been able to target the resulting market segments across international boundaries. The success of Benetton and Stefanel illustrate this international activity. Thus, while consumption patterns have increased in diversity, cross-country comparisons reveal fundamental similarities.

As society and consumer groupings have changed, so, too, has the structure of personal expenditure. Most significantly, the proportions of consumer expenditure accounted for by products (such as clothing, household equipment, food, drink and tobacco) have fallen, while service-based consumption (e.g., leisure, education and health-related services) has grown considerably. Trends in France (Table 9.2) illustrate the general Western European experience. In most nations, as expenditure on services has grown, retail sales as a proportion of consumer expenditure have fallen, typically from around 55 per cent in the early 1970s to 45–50 per cent in the late 1980s. Excluding the UK, where widely available credit has played a role, the volume of retail sales has been static, or has even declined, in most Western European markets over the past decade (Table 9.3).

145

Table 9.2 *Expenditure patterns, France, 1963–90 (per cent)*

Expenditure class	1963	1973	1984	1990* Low variant	1990* High variant
Food	30.5	24.5	21.0	20.2	19.5
Clothing	9.5	8.2	6.4	6.1	6.0
Housing	12.9	14.7	16.7	16.7	16.7
Household equipment	10.3	10.8	9.2	9.1	9.1
Health	7.9	10.7	15.7	16.6	16.7
Transport	10.0	12.5	12.3	12.8	13.4
Leisure/culture	5.8	6.4	7.8	8.8	8.9
Other	13.1	12.2	10.9	9.7	9.7

*Estimate
Source: INSEE (1985).

Table 9.3 *Index of retail sales volume, 1980–87*

	1980	1984	1987
Belgium	100	95.2	96.4
Denmark	100	105.8	106.2
France	100	92.0	90.9
Greece	100	96.2	96.2
Ireland	100	89.4	89.3
Luxembourg	100	101.1	108.6
Netherlands	100	89.3	94.6
UK	100	111.3	129.8
West Germany	100	96.6	102.3

Source: Derived from Eurostat data.

Within this static total retail market, changes in consumer behaviour — especially those caused by shifts in consumer values and attitudes — have provided growth segments which have been targeted by retailers. Fragmentation of the mass market has required organisational and operational adjustments by retail companies with, as is shown below, far-reaching implications for corporate structures, the presentation of products and, not least, the location of retail facilities.

Changes in corporate structure

A striking feature of the retail industry has been its transformation from one dominated by small family firms to one in which large corporate organisations are pre-eminent. Companies involved

Table 9.4 *Major retailers based in the European Community*

Company	Total turnover, 1988–89 (million ECU*)	Country of origin
Metro	17,109	Germany
Tengelmann	16,872	Germany
Leclerc	10,517	France
Intermarché	9,522	France
Carrefour	9,214	France
J. Sainsbury	8,885	UK
Albrecht	8,700†	Germany
Rewe-Leibbrand	7,857	Germany
Marks and Spencer	7,714	UK
Vendex International	7,497	Netherlands
Tesco	7,105	UK
Asko-Massa	7,038	Germany
Karstadt	6,894	Germany
Gateway	6,807	UK
Promodès	6,564	France
Ahold	6,543	Netherlands
Otto Versand	6,412	Germany

*1 million ECU equalled £0.664 million in 1988.
†Estimate.
Source: Institute for Retail Studies from Annual Reports/Trade Press.

in retailing, such as those listed in Table 9.4, are now among the largest concerns in Europe. These large retailers have been required to look beyond short-term tactical decisions and respond to longer-term developments in their market environment. This longer-term and wider view of business horizons has entailed the construction of formal strategic plans, typically based on a policy of substantial investment growth and often envisaging international operations. This level of expansion normally requires companies to use external capital to implement their plans, the usual source being institutional finance obtained either through investment in corporate equity or by loans. An important consequence of this development is that new demands have been placed on retail companies, which must achieve improved financial performances to ensure the continued support of the financial institutions. Thus retailers have become part of the system of capital. In the manufacturing sector this adaptation took place some time ago, but it is now well under way in the retail sector, and many of the structural and spatial adjustments currently

GENERAL STRATEGIES (after Ansoff, 1988)

	MARKETS	
PRODUCTS	Existing	New
Existing	Market penetration	Market development
New	Product development	Diversification

RETAIL STRATEGIES

	MARKETS	
PRODUCTS	Existing	New
Existing	Product market dominance / Increase market share	National geographical expansion / International expansion
New	Product range development -addition of products/ services	Entry into new product and service markets

Figure 9.2 *The product market matrix*

occurring in retailing reflect this change in capital structure.

In practical terms, the formal strategy options available to retailers at the corporate level are relatively few. Ansoff (1988) relates strategic opportunity to market and product characteristics (Figure 9.2), providing a matrix that has been applied to retailing by Knee and Walters (1985). Each of these corporate strategies has its own particular spatial implications, and all have been pursued by companies in the Western European retail sector since the early 1980s.

Market-dominance strategies

The strategy of market dominance seeks corporate growth through increases in market share. Generally, as a company pursues this strategy it seeks to control a steadily larger and larger proportion of stores, floorspace and, in consequence, retail sales within the retail product

market concerned. Some retailers pursuing this strategy have invested heavily in new stores and additional floorspace. Asda's development in the UK illustrates this approach. But incremental growth, store by store, can be costly, and development costs rise as the easiest sites are used first and the remaining good sites become fewer and more expensive to develop.

A more popular means of achieving market dominance has been acquisition. Marenco (1984) has shown that, between 1963 and 1980, two-thirds of the 50 largest food chains in France were absorbed by competitors; indeed, five companies were responsible for the disappearance of 26 of these chains. The pace of growth that may be achieved through acquisition is often impressive. In the UK, Argyll had almost 10 per cent of the grocery market in 1988, despite having had a negligible interest in this sector before 1982. Similarly Ratners had 15 per cent of the UK jewellery market in 1988, compared with a 1980 figure of under 3 per cent. Asko in West Germany and Docks de France in France have achieved comparable growth patterns. The traditional regional nature of retailing in much of Western Europe has provided many of the targets for these acquisition policies, and the main geographical consequence has been the emergence of national chains which are represented in all major towns. Figure 9.3 provides an example of this process in action. In this instance a previously regional company, Docks de France, has expanded throughout the country by making three major acquisitions in the 1980s. Here we may note a potential disadvantage for areas affected by acquisition. If operational decisions throughout a retail chain are made at a central head office, this process reduces levels of regional independence and alters the nature of management in the regions. Unless a decentralised management structure is in operation, management at the regional and local levels becomes essentially supervisory, while key business decisions are made at national headquarters. The problems are analogous to those produced by branch-plant development in the manufacturing sector.

The creation of more informal chains of stores is a further mechanism used by firms to increase market share and achieve sector dominance. Spar operates internationally but is organised more formally on a national basis. In national markets, retailer- or wholesaler-led buying groups such as

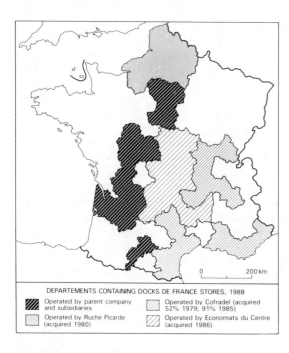

DEPARTEMENTS CONTAINING DOCKS DE FRANCE STORES, 1988

Operated by parent company and subsidiaries

Operated by Cofradel (acquired 52% 1979; 91% 1985)

Operated by Ruche Picarde (acquired 1980)

Operated by Economats du Centre (acquired 1986)

Figure 9.3 *Docks de France: national expansion through acquisition*
Source: Based on Company Reports.

Intermarché (France), and Rewe and Edeka (West Germany) are amongst the largest retail concerns. Such buying groups generally comprise associations of smaller companies which seek benefits of scale through group action. Buying groups for multiple food retailers, as opposed to small retail companies are also present in France and West Germany; in 1987 the Paridoc group in France had a turnover of £6.5 billion. On the one hand, consumers benefit from these retail alliances as a result of bulk purchasing economies. On the other hand, however, it is arguable that the penalty paid for this central co-ordination of the activities of small firms may be reduced product variety throughout the system. The question is again one of external control. Consumer preferences for regional products may be overlooked as retailers' buying decisions are made outside their region of operation.

Product-range development

A second strategic option to attain corporate expansion within existing markets is to develop new products and complementary services. This can be readily achieved through the addition of new products in existing stores. Consumer demand for one-stop shopping for certain types of product, in conjunction with acceptance of mass-merchandising management techniques, has allowed retailers to extend their product ranges into bulky non-food products. This is particularly relevant in the food sector, in which the product range of existing self-service stores has been diversified through the addition of specialist service counters selling, for example, delicatessen items, wet fish or bakery products. French hypermarket schemes illustrate the implementation of this strategy extremely well. Here there is often a core of mass-merchandising activity and a supplementary range of facilities providing more specialist products, often with a personal service element.

In addition to restructuring facilities within their stores, retailers have taken a greater role in new product development. This has been made possible through better information systems which have collected data on individual purchases made by consumers at the checkout. The ownership of consumer information is a source of power for retailers in their dealings with manufacturers, as Piercy (1983) recognises. Use of this power can result in co-operation and co-ordination between retailer and supplier within the distribution channel, and an important aspect of this co-ordination lies in the potential for retailers and manufacturers to work together to develop and test new products.

One advantage of goods developed in this way is that they allow product-range extensions which are unique to a particular retailer. In many cases these new products are the retailer's 'own-brand' products and their growth reflects the new balance of power in the distribution channel (Thil and Baroux, 1983). Improved margins and cost control motivated many initial moves into own-brand labels (Dawson *et al.*, 1986), but this strategy is now frequently driven by the goals of image projection and extension of the product range. Geographically, their primary significance lies in their effects on manufacturers. The centralisation of management by retailers often means that manufacturers at a distance from a retailer's corporate head office are disadvantaged by their relative remoteness. This handicap is more significant for small manufacturers than large ones, who are able to maintain contact with

Table 9.5 *Diversification into DIY retailing: major companies and parents*

Company	Turnover 1988 (£ million)	Country	Parent company*	Main retail activity
B & Q	779.8	UK	Kingfisher	Mixed
Castorama	622.8	France	Carrefour (33%)	Food
Leroy Merlin	471.8†	France	Auchan	Food
Texas Homecare	440.3	UK	Ladbroke	Non-retail
OBI	415.5	Germany	Tengelmann (50%)	Food
Asko Group	384.1	Germany	Asko	Food
Bauhaus	304.1†	Germany	–	–
Do-It-All	238.0	UK	W.H. Smith	Stationery/books
Payless	230.0†	UK	Ward White	Mixed
ANPF/M Bricolage	196.7	France	–	–
Massa	192.1	Germany	Asko (49%)	Food
Homebase	180.3	UK	J. Sainsbury (75%)	Food
			GIB (25%)	Mixed
Plaza	179.3	Germany	Coop AG	Food
Metro Group	176.1†	Germany	Metro	Mixed
Stinnes	164.9	Germany	–	–
Bahr	160.1	Germany	–	–
Brico GB	151.4	Belgium	GIB	Mixed
OBI France	150.9	France	Casino (65%)	Food
			GIB (35%)	Mixed

*Shareholding shown in brackets if less than 100 per cent.
†Estimate
Source: Institute for Retail Studies from Trade Press.

the head office more easily. Small firms are further disadvantaged by the size of order being placed by retailers: small manufacturers often lack the production capacity to meet retailers' demands.

Diversification into new product and service markets

Going beyond the development of new products in existing markets, a common theme in the strategies of retailers since the early 1980s has been diversification into new product and service markets. This has generally meant that retailers have set up new chains of shops to sell different products, or have acquired companies in other retail sectors. In parallel there is also evidence of consumer service companies moving into the retail sector. An example of these trends is investment in DIY retailing by retailers based in other sectors (Table 9.5). Only four of the leading 17 DIY retailers in Western Europe are not at least partly owned by other companies.

Acquisition has been the preferred means of undertaking diversification strategies. This involves the purchase of an established, but not necessarily efficient, management team and chain of existing stores. One advantage to this solution is that it provides rapid entry into the new market. For example, in the 15 months prior to June 1988 Hertie (a West German department store chain) acquired Wehmeyer (clothing), Schumann (electrical goods), Schaulandt (leisure electrical goods), WOM (records) and Vamos (mail-order footwear). In this way, chains of stores operating under different names become controlled by the one company. Although it may not be apparent to the consumer, the diversity of ownership in the High Street and in other locations is reduced as more outlets become owned by a small number of companies.

Diversification has been sought also through co-operation and collaboration between companies, the aim being to exploit complementary expertise to allow entry into new markets. In France, for

Table 9.6 *Vendex International activities, 1988–89*

Retail trade	No. of subsidiary companies	Service activities	No. of subsidiary companies
Department Store Group	4	Information Group	5
Home Furnishing Group	4	Maintenance Services Group	9
Fashion Group	5	Temporary Help Services Group	8
Hard Goods Group	7	Financial Services Group	2
Food Group	7	Leisure Group	2
Mail Order Group	7	Miscellaneous Services Group	10
Minority Holdings in the Netherlands	3		
Holdings outside the Netherlands			
– USA	7		
– Brazil	7		

Source: Vendex International (1989).

example, joint ventures have been established between petrol station chains (which possess sites and expertise in locational assessment) and food retailers (offering marketing expertise) to develop convenience stores. These have long opening hours and sell a range of basic foods and other essential products at particularly accessible sites (Libre Service Actualités, 1987). The Marché Minute chain is a case in point. In other instances collaborative ventures have allowed the creation of unique store formats such as the ABC stores in the Netherlands; these involve Ahold (food), Blokker (household goods) and C & A (clothes). Similarly in the UK, until 1989, the Savacentre hypermarket chain was a joint venture between J. Sainsbury and BHS (itself part of the Storehouse group).

Some diversification has gone beyond other product groups into, for example, financial activities such as banking or insurance. The West German retailers Schickedanz, Kaufhof and Otto Versand are among those who have entered into banking, as are the French food retailers Auchan, Cora, Intermarché and Leclerc. In the same vein, Carrefour and Benetton have developed insurance services. Service retailing — in the form, for example, of travel agencies and opticians — has also attracted investment, as have a wide range of business services. In this respect Vendex International, which grew out of the Vroom and Dreesman department store chain in the Netherlands, has an impressive portfolio of activities, ranging from traditional and multi-sector retailing to an equally wide range of business services (Table 9.6).

With these diversification strategies a large European retail company may be operating in several retail sectors, in a variety of service sectors and in a number of countries. The spatial strategies of these firms become very complex as attempts are made to optimise activity within a set of retailing operations. The one company may operate a range of different store types in the same area, and these distinct operations may be managed by separate subsidiary companies.

Internationalisation

The extension of operations into foreign markets is a further strategic option for retailers. Expansion from a regional base into the national market is a well-established process for retail companies, but the move into foreign markets is now becoming increasingly common. Opportunities in foreign retail markets may appear attractive if, relative to domestic markets, they are growing or are less competitive. Moreover, this attraction is frequently increased by the potential benefits to be gained by transferring a particular store format or managerial style to another country. Several international investment strategies have been identified (Kacker, 1988; Treadgold, 1988; Tordjman and Salmon, 1989), ranging from investment in an overseas retailer to the creation of global retail concepts which may be standardised and operated worldwide, of which Benetton is an example.

The spatial process most closely associated with the internationalisation of Western European retailing has initially entailed entry into

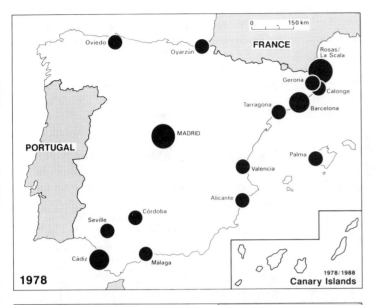

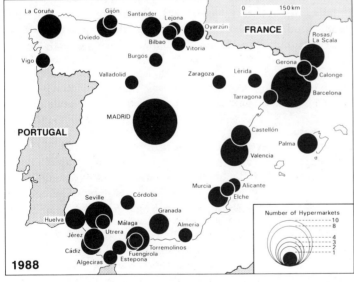

Figure 9.4 *Hypermarkets in Spain, 1978 and 1988*
Source: Based on Distribucion Actualidad data.

geographically or culturally close markets. For example, French retailers dominate the hypermarket sector in Spain; in 1988, 50 of the 79 units operating were under French ownership. Similarly, a number of Dutch retail chains have moved into Belgium, while British retailers have generally favoured investment in America.

As with other strategies, internationalisation may be achieved by a variety of mechanisms. If the source of competitive advantage for a company is closely associated with the company brand or image — for example, Benetton, Laura Ashley or Ikea — then strict managerial control will be sought over any expansion. Because in these circumstances growth takes place in a steady series of incremental steps, the pace at which international expansion can be undertaken is limited. This problem may be overcome, however, by franchising agreements. These allow tight central control by the franchisor to be associated

151

BASIS OF ADVANTAGE

		Cost control	Uniqueness
TARGET/SCOPE	Broad (industry)	Cost leadership	Differentiation
	Narrow (segment)	Cost focus	Differentiation focus

Figure 9.5 *Sources of competitive advantage*
Source: Porter (1980).

with limited operational decision-making by the franchisee.

An alternative to incremental growth is, once again, acquisition. Acquisition at the international scale has usually been undertaken by retailers from the smaller European countries, where domestic growth has been limited by the size of national markets and also by legislative controls. Delhaize le Lion, the Belgian food retailer, has invested heavily in the USA and Portugal; altogether, almost two-thirds of the group's food stores are now located in countries other than Belgium, and 60 per cent of sales are achieved in the USA. Similarly, Ahold of the Netherlands has expanded its American portfolio steadily by acquiring chains in the USA in 1977, 1981 and 1988.

Primarily because of the high risks of operating in an unknown foreign environment, co-operative forms of international expansion have developed. These range from operational joint ventures to contractual or franchise agreements concerning the use of management systems and brand names. Spar operates in 17 countries; throughout its system small member retailers are provided with the benefits of group buying by wholesalers, and in each country Spar is heavily dependent for success on its members' knowledge of the local market. Another major exponent of joint venture arrangements has been the Belgian GIB group; this organisation operates companies in conjunction with local retailers in the DIY markets of the UK (Homebase), France (OBI) and Spain (Alba). In addition, GIB is also active in the fast food sectors in France and Italy, where it trades as Quick.

As more retail companies become international, the scope for repetition, if not standardisation, of shop types and management techniques throughout Western Europe is increased. The spread of the hypermarket concept in Spain (Figure 9.4) is an example of a retail technique 'imported' into a country through internationalisation. Cultural differences may necessitate adaptations to product ranges and trade names, yet broad retailing concepts are often transferable. The types of shop and retailer operating in a shopping centre in Cologne or Lyon will be similar to those found in Manchester or Odense.

Changes at the establishment level

Changes in the consumer environment, and the strategic responses to them, have necessitated related adaptation at the establishment or store level in retailing. As we have seen, the development of new products and markets in response to changing consumer tastes has involved retailers in new types of store-based retailing and new approaches to corporate management.

Other environmental factors such as legislative controls, particularly those associated with land-use planning, have shaped the nature and location of store-based retailing. In all national markets there is legislation which influences, to varying degrees, the establishment of stores, their location and the nature of their operation (Davies, 1979; Commission of the European Communities, 1985). Within this regulatory framework, and within product and geographical markets, retailers seek to establish and maintain some form of advantage over competitors. According to Porter (1980), the basis of competitive advantage at the operational level is to be found in three generic strategies (Figure 9.5), each of which has different spatial implications.

Cost leadership requires continuous attention to the minimisation of costs in stores, relative to those of competitors. This may be achieved through strict cost-control procedures, through seeking cost economies of scale or from the application of specific skills, technologies or systems. One effect of this strategy is to generate large mass-merchandising stores which gain scale-related benefits and, in terms of their spatial impact, attract large numbers of consumers from a wide area. Another is to seek locations which minimise operating costs, for example in regions with low land costs or at sites with low rents.

Differentiation requires the company to create a store concept which is perceived as unique by its users. This uniqueness may take many forms, including location, but it generally arises from a variety of complementary factors. Opening hours, price, quality, store location and image form a 'package' which is perceived as unique by the consumer. Location is particularly important to convenience stores and local corner shops. Close proximity or accessibility for consumer groups allows these types of retailing to offer the convenience which is the basis of their competitive advantage. Similarly, out-of-town locations for hypermarkets couple low land costs with the space to provide large sales areas and extensive car parks. The outcome is that these shops offer the unique advantage of competitive prices combined with one-stop shopping.

Focus, the third strategy, encompasses both the cost-leadership and the differentiation strategies discussed above. Its distinctive feature is that it concentrates on a particular segment of the market, such as a user group or a geographical area. The strategies pursued by, for example, Tie Rack and Next in the late 1980s fall into this category. The key geographical dimension is the need to match each outlet's location to the particular consumer group that is targeted. Hence, Tie Rack is found in locations with a high consumer throughput, such as transport termini, while Next shops are located in the controlled retail environments of shopping centres or in prime property on the High Street. The importance of new housing areas, and housing renovation areas, for DIY operators provides another example of the locational dimension of this focus strategy.

Changes in store characteristics

As these options for competitive advantage have been pursued, changes have been imposed on both the range and the types of retail store operating in Western Europe. The total number of retail outlets has declined for several decades, primarily because high opening rates have been more than offset by even higher closure rates. In Western Europe as a whole, by the late 1980s there were approximately 2.6 million shops, compared with about 3.5 million in 1955. This downward trend has naturally varied from sector to sector and, as the earlier discussion implies, in some specialist markets numbers have risen. In general, however, the greatest net loss in store numbers has been

Table 9.7 *Hypermarkets and superstores, 1975–87*

	1975	1978	1981	1984	1987
		Number			
Belgium	70	76	79	82	88
Denmark	5	n/a	n/a	12	13
France	291	370	433	527	639
Italy	3	8	12	20	28
Luxembourg	3	3	3	4	4
Netherlands	30	38	39	35	35
Spain	4	21	31	45	60
UK	102	176	279	372	457
West Germany	627	778	821	907	956

n/a: not available.
Source: Derived from International Self Service Organisation *Biannual Reports.*

experienced among food outlets. Also, recent changes in consumer preferences — particularly those that have fragmented mass markets — have accelerated the decline of general store types and the growth of certain specialist outlets.

Although the number of shops has fallen, the amount of retail floorspace has risen in most countries. There are now fewer but larger outlets, especially in the food trades where superstore and hypermarket growth reflects the adoption of mass-merchandising techniques by retailers (Table 9.7). As companies have diversified into new markets, relatively standardised retail techniques have been transferred into the sale of a wide range of products, such as DIY materials, furniture and electrical goods. The specialist mass-merchandise outlet — the *Fachmarkt* and *grandes et moyennes surfaces specialisées* — are the new large-scale outlets. The site requirements of these large units often mean that out-of-town locations are necessary, and thus a new competition is emerging in many countries between established town centres and the out-of-town specialist superstores.

While it is important to recognise the significance of the emergence of large stores, investment in highly specialised small shops must not be overlooked (Dawson, 1985). From the preceding discussion, it will be apparent that this investment is strongly associated with the strategy of targeting specific consumer groups.

Various widely used small-shop formats can be recognised: discount stores designed to meet the low price requirements of certain consumer groups; convenience stores catering for time constraints on behaviour; specialist stores

Table 9.8 *Targeted small shop formats: core characteristics*

Store format	Target consumer requirement	Good/product elements	Store/environment elements	Examples
Discount stores	Low prices	Discount prices; relatively wide product range with few lines; variable- to low-quality goods.	Relatively large size; located in surburban/ secondary High Streets; décor sparse, providing frugal ambiance.	Aldi Ed Kiabi Poundstretcher
Convenience stores	Time/convenience	Premium prices; wide range but few lines; high-quality products, usually leading brands.	Typically 100–200 m²; located in suburban sites or traffic routes; design and décor stress efficiency.	7-Eleven Marché Minute Sperrings
Specialist stores	Choice	Premium or competitive prices; narrow or very narrow range of products but depth of lines; product quality variable to high.	Can be very small if ultra-specialist; high visibility sites in shopping centres/on High Street/in store; décor and design stress the specialist nature of the offer.	Athletes Foot Pingouin Tie Rack Oddbins
Style shops	Design/originality/ exclusivity	Premium/high prices; very narrow product ranges with very few lines; product quality is high.	Size range around 50–200 m²; High Street/targeted shopping centre locations; design stresses quality and exclusivity.	Benetton Next Bally Laura Ashley
Branded goods shops	Assurance/ reputation	Premium prices; very limited product range, narrow with few lines; high-quality merchandise.	In-store locations, or very small stores, 50– 100 m²; ambiance created by store design is quality-based.	Yves Rocher Estée Lauder
Service shops	Added value service	Premium/competitive prices; limited service offer; quality variable but often intangible.	Small outlets; High Street and in-store locations; décor and design are functional.	FNAC Service Mr Minit Supa-Snaps Sketchley
Locality-specific shops	Impulse/emotion	Prices are variable, as is quality of products but usually either high or low; product range is narrow and activity or theme related, but choice within the range.	Location related to theme; as are design and décor, which often provide a 'bazaar'- type ambiance.	Tourist shops National Trust shops

Source: Burt and Dawson (1988).

providing choice in a particular range; style shops based on the value placed on originality, design and exclusivity; branded goods shops that are dependent on brand reputation; the service shop which reflects the consumer's wish to consume services and not just products; and the locality-specific outlets — such as souvenir shops — which enable consumers to make impulse purchases following a particular experience. Details on these small-shop formats are provided in Table 9.8. As this table makes clear, each format has its particular type of optimal location. Several succeed best in town centres, often in the purpose-built shopping centres that are part of city-centre development schemes. Others require sites which are geographically closer to their consumer groups. The key is that location, and the other factors identified, combine to meet the specific needs of the consumer.

Management systems

Advances in information technologies, particularly data collection at the point of sale, have resulted in a considerable improvement in the quality of data used in management systems. Important consequences are that it has become possible to manage more effectively large chains of stores and also to respond to local differences in consumer demand (Dawson and Sparks, 1986). The integration of information produced at the point of sale with spatially based demographic data, for example from population censuses, has enabled companies to adapt store operation to regional and even local conditions.

Laser scanning at the point of sale is a key technology in the move to electronic data collection. The number of scanning stores in Western Europe has risen steadily over the last decade and a number of countries, most notably France, are now in the rapid growth stage of the adoption curve (Table 9.9). Investment in scanning has been greatest in the food sector, where bar coding of products has been most widely adopted. A second significant technology which must be noted is Electronic Data Interchange (EDI). In the late 1980s investment took place in inter-organisational communication networks, creating systems such as TRADANET in the UK, TRANSNET in the Netherlands, SEDAS in West Germany, and ALLEGRO in France. These networks allow the

Table 9.9 *Innovation dispersal in store management: EAN-system scanning equipment, 1981–88*

	1981	1983	1985	1988
		Number of stores		
Belgium	0	12	115	356
Denmark	0	0	14	416
France	2	37	420	2,300
Greece	0	0	0	0
Ireland	0	0	n/a	14
Italy	9	13	20	618
Luxembourg	0	n/a	n/a	18
Netherlands	1	36	134	580
Portugal	0	0	0	81
Spain	0	2	36	422
UK	7	42	160	1,152
West Germany	23	69	290	1,544

n/a: not available.
Source: Derived from International Article Numbering Association, *Annual Reports.*

automated transmission and processing of orders and invoices between suppliers and manufacturers. For each institution in the distribution channel, EDI systems therefore reduce the effects of space by speeding up the transmission of information between the scattered set of stores, head office and factories.

While the costs of technology are now considerable in retailing, labour costs remain the single largest item of expenditure after the purchase of stock: wages and salaries account for at least 50 per cent of store operating costs in most Western European countries. Because of this, retailers have moved towards flexible labour scheduling, using cheaper part-time labour in unskilled and semi-skilled occupations, to match daily and even hourly fluctuations in store sales. This trend is of considerable significance in the context of the spatial restructuring of retailing. In particular, the growth of large stores in suburban locations has changed the spatial structure of the retail labour market and has provided new, often part-time, jobs which are easily accessible to women whose family commitments place a premium on flexibility and short commuting times.

Locational change in retailing

Store location influences costs directly through land prices, rents and property taxes. Location also affects sales volume, and therefore revenue.

Thus the right location to meet consumer needs continues to be a critical success factor in retailing, and across Western Europe several common patterns of retail locational change can be identified. These include a suburbanisation of investments as stores are developed in locations closer and more accessible to catchment populations; the redevelopment of central areas of towns and cities in response to the suburbanisation of retailing; a decline in rural retailing as rural populations have declined or become more mobile; and the creation of concentrations of specialist retailing designed for specific consumer groups. Each of these has naturally evolved within a framework of planning legislation operated by national and local government agencies. In this section we examine trends relating particularly to urban structures.

Retail location and settlement structure

The major focus of debate in retail location over the past decade has been the changing relationship between in-town and out-of-town retailing. The suburbanisation of retailing, particularly large stores, has been a common feature throughout Western Europe. Dawson (1974) and Kivell and Shaw (1980) identify several underlying, and by now well-known causes: the spatial decentralisation of population and consumer demand; an increase in personal mobility based on widespread car ownership; the lack of suitable space for large retail stores in city centres; problems of land availability in town centres; and government policy in relation to out-of-town sites. These reasons have not changed substantially since the mid-1970s.

Developments outside the town centre, the traditional location of retailing, have taken several forms. Initially they generally comprised stand-alone hypermarkets or large supermarkets. As development trends progressed, it became more common for large units to be grouped together in 'retail parks'. Today there are free-standing units such as hypermarkets, clusters of superstore operations in retail parks, and several other types of purpose-built shopping centres. Major traffic routes have played a key role in determining the location of all these forms of suburban retailing. The common picture in larger towns and cities in most of Western Europe is the emergence of a

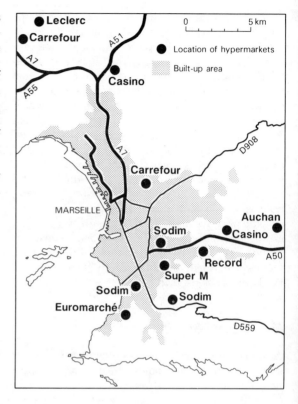

Figure 9.6 *Hypermarkets in Marseille, mid-1980s*
Source: Based on LSA data.

discontinuous ring of large-scale retail units on arterial routes in the suburbs. Developments in Marseille (Figure 9.6) illustrate this phenomenon.

The perceived threat of out-of-town retailing to the town centre has resulted in a variety of attempts to maintain or revitalise retailing in central areas. Most of these schemes have involved some form of renovation of town-centre environments. Renovation may entail full-scale redevelopment of parts of city centres, as in Utrecht and Newcastle, or it may involve a more piecemeal approach including the creation of traffic-free zones, as in Bordeaux, Cologne and Bruges. In-town redevelopment schemes such as City 2 in Brussels, the Forum les Halles in Paris (Chapter 8) and Part Dieu in Lyon (Tuppen, 1977) are other well-known examples of this response to the perceived threat from out-of-town developments. Taking a broader view, Davies and Bennison (1979) comment on large-scale redevelopment in Britain, while Knee (1988) reviews such developments on a wider European

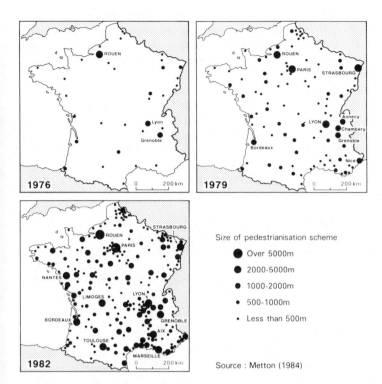

Source : Metton (1984)

Figure 9.7 *Pedestrianisation schemes in France*
Source: Metton (1984).

scale. Even though many of these projects in major cities have often received considerable publicity, however, as a rule redevelopment has only affected small sections of the urban core.

A common feature in smaller Western European towns has been the pedestrianisation of the core to improve the retail environment. Naturally, the speed with which this innovation was adopted varied considerably from country to country. For example, in France pedestrianisation did not really come of age until the early 1980s, as Metton (1984) has illustrated (Figure 9.7). Local retailers initially resisted pedestrianisation for fear that passing trade would be lost, but in most instances they have ultimately benefited from the schemes and opposition has disappeared. Indeed, early worries have given place to demands by retailers to extend pedestrian areas.

The development of controlled retail environments

Many of the changes in town-centre retailing reflect the desire to have more control over the retail environment. For the retailer this will provide an additional attraction to consumers, and from the planners' viewpoint such control can be used to maintain the vitality of the urban core. The purpose-built shopping centre, in all its forms, is one mechanism of achieving this by providing an enclosed retail area. Such centres also allow influence to be exerted over the shopping environment through management of the mix of shop types, usually termed the 'tenant mix'. The number of purpose-built shopping centres developed in Western Europe over the past decade bears testament to the benefits to be gained from this type of control. In France, for example, the total rose from 297 in 1977 to 478 in 1989; in the UK over 10 million square metres of retail space were provided in these new centres between 1965 and 1988. The development and changing nature of shopping centres has been well documented at a variety of scales. For example, Delobez (1985) has detailed their emergence in the Paris region, while Shaw (1983) examines their development in West Germany.

Established classifications of shopping centres have been based on size, and rely on the concepts

157

of neighbourhood, community and suburban regional centres that have been transferred from the United States. Dawson (1983) has extended this classification to take into account town-centre schemes and new types of shopping-centre development such as the focused centre and the specialised centre. The focused centre, in which a main store is supported by a limited number of smaller shops, has been a common development in Western Europe, where most are based on a hypermarket. Hypermarket schemes in France, West Germany and Spain have moved away from dependency on a single free-standing store towards these focused centres.

The development of specialist shopping centres is a response to retailer interest in the targeting of specific consumer groups. The shopping centre takes on a theme which is reflected in the environment created by the centre. Specialist centres such as Princes Square in Glasgow, Covent Garden in London and the Kö-Gallerie in Dusseldorf provide a designer environment and a general ambiance to appeal to style-conscious consumers. Not all these developments are new: the arcades in Brussels are a form of this type of centre and many date from the last century; London's Burlington Arcade originated in 1818. In a sense, therefore, recent investment following this pattern represents a strategy to increase the competitiveness of city-centre retailing by modernising a well-tried formula. Another form of specialist centre in the urban core, prevalent in the USA and now appearing in Western Europe, is the service centre established to provide a range of business and personal services for a different type of consumer — small businesses in a commercial business district. In France and Belgium, meanwhile, discount centres have recently appeared (Bondue, 1987). These comprise retailers who are either direct branches of manufacturers or who are providing branded goods at discount prices within a suitable environment.

While most of these centres are purpose-built, some department-store companies are reducing the number of departments in the store, while simultaneously accentuating the distinctiveness of each department and increasing the number of specialist in-store concessions. In effect they are providing a speciality shopping centre in the tightly controlled environment of an existing department store building which, in most instances, has the advantage of occupying a prime retail location. The development of this galleria concept is seen in the Debenhams stores refurbished since the mid-1980s and in the similarly redesigned Horten stores in West Germany.

Implications and future trends

The strategic, operational and locational changes currently taking place within retailing have substantial implications for the future structure and development of the industry in Western Europe. At a general level a dual structure is developing: a small number of very large enterprises is emerging, yet at the same time there remains a large number of small firms. It is the medium-sized companies that are under the most pressure. As these reach a critical size, measured either by sales or store numbers, they may attract the attention of larger predator companies, they may be unable to compete with the larger companies, or they may lose the benefits previously provided by the market responsiveness inherent in their small size. The spatial implications of these structural trends are, first, the loss of regional firms and their replacement by centralised national companies and, second, greater uniformity in the mix of retailers found in out-of-town locations and in urban shopping centres.

A major consequence of the changes occurring over the past two decades has been market concentration. In Belgium, for example, the share of all retail sales taken by corporate chains has risen from under 9 per cent in 1960 to around 25 per cent in 1990, while that of the unaffiliated independents has fallen from over 80 per cent to under 66 per cent over the same period. This pattern of increasing market dominance by integrated retailers is even more acute at the regional level and in individual product sectors. Such concentration is most advanced in the food sector: data collected by the market research firm A.C. Nielsen show that, with the exception of the southern European states, multiple chains and consumer co-operatives account for at least 50 per cent of food sales in the European Community, and in the UK the figure is no less than 83 per cent.

A further outcome of the trends identified is that individual companies and affiliated groups have major shares of national product markets, as Table 9.10 shows for food sales in the UK, West

Table 9.10 *Market shares in food retailing by major companies, 1987 (per cent)*

UK		West Germany		France	
J. Sainsbury	13.7	Aldi	10.6	Leclerc	9.0
Tesco	12.9	Metro Group	10.4	Carrefour	6.5
Gateway	11.3	Rewe-Leibbrand	8.3	Intermarché	5.7
Argyll	9.5	Tengelmann	6.8	Casino	4.7
Asda	7.4	Asko Group	6.2	Auchan	4.0
CRS	3.7	Coop-Bolle	5.7	Euromarché	4.0
Top six companies	58.5		48.0		33.9

Source: Trade press coverage of market research reports (primarily by AGB/Mintel, Glendinning & Lehning and Libre Service Actualités).

Germany and France. Similar patterns of corporate dominance also exist in the non-food product markets, particularly in the UK. Here, estimates for 1987 show that Dixons has 16 per cent of the electrical goods market, while Marks and Spencer, the Burton Group and Sears together account for 39 per cent of the menswear market. This restructuring raises issues of consumer access to essential facilities as the number of small, local shops falls. Also important are changing relationships between manufacturers and retailers; the power of the latter continues to increase as large retailers extend their market dominance and exhibit a growing propensity to buy from foreign manufacturers (Morgan, 1988).

Finally, it is necessary to consider this dynamic industry in the context of the European Community and, in particular, the Single European Act. The removal of many of the internal barriers between countries in the European Community as the single market initiative takes effect will naturally have implications for retailers in the 1990s. As product specifications are standardised and restrictions on the movement of goods are lifted, the opportunities for buying goods from further afield will increase. It is also likely that the internationalisation of the industry, discussed above, will gather pace simply because the media coverage of 1992 has brought a geographically larger market to the attention of retailers. However, although change will come, we believe that the likely impact of the single market on retailing may be easily overestimated. The probability is that the variety of legislation controlling land use, opening hours and other operational aspects of retailing, when coupled with deep-seated differences in consumer demand and behaviour among the various countries of Western Europe, will ensure that retailing will continue to exhibit great international variation. Although general forms of retailing will be common (for example, the convenience store, the supermarket, the pedestrian zone or the specialist shopping centre) there will be regional and national adaptations in each of these forms. The single European market after 1992 will not lead inexorably towards homogeneity in retailing.

References

Ansoff, I. (1988) *New Corporate Strategy*, John Wiley & Sons, New York.

Bell, D. (1974) *The Coming of Post-Industrial Society*, Heinemann, London.

Bondue, C. (1987) L'essor des centres commerciaux de magasins d'usines en France. In B. Merenne-Schoumaker (ed.), *Le Commerce de Detail Face aux Mutations Actuelles. Les faits et leur analyse*, Papers from IGU Conference, Liège, pp. 379–86.

Burt, S.L. and Dawson, J.A. (1988) *The Evolution of European Retailing*, ICL, Slough.

Commission of the European Communities (1985) *Measures Taken in the Field of Commerce by the Member States of the European Communities*, Commission of the EC, third edn, Brussels.

Dahrendorf, R. (1975) *The New Liberty*, Routledge & Kegan Paul, London.

Davies, R.L. (ed.) (1979) *Retail Planning in the European Community*, Saxon House, Farnborough.

Davies, R.L. and Bennison, D.J. (1979) *British Town Centre Shopping Schemes*, Unit for Retail Planning Information Report U11, Reading.

Dawson, J.A. (1974) The suburbanization of retail activity. In J.H. Johnson (ed.), *Suburban Growth*, Wiley, London, pp. 155–75.

Dawson, J.A. (1982) *Commercial Distribution in Europe*, Croom Helm, London.

Dawson, J.A. (1983) *Shopping Centre Development*, Longman, Harlow.

Dawson, J.A. (1985) Structural change in European retailing: polarisation of operating scale. In E. Kaynak (ed.), *Global Perspectives in Marketing*, Praeger, New York, pp. 211–29.

Dawson, J.A., Shaw, S.A., Burt, S.L. and Rana, J. (1986) *Distributor Brand Packaged Groceries in the European Community*, FAST Occasional Paper 104, Commission of the European Communities, Brussels.

Dawson, J.A. and Sparks, L. (1986) New technology in UK retailing: issues and responses, *Journal of Marketing Management*, 2(1), 7–29.

Delobez, A. (1985) The development of shopping centres in the Paris region. In J.A. Dawson and J.D. Lord (eds), *Shopping Centre Development*, Croom Helm, London, pp. 126–60.

Distributive Trades Economic Development Council (1973) *The Distributive Trades in the Common Market*, HMSO, London.

Dubois, B. (1987) *Cours: comportement du consommateur, le style de vie*, Centre HEC-ISA, Jouy en Josas.

Eurostat (annually) *Demographic Statistics*, Statistical Office of the European Communities, Luxembourg.

Eurostat (monthly) *Trends in Distributive Trades*, Theme 7, Series B, Office of the European Communities, Luxembourg.

Europanel and CCA (1989) Euro-styles, *Marketing Journal*, April/May, 106–11.

Fielding, A.J. (1982) Counterurbanisation in Western Europe, *Progress in Planning*, 17, 1–52.

Ghosh, A. and McLafferty, S.L. (1987) *Location Strategies for Retail and Service Firms*, Lexington Books, Lexington, Mass.

INSEE (1985) *Previsions économiques à l'horizon 1990*, Institut National de la Statistique et des Etudes Economiques, Paris.

International Article Numbering Association (annually) *Annual Report*, International Article Numbering Association EAN, Brussels.

International Self Service Organisation (biannually) *Self Service Report*, International Self Service Organisation, Cologne.

Jefferys, J.B. and Knee, D. (1962) *Retailing in Europe: present structure and future trends*, Macmillan, London.

Jones, K. and Simmons, J. (1987) *Location, Location, Location*, Methuen, Toronto.

Kacker, M. (1988) International flow of retailing know-how: bridging the technology gap in distribution, *Journal of Retailing*, 64, Spring, 41–67.

Kinsman, F. (1987) *The Telecommuters*, Wiley, Chichester.

Kivell, P.T. and Shaw, G. (1980) The study of retail location. In J.A. Dawson (ed.), *Retail Geography*, Croom Helm, London, pp. 95–155.

Knee, D. (1988) *City Centre Retailing in Continental Europe*, Longman, Harlow.

Knee, D. and Walters, D. (1985) *Strategy in Retailing*, Philip Allen, Oxford.

Laulajainen, R. (1987) *Spatial Strategies in Retailing*, Reidel, Dordrecht.

Lawson, R.W. (1988) The family life cycle: a demographic analysis, *Journal of Marketing Management*, 4 (1), 13–32.

Libre Service Actualités (1987) Les convenience stores à la française, *LSA* 1076, 26 June, 16–20.

Marenco, C. (1984) The economies and rationality of concentration in food retailing in France. In CESCOM/IRM, *The Economics of Distribution*, Franco Angeli, Milan, pp. 159–89.

McNulty, C. (1989) *Future of the UK 2010*, Applied Futures, London.

Metton, A. (1984) L'Expansion des espaces piétonniers en France. In A. Metton (ed.), *Le Commerce Urbain Français*, Collection Université d'Orléans, Presses Universitaires de France, pp. 61–76.

Meyrowitz, J. (1985) *No Sense of Place*, Oxford University Press, New York.

Mitchell, A. (1978) *Consumer Values: a typology*, Stamford Research Institute, California.

Morgan, A. (1988) *British Imports of Consumer Goods*, Cambridge University Press, London.

Murphy, P.E. and Staples, W.A. (1979) A modernised family life cycle, *Journal of Consumer Research*, 6, June, 12–22.

Nielsen (annually) *International Food and Drug Store Trends*, A.C. Nielsen, Northbrook, Il.

Ochel, W. and Wegner, M. (1987) *Service Economies in Europe*, Frances Pinter, London.

Piercy, N. (1983) Retailer information power — the channel marketing information system, *Marketing Intelligence and Planning*, 1, 40–55.

Porter, M. (1980) *Competitive Strategy*, Free Press, New York.

Shaw, G. (1983) Trends in consumer behaviour and retailing. In T. Wild (ed.), *Urban and Rural Change in West Germany*, Croom Helm, London, pp. 108–29.

Stern, W. and El-Ansary, A.J. (1977) *Marketing Channels*, Prentice Hall, Engelwood Cliffs, NJ.

Thil, E. and Baroux, C. (1983) *Un Pavé dans la Marque*, Flammarion, Paris.

Tordjman, A. and Salmon, W.J. (1989) The internationalisation of retailing, *International Journal of*

Retailing, 4 (2), 3–16.

Treadgold, A. (1988) Retailing without frontiers, *Retail and Distribution Management*, November–December, 8–12.

Tuppen, J. (1977) Redevelopment of the city centre: the case of Lyon — la Part Dieu, *Scottish Geographical Magazine*, 12, 151–8.

Tynan, A.C. and Drayton, J.L. (1985) The Methuselah Market, *Journal of Marketing Management*, 1(1), 75–85.

United Nations (annually) *Demographic Yearbook*, United Nations, New York.

Vendex International (1989) *Annual Report 1988/89*, Vendex International NV, Amsterdam.

Walters, D. and White, D. (1987) *Retail Marketing Management*, Macmillan, London.

10

The social and economic geography of labour migration: from guestworkers to immigrants

Russell King

Introduction: the context of international labour migration in Western Europe

Since the end of the Second World War, Western Europe has been a major theatre of international migration. The main transfers of people have been from the countries of southern Europe and the Mediterranean Basin to the industrialised countries of northern and western Europe, with more local movements from Ireland to the United Kingdom and from Finland to Sweden. A separate set of migration flows has developed from former colonial countries in Asia, Africa and the Caribbean; these flows have been mainly directed towards Britain, France and the Netherlands.

There are three main demographic components in this massive international transfer of people (Kirk, 1981). The core factor is the large number of young adult workers, mostly male, who moved from Italy, Spain, Portugal, Greece, Turkey, Yugoslavia and North Africa into France, West Germany, Switzerland, Belgium, the Netherlands and other countries during the 1950s, 1960s and early 1970s, most of them seeking work on a temporary basis. This group came to be called 'guestworkers', after the German *Gastarbeiter*. The second component is that of wives and families joining workers already established abroad. By about 1975, the second component had grown to equal the first in terms of crude numbers living abroad. Thirdly, while family-reunion migration has continued into the 1980s, it has been joined by a new movement of refugees and those seeking political asylum. Indeed, if one were to regard the 1960s as the decade of the guestworker and the 1970s as the decade of family

reunification, then the 1980s seem destined to be known as the decade of the political refugee.

The evolving interplay of migration, economic change and social problems has left its imprint in a variety of ways across the continent.[1] The various flows of migrants have been a major component in the dynamics of regional development, urban change and rural decline (Knox, 1984). Within the urban setting in which most of the foreign guestworkers have been located, the process has brought new complexities of population composition, new social structures and new subtleties in the social organisation of urban space (White, 1984).

Geographers in Europe have a fairly well-established tradition of studying minority groups in their own respective countries. These studies are far too numerous to be cited here, for in addition to the many investigations of immigrant minorities in the UK, there is a rich vein of literature in French, German and Dutch which explores the experience of individual countries, regions and cities. Perhaps more important to note here is the limited number of geographical studies which take a European-wide or comparative perspective on the theme of guestworker migration: the collection of essays edited by Salt and Clout (1976), regrettably not updated, and work by Kayser (1977), King (1976a; 1989), Kirk (1981), Knox (1984), pp. 32–47), Salt (1981), van Amersfoort *et al.* (1984), and White (1984; 1986).

The economic and political background to European labour migration

Post-war international migration in Western

Europe has been overwhelmingly conditioned by economic factors. The migrants are 'employment-seeking' and have tended to flow from countries with weak economies to those with strong economies. While the push factors of unemployment, poverty and poor career prospects were important, the movement was also demand-driven as labour-hungry economies in northern Europe sought deliberately to attract labour from the continent's southern and colonial peripheries. The causes of labour shortage in the more advanced economies included the lack of further potential for labour-shedding from agriculture, fertility decline, increasing rates of participation in higher education, a rapid rate of economic growth creating full employment, and the upward mobility of indigenous populations. This last trend led to labour-supply difficulties in jobs that were perceived as undesirable.

These low-status jobs — unpleasant, insecure, menial and poorly-paid — nevertheless presented welcome opportunities to many of the un-employed and destitute workers of Mediterranean Europe and the more distant realms of former European colonies. As guestworkers poured in, the native workers of countries such as West Germany, the Netherlands and Switzerland were relieved of the necessity of doing such jobs and thus enjoyed a measure of *embourgeoisement* or social uplift. Furthermore, economists, manpower planners and employers in the more-prosperous countries realised that temporarily employed foreign workers might provide a cushion for the indigenous labour force against the effects of economic cycles — the so-called *Konjunkturpuffer* philosophy.

The temporary nature of post-war European international migration was seen as one of its essential features, at least during the 1950s and 1960s. Böhning (1974) coined the term 'polyannual' to define the likely length of stay of the migrants — typically two or three years. West Germany and Switzerland — the most ruthless exponents of *Konjunkturpuffer* — kept their guestworkers in a system of constant rotation by employing them only on short-term contracts. A temporary stay was also probably the intention of most of the migrants, at least initially. Their aim was to work for a few years, accumulate some savings, and return home, hopefully to a better life. However, by the early 1970s, as competition between countries to obtain migrants intensified,

and as the longer-term need for migrant labour was finally accepted, the 'rotation' systems were liberalised, allowing more permanent settlement and the reunification of families by the arrival of dependants. As a result of these, and other, processes, labour migrants are now a permanent structural feature of north-west European labour markets and societies.

The accumulation of labour migrants in northern European cities was structured around the threshold of 1973–4 when the first oil-price crisis changed the basic patterns of economic growth — and therefore the demand for labour — almost overnight. The onset of the deep general recession late in 1973 brought a dramatic check to the recruitment of guestworkers. West Germany shut down its recruitment offices in seven Mediterranean countries. Restrictions on the admission of non-EEC immigrants began in West Germany in November 1973 and within a year France, Belgium and the Netherlands had followed suit, clamping down on new immigrants and enforcing stricter controls on clandestine immigration and on the illegal employment of foreign labour. The reason for this dramatic change in attitude was the rising tide of home unemployment, particularly among young workers and school-leavers seeking their first job. As a result the emigration of workers from Yugoslavia, Greece, Turkey, Spain and Portugal virtually halted during the mid-1970s. Return migration became a much-publicised feature of the European international migration scene (Böhning, 1979). In West Germany 650,000 foreign workers left as the *Konjunkturpuffer* approach took effect. In France, after 1977, a free travel ticket home and a 'return bonus' of £1,100 were offered as an inducement to migrants to repatriate. The original migrant-exporting countries became, at least for a time, importers of their own labour. The reduced numbers of emigrants still leaving were mainly made up of family dependants joining workers remaining abroad.

The end of the great migration boom period of *c*.1950–73 has resulted in a complex and confused situation. The agreements on the free movement of labour enshrined in Articles 48 and 49 of the Treaty of Rome and codified in legislation during 1961–8 are not affected, nor are the arrangements for labour migration among the Scandinavian countries. But, of the nine senior members of the EC, only Italy and Ireland are major emigrant countries; and Greece, Spain and Portugal are still in a transitional stage as far as free labour movement is concerned. All EC countries have, however, stopped or severely curtailed 'third country' immigration, except in some cases for humanitarian or political asylum reasons. There remains a certain level of labour circulation among EC states, but this is now composed of labour of a range of types, highly skilled as well as in the traditional low-skill bracket.[2]

Most recently, it has become clear that certain southern European countries are starting to experience the arrival of international labour migrants from North Africa and the Third World. The Italian case is the best documented (Arena, 1982; King, 1985). During the 1970s female domestic servants started arriving to work in Rome and other major Italian cities. They came from the Philippines, Ethiopia, Somalia and Cape Verde. Male migrants from these countries work mainly in low-grade and non-unionised employment in hotels and restaurants. Other cases can also be cited: Egyptian building workers in Athens, Moroccan harvest workers in the *huertas* of Spain, Tunisian fishermen and farm labourers in Sicily. Such labour has some parallels with the classic guestworker types of northern Europe but there are differences, notably the high percentage of females in some types of job, the dispersed character of the migrants and their even greater economic and socio-cultural marginalisation (King, 1984).

Theoretical frameworks for interpreting Western European labour migration

Given the importance of migration to an understanding of the social geography and political economy of contemporary Western Europe, it is rather surprising that so little guidance as to the essential character of the migration process has been given, either by migration theory or by the multitude of mainly small-scale empirical investigations of migrants in different countries. Fielding (1985), following Zelinsky (1983), points to the 'impasse in migration theory' in Western Europe and suggests that labour migration is a 'chaotic concept' which needs to be 'unpacked' in order that it can be interpreted within its proper social and historical context.

In their discussions of the relationship between migration and social well-being in Western Europe Knox (1984) and Fielding (1985) explore a range of interpretations extending from neo-classical to Marxist. Traditionally, labour migration has been seen as an equalising mechanism, moving workers from surplus to deficit regions in response to differentials in income levels and unemployment rates. According to equilibrium theorists, these movements of labour eventually produce a narrowing of wage and unemployment levels, leading in turn to a convergence of inter-regional and international well-being. Such arguments have been advanced for instance by Lutz (1961) for southern Italy and by Kindleberger (1965) for emigration from Greece. Fielding (1985) accuses these neo-classical economic arguments of being the worst offenders in propagating a falsely ahistorical and asocial account of migration. The creation of a race of fictional 'rational economic men' who determine the workings of the spatial economic system through their location decisions based on well-informed individual preferences and cost-benefit analyses is now seen by most analysts as naive and misleading. Such an approach vastly exaggerates the range of choices open to most individuals and involves a highly simplistic view of the European labour market. This view is that the market is unsegmented, and that movement between jobs is unconstrained either by institutional arrangements, such as union opposition or international quotas, or by the socio-demographic characteristics of the migrants themselves, such as their ethnicity and information fields.

Fielding is almost equally critical of the neo-Marxists who, while being most sensitive to the class structuration of society, to the constraints upon individual action and to the segmentation of the labour market, have nevertheless created their own narrow orthodoxy about contemporary European migration, seeing potential migrants in backward regions as a 'reserve army of labour' available to 'capital' to draw on at will (cf. Carney, 1976; Castles and Kosack, 1985; Marshall, 1973; Ward, 1975a; 1975b). There is no doubt, however, that this neo-Marxist interpretation, while crude, does contain an element of truth. The Western European experience has indeed shown that at times of economic expansion the labour power of potential migrants residing in the Mediterranean and Atlantic peripheries has been tapped by importing low-paid and poorly protected workers into the industrial cities of north-western Europe. By raising the 'ceiling' on full employment, such 'cheap' migrant workers have permitted capital accumulation to continue at a time when labour would otherwise have threatened to recoup through wage increases a larger share of the 'value' produced. Also, being foreign, they have tended to fragment the working class, undermining its solidarity in its relationship with capital and employers. At times of economic depression, such as occurred in West Germany in 1966–7 or in Europe as a whole in 1973–4, unemployed foreign workers have been exported back to their countries of origin — the classic *Konjunkturpuffer* mechanism.

Although Fielding (1985) acknowledges that the above model has some value as a general portrait of post-war European international labour migration, he criticises it for being too overtly functionalist and, again, ahistorical (cf. Shanin, 1978). It is not enough, he maintains, to say that migration is somehow necessary for the survival of capitalism in Western Europe; one must elucidate the specific historical and social circumstances which led to the movement of people from certain origins to certain destinations. Given the extraordinary multiplicity of chain migrations linking individual emigrant villages with specific destinations this is an almost impossible task: if it could be done, it would take an army of historical geographers. More fundamental is the question of whether it is any longer true to say that migration from the European periphery is necessary for the maintenance of capital accumulation in the industrial heartlands of the north. A reverse movement of capital to the periphery has also taken place (Paine, 1979), although this has not proceeded without undesirable effects in terms of increased dependency and spatial deformation (see, for example, Hudson and Lewis, 1985).

Between the theoretical poles of neo-classicism and neo-Marxism are the frameworks offered by writers such as Hirschmann (1958) and Myrdal (1957), who emphasise that migration often serves to maintain or exacerbate the spatial economic disparities of which it is initially a function. Migration is seen as one of the key motors of 'cumulative causation' whereby the initial advantage held by more prosperous nations and regions is progressively reinforced by inflows of capital, labour and enterprise. At the same time, the loss

of these factors of production launches the disadvantaged areas into a degenerative downward spiral. King (1982) views international labour migration as the human bond linking the prosperous core of Europe with its periphery: one of the essential expressions of the centre–periphery structure and crucial to the evolution of broad patterns of spatial inequality in Western Europe. While it is true that the great exodus of low-grade migrant labour from the rural peripheries has now ended, at least for the time being, the core–periphery divide persists. The dual processes of factor accumulation in the urbanised core and rural depopulation on the periphery have left an indelible mark on the social and economic geography of Europe.

Each of the above frameworks recognises, in one way or another, that migration is both a response to the dynamism of economic development and a catalytic stimulus for further economic and social change. There are three major aspects of this change: the well-being of the migrants themselves; the impact on the recipient communities, mainly urban-industrial settings; and the welfare of the sending areas (Knox, 1984, p. 32). Each of these impacts involves a range of costs and benefits for which, however, a simple balance sheet cannot easily be drawn up; the effects are multi-dimensional and may change over time as well as varying in space. The impact of emigration on sending countries, for example, involves both positive and negative aspects. On the positive side it has been argued that emigration successfully removes surplus labour and reduces unemployment in rural areas of the periphery (Griffin, 1976). Another positive effect, from the perspective of the ruling classes of the sending area, is emigration's stabilising role of channelling out the frustrated, thus defusing political and social tension; Dench (1975) has identified this mechanism within Maltese emigration. On the other hand, emigration has been highly selective in taking out disproportionate numbers of young, ambitious, economically active adults, including many of the better qualified with trades and skills. The residual community is thus deprived of these important human elements. Nor does subsequent return migration seem to restore the balance, for those who return are the less skilled of the emigrants, whose impact on the communities in which they resettle is generally minimal, beyond the building of numerous new houses and the establishment of some small and often economically marginal service enterprises (King, 1979).

The main concern of this chapter, however, is not the sending areas from which the migrants have departed and to which they may or may not return, but the destination environment. Here the economic multiplier effects of immigration have to be balanced against the social tensions and welfare needs generated by the guestworkers and their families. Accordingly the rest of this chapter is structured as follows: first the statistical and demographic profile of the post-war migration phenomenon is sketched; second the employment characteristics of the guestworkers are briefly described; and finally some of the socio-cultural patterns and problems are discussed.

Flows, stocks, workers and migrants

The totals of foreign workers residing in host countries are a function of three factors: the annual inflow of migrants; the mean length of stay; and the rates of return to the home country. (Given the temporary nature of much European guestworker migration, there has always been a significant return factor, even in the years of rising immigration.) Many data sets relating to these factors are available, yet these international migration statistics are notoriously inaccurate, although they may superficially appear accurate because of their detail.[3] Sometimes the differences between estimates of the same flow that are produced by host and sender countries are enormous. It is necessary, therefore, to consider the nature of the problems inherent in migrant data.

Data problems

Kirk (1981, p. 61) has summarised the data problems for European international migration as follows. A first and major difficulty is the lack of comparability between the criteria adopted by different countries for defining who exactly is an international migrant. Some nations' records relate to 'total foreign residents', others are limited to 'migrant workers'; some are based on census totals, others on registered 'aliens'. Some figures derive from persons crossing frontiers for any purpose and any period, some refer only to long-

term residents, and others refer only to those registered for employment. Switzerland classifies migrant workers into four types: frontier, seasonal, annual and permanent. A particular problem with UK statistics on immigration is that census and other official figures refer only to 'foreign-born', which is no guarantee of a particular nationality. On the other hand, nationality creates its own inconsistencies: in the Netherlands Surinamers have Dutch nationality, and in France the residents of overseas *départements* such as Guadeloupe and Réunion have French citizenship. In certain countries a significant number of foreign migrants have become legally naturalised or have taken up citizenship of the host country. Clandestine migration is a significant problem for many countries, such as France and Italy, adding further imprecision to estimates of foreign residents. Finally, return flows are particularly poorly monitored.

In spite of these grave difficulties, several estimates of Western Europe's immigrant population have been made. During 1970–5 these estimates generally ranged between 12 and 13.5 million, the majority of whom were southern Europeans. Ten years later, in the early 1980s, estimates generally hovered around 14 million and, by the late 1980s, we may suppose that the average collective estimate has topped 15 million, with an increasing proportion of non-Europeans involved in the migration streams.

The general pattern

These brief numerical snapshots hide a more complex reality of a constantly evolving situation with varying numbers of migrants going to a range of different destination countries. Moreover, these migrant flows expand and contract with time, and different rates of return and mean length of stay result in different behaviour of the stocks of individual nationalities in different countries. The general spatio-temporal pattern of post-war European international migration can therefore be described only in its barest outlines.

As the regional and national flows portrayed in Figure 10.1 show, there is a broad spatial symmetry to the geographical pattern of migrant movement. France and the Benelux countries have taken guestworkers mainly from rural south-

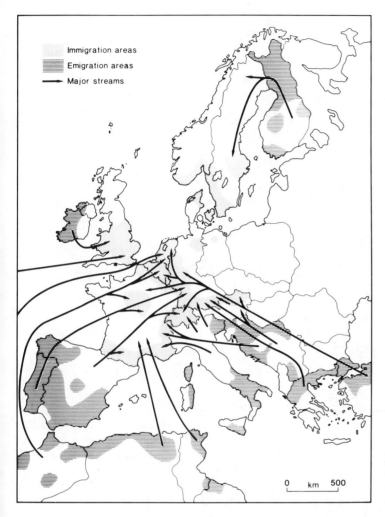

Immigration areas
Emigration areas
Major streams

Figure 10.1 *Major international labour migration flows in post-war Europe*
Source: Hammar (1983), 3.

0   km   500

western parts of the Mediterranean Basin (Spain, Portugal, Algeria, Morocco); Switzerland has recruited predominantly from Italy in the central Mediterranean; and Austria and West Germany from the south-eastern European countries (Yugoslavia, Greece, Turkey). Moving north from the main Mediterranean sources of supply, the spatial coherence of the periphery-to-centre pattern is maintained by the movement of large numbers of Irish into the United Kingdom and of Finns into Sweden.

The migration scene within Western Europe has been dominated by the rise and fall of Italy as a major supply country (King, 1976b). In the immediate post-war years, Italy provided the first

migrant labour on any scale, particularly to Switzerland, France and Belgium. West Germany started taking Italian workers in the late 1950s. Most Italian guestworkers came from the over-populated and poverty-stricken south, including Sicily. However, during the 1960s southern Italians came to be increasingly in demand for the growing industries of northern Italy, so their need to emigrate tailed off. Thus, by the late 1960s, after the short-lived German recession of 1966–7, Yugoslavs and Turks began to replace Italians as the main labour migrants to West Germany, while in France the Portuguese and Maghrebins took over from the Italians and the Spanish.

For most countries, the migration transfers

167

Table 10.1 *Distribution of foreign workers by nationality, selected countries, 1976–86 (per cent)*

West Germany	1976	1986	France	1976	1984*
Turkish	27.2	33.0	North African	35.1	33.0
Yugoslav	20.1	17.6	Portuguese	22.8	28.0
Italian	14.3	12.1	Spanish	12.9	8.4
Greek	9.2	6.2	Italian	12.6	8.4

Netherlands	1976	1986	Belgium	1976	1986
EC**	30.1	38.8	Italian	32.0	22.3
Turkish	21.1	21.5	Spanish	10.0	9.2
North African†	16.6	15.6	North African‡	9.3	13.9
Spanish	8.8	4.5	Turkish	5.0	8.3

Switzerland	1976	1986	Sweden	1976	1986
Italian	47.7	40.5	Finnish	43.7	38.8
Spanish	11.5	12.4	Danish and Norwegian	13.3	14.4
French	10.7	4.9	Yugoslav	11.5	9.6
German	10.1	8.4	Greek	5.3	2.2

Notes: *Most recent French data are for 1984, not 1986.
      **EC refers to the Nine.
      †Virtually all are Moroccan.
      ‡Mainly Moroccan.
Source: SOPEMI, *Annual Reports* (Paris, OECD, various years).

Table 10.2 *Evolution of stocks of foreign workers and foreign residents, selected countries, 1970–85 (thousands)*

	1970	Foreign workers 1975	1980	1985
West Germany	1,839	2,170	2,116	1,824
France	1,140	1,900	1,458	1,658*
Netherlands		176	188	166
Belgium	190	203	333	396
Switzerland	593	552	501	549
Sweden	209	225	234	216

	1970	Foreign residents 1975	1980	1985
West Germany	2,725	4,090	4,453	4,379
France	3,140	3,442	3,680†	
Netherlands		345	521	552
Belgium	696	805	903	898*
Switzerland	1,080	1,010	893	940
Sweden	411	410	422	389

Notes: *1984.
      †1982.
Source: Kayser (1977); SOPEMI, *Annual Reports* (Paris, OECD, various years)

gathered momentum during the two decades between the early 1950s and the early 1970s. The peak year for this predominantly south-to-north migration was 1969 when about 700,000 moved; this excludes Algerian migrants to France (Kirk, 1981, p. 62). Tables 10.1 and 10.2 present some relevant figures for six host countries since the 1970s. It must again be emphasised that, while these figures are the best available, they may be subject to quite significant errors.

The composition of the immigrant workforce varies both between countries and in the same country through time. Turks predominate in West Germany, Algerians in France, Italians in Switzerland and Finns in Sweden. For most countries, the proportion of Italians, Spanish, Greeks and Yugoslavs declined over the 1976–86 period as the relative importance of Turks and North Africans increased (Table 10.1). Table 10.2 documents the evolution of both migrant workers and total foreign residents for the period 1970–85 for the same selected destination countries. The stagnation or decline in the numbers of workers over the period 1975–85 (except in Belgium)

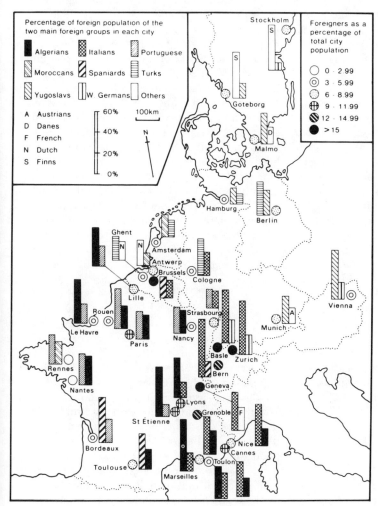

Figure 10.2 *Foreigners in selected European cities*
Source: White (1984), 104, 109.

contrasts with a continuing pattern of growth in total population numbers, at least until 1980, for most countries. This trend towards an increasing number of family dependants joining the migration streams during the 1970s is clearly shown by the annual flow data. These are too cumbersome to be tabulated here, but a simple example will illustrate the point. In 1970 France took in 174,000 foreign workers and 81,000 foreign family members; in 1980 the respective figures had become 17,400 and 51,000. Thus the ratios had been reversed from 2:1 to 1:3.

These national pictures ignore the spatial distribution of migrants within countries and the fact that, with a few exceptions such as harvest migrants (e.g., for the grape harvest in France) and some tourist employees (e.g., in Alpine ski resorts), guestworkers are overwhelmingly urban in their location. In his masterly survey of European urban social geography, White (1984) has compiled interesting comparative data of foreigners in European cities during the 1970s. These data are synthesised and presented in Figure 10.2. Among cities with more than a tenth of their population made up of foreign citizens are Geneva (34.0 per cent foreign), Brussels (16.1 per cent), Grenoble (12.8 per cent), Paris (11.9 per cent), Lyon (11.8 per cent), Nuremburg (11.8 per cent), and Stuttgart (11.5 per cent). Most of these percentages relate to the period 1970–5. In spite

169

of the decrease in foreign worker entry, many cities increased their foreign populations dramatically during the decade of the 1970s as family members flocked in to join the guestworkers. West Berlin's foreign population grew by 63.3 per cent, Munich's by 75.5 per cent, Frankfurt's by 80.4 per cent, Stuttgart's by 26.2 per cent and Vienna's by 48.8 per cent.

Regarding the nationalities present in different cities, certain international contrasts can be noted (Figure 10.2). In France the two leading places are almost everywhere taken by either Algerians, Portuguese or Italians (Algerians and Portuguese in the north, Algerians and Italians in the south). In Vienna and the limited number of West German cities for which comparable data are available, the predominant nationalities are Yugoslavs and Turks. Italians are the most numerous group in Swiss cities, and Finns dominate in urban Sweden. Dutch and Belgian cities have more variable mixes: Moroccans, for instance, are the biggest foreign group in Amsterdam. Again, however, variation through time must be stressed. In Stuttgart, for example, Italians were the dominant group until 1970 when they were replaced by the Yugoslavs; by 1975 the Greeks had pushed the Italians into third place; by 1980 the Italians were in fourth place, after Yugoslavs, Turks and Greeks (Kolodny, 1982, p. 16).

Demography of migrants

Demographically, the migrant labour force of Western Europe differs from the host society in many respects (Kirk, 1981, pp. 69–74). Currently there is a striking weighting of the migrant populations towards the younger adult age groups, with a growing number of young and teenage children. By the 1990s this will manifest itself increasingly in terms of middle-aged migrants with working-age children. For the early post-war immigrants, such as the Italians in France and Belgium or the Irish in Britain, this stage is already past, with the third generation now being born.

An uneven age–sex balance inevitably influences patterns of migrant nuptiality and fertility. Before about 1970 the age structure favoured high marriage rates, but the sex balance worked against this. In 1971, for instance, both France

and Germany had twice as many immigrant males as females in the 20–50 age bracket. This in turn has tended to create pressures for mixed marriages, usually between migrant males and local females. Linke (1976) discovered that among marriages involving foreigners, mixed marriages outnumber those in which both spouses are of the same nationality. The significance of this is the contribution that mixed marriages probably make to the more rapid cultural adjustment of foreigners and their easier assimilation into the host population.

A third element of the demography of immigrant populations is their different fertility from that of the host populations. This distinction generally derives from the notion that migration will tend to take place from high-fertility to low-fertility populations, partly because of the surplus of people and partly because of the historic inverse relationship between economic growth and fertility which has associated high birth rates with low living standards (Kirk, 1981). One of the most critical aspects of migrant fertility is the contribution of immigrant births to the national total. This depends partly on the natural increase or decrease characteristics of the indigenous population, and in the long term it also depends on the propensity of migrant families to return home, taking their foreign-born children with them. West Germany has in recent years recorded deficits in births over deaths for the German population; excess foreign births work against the national trend but do not completely compensate, with the result that there has been a small drop in total population for most years since 1974. Generally it seems to be the case that immigrant families are larger than those of the host society; but host society influence can be felt to the extent that immigrants' family sizes are smaller than they are in the country of origin. The balance between births and deaths among the immigrant communities is heavily influenced by the age structure of the communities concerned. Turkish data are very revealing here: in 1986 the Turkish emigrant communities in Europe yielded 45,000 births but only 2,282 deaths.

Clear evidence therefore exists that the migration flows of the past 30–40 years have added not only absolute numbers but also (at least for a time) considerable demographic growth potential to the host countries of northern and Western Europe. A crucial element in all these calculations is the inevitably low death rate among the migrant communities. As these communities mature into settled ethnic groups, however, the death rates will rise, the birth rates will fall and the demography of the immigrants will gradually approximate to that of the host societies.

Employment, occupational specialisation and diversification

At the heart of the influx of foreign-born workers in the more prosperous countries of Western Europe lie the labour needs of these more developed countries. These needs have been reinforced by low birth rates in the 1930s and 1940s, leading to a slow post-war rate of growth in the indigenous labour forces of the industrial countries. An increasing trend to prolong education in these countries has had two further effects: delaying the entry on to the labour market of college and university students, and orienting an increasing number of indigenous workers to high-status rather than low-status jobs. This background explains the immigrants' role as 'replacement labour' for the low-paid, assembly-line and service-sector jobs vacated by the upward social mobility of the indigenous population (Salt, 1976). Foreign workers have been seen, especially by many big employers, as cheap labour which will tolerate low wages and poor working conditions, accepting also the deskilling of jobs through more detailed divisions of labour, and often working outside the trade-union framework which has generally acted to preserve the interests of indigenous labour. Most migrants do not have the education, skills or language to compete for higher-status jobs. However, as Castles and Kosack (1985) point out, most migrants are *kept* at the foot of the employment ladder by a combination of official restrictions and social discrimination.

The most detailed published data on migrant employment in Europe are those assembled by Castles and Kosack (1985, pp. 57–115). Unfortunately these comparative figures refer to the 1960s and not the 1980s, and the parameters surrounding the use of guestworker labour have shifted in the meantime. As the European and world economies moved into a phase of stagnation and restructuring, labour-intensive production processes were relocated to Third World

countries where labour was even cheaper (Chapter 3; see also Fröbel *et al.*, 1980). This 'new international division of labour' resulted in a decline in the demand for jobs available to *Gastarbeiter* in most forms of manufacturing industry. Less affected were low-grade service jobs such as street cleaners and hospital orderlies.

In spite of these structural economic changes, migrant workers have tended to remain concentrated in three main employment sectors. First, the building and construction industries are very dependent on migrant labour. The insecure, geographically mobile nature of construction work, and the widespread subcontracting operating in this sector, make building work especially attractive for those who are trying to evade official systems of registration and control. The second key employment sector for foreign migrants is manufacturing industry, where the foreigners are almost always found in unskilled or semi-skilled capacities and are disproportionately concentrated in higher-risk processes involving products such as rubber or chemicals. Third, foreign workers cluster in low-grade tertiary occupations such as refuse-collecting, office-cleaning and hotel kitchen work.

Within these broad occupational groups a hierarchy has tended to evolve. Non-European workers have become concentrated at the very bottom of the socio-occupational ladder. For instance, a study of Algerian workers in the Paris region found that 86 per cent of them were unskilled labourers and that 88 per cent had no occupational qualifications whatsoever (El Gharbaoui, 1971). On the other hand, the migrants from the more 'mature' migration streams, such as the Italians and the Spanish, have progressed further up the ladder. These nationalities are often found as building-site foremen or on the factory floor as workgroup leaders (King, 1989).

Nevertheless, certain national specialisations within the guestworker labour market have evolved which are independent of the hierarchy noted above. One of the best known is the Irish domination of casual construction work in the United Kingdom, an occupational association that dates back two centuries when Irish 'navvies' were imported to dig the canals and later build the railways. Salt (1976, pp. 110–14) gives examples from West Germany where employers have tended to recruit from single nationalities in order to minimise language and job-instruction problems:

Table 10.3 *Occupational status of foreign employees in French industrial and commercial firms with more than ten employees (per cent)**

	1971	1973	1976	1979
Supervisors	1.1	1.3	1.7	2.1
Technicians and foremen/women	1.7	1.7	2.1	2.6
Clerical employees	3.6	5.1	6.7	9.2
Skilled manual	20.8	30.5	34.8	37.9
Semi-skilled manual	46.1	42.6	38.2	34.5
Unskilled	26.7	18.8	16.1	13.4

*Columns do not all total to 100 due to rounding errors.
Source: SOPEMI (1983) *Annual Report 1982* (Paris, OECD).

in Cologne this explains the dominance of Turks in the Ford automobile plant and of Spaniards in the postal service. In West Germany as a whole, Yugoslavs are found particularly in construction, Greeks in manufacturing, and Moroccans in mining.

The ending of labour migration and the shift towards more permanent migrant settlement has not changed basic patterns of migrant employment (Castles and Kosack, 1985, pp. 495–9). Comparative figures for 1972 and 1980 for West Germany, for instance, show that foreign workers are still very highly concentrated (about 80 per cent) in the secondary sector — manufacturing and construction. Female migrant employment has shifted somewhat from manufacturing (74 per cent in 1972, 56 per cent in 1980) to services (21 per cent in 1972, 40 per cent in 1980) but this does not necessarily represent an upward change in employment status. Taken as a whole, 81.1 per cent of foreign workers in West Germany were still manual workers in 1978 (compared with 39.7 per cent of Germans), 14.5 per cent were non-manual workers (Germans 37.5 per cent), 3.8 per cent were employers, self-employed or civil servants (Germans 18.5 per cent). French data do show some status improvement for foreign workers during the 1970s (Table 10.3), but the dominance of manual workers remains over-whelming. Eighty-eight per cent of the foreign labour force were employed in industrial and commercial firms with more than ten employees in 1971, and 85.8 per cent in 1979.

A longer-term trend has been the growth of

ethnic entrepreneurship. The migrant communities themselves have created a seed-bed for mainly small-scale ethnic businesses in such fields as retailing, wholesaling, restaurants and other minor professions to serve the needs of the ethnic communities. Some examples, such as Chinese restaurants, are fairly long established. Others are more recent. Gentileschi (1978) has documented the rise of the 'independent migrants' from the ranks of the Italian labour migrants in Stuttgart, and Palmer (1984) has described the special circumstances surrounding the occupational specialisation of Italian restaurateurs and snack-bar owners in London. The rise of Asian and Cypriot businesses, in both catering and textiles/clothing, is also worthy of note in the British post-war economic scene.

Housing and the residential distribution of migrant workers

The residential segregation of foreign workers is an important dimension of urban structure in north-western Europe. Their concentration in areas of low-quality housing is conditioned by a number of factors. The first and most important has to do with the types of work available to the immigrants; as we saw from the previous section, these are almost exclusively low-paid and insecure. Because of their weak competitive power in the housing market, immigrants have inevitably concentrated in the worst housing and the most run-down neighbourhoods, thereby initiating a spiral of social problems and deprivation. The second source of disadvantage concerns the public housing sector, in which allocation policies often place recent immigrants at the bottom of the waiting list. Single migrants are often ineligible for public housing, yet family members are often denied entry to the country unless satisfactory accommodation has been secured — a classic Catch-22 situation. Racial discrimination constitutes a third arena of disadvantage, with landlords either refusing to take migrants or, if they do, accommodating them in inferior conditions and charging them excessive rents, exploiting their ignorance of the language and of their rights, and perhaps also their illiteracy and their illegal status (for a wealth of evidence, see Castles and Kosack, 1985, pp. 240–317). Finally, it should be borne in mind that migrants themselves often opt for cheap (and therefore poor) housing in order to maximise remittances and savings.

Clusters of migrants in European cities generally exhibit a consistent pattern: the overriding need is for low-priced accommodation near to their jobs. Thus they are found in inner-city districts (for easy access to city-centre service employment) or, further out, near industrial areas (for factory jobs). Once established in these areas, immigrant concentrations have tended to persist, partly because of ethnic cohesiveness and partly because of the diminishing attractiveness of immigrant areas for indigenous residence, except sometimes at a later phase when gentrification or commercial expansion may reverse the trend.[4]

Concentration of immigrants near industrial zones derives in part from the operation of the company hostel system, which has been particularly prevalent in Switzerland and West Germany. Many large companies in these countries have had to build hostels before being allowed to recruit guestworkers (O'Loughlin, 1980). These single-sex dormitories, the ultimate in residential and social segregation, have often been built within the factory gates; Turks and Yugoslavs are particularly overrepresented in this form of accommodation (Rist, 1978). Movement out of the hostels has taken place, but the migrants' narrowly constrained mental map of the city into which they are thrust tends to confine them to overcrowded tenancies in nearby slums (Clark, 1977).

Company housing for migrant workers has been less common in other countries; in France it has had some importance in the building industry where construction gangs have often been housed in caravans and huts on building sites, and in the company-owned *foyers-hôtels* organised by firms for their workers (White, 1984, pp. 116–17). More common in France have been the *garnis* or *hôtels-meublés* (privately owned, inner-city 'hotels' for guestworkers) and the peripheral squatter settlements known as *bidonvilles*. The growth of these shanty towns round major French cities was a feature of the 1950s and 1960s, corresponding to the massive influxes of Portuguese and North African labour. Although a law was passed in 1964 aiming at the reabsorption of *bidonville* inhabitants into 'normal' housing, progress has been slow in some cities, notably Marseilles (Jones, 1984; Kinsey, 1979).

A very considerable geographical literature — too extensive to be surveyed even briefly here — already exists on the Asian and Caribbean subareas of British cities. The literature on the emerging social geography of immigrant groups in the cities of continental Europe reveals that, in spite of the different mixes of nationalities in different cities (Figure 10.2), the immigrants' role in the dynamics of the housing market and in the demography of inner-urban areas is fairly consistent and, moreover, exhibits an interesting parallel with their role in the labour market noted earlier. Just as the guestworkers have replaced the lower echelons of the indigenous population in the occupational hierarchy, so they are acting as a partial replacement of the declining native population in the older neighbourhoods of privately rented property in the inner-urban areas (see, for example, Drewe *et al.*, 1975; Geiger, 1975; Jones, 1983; Mik, 1983; O'Loughlin, 1980).

Social geographers have also revealed that there is a high degree of social segregation *within* immigrant areas in European cities. Individual apartment buildings are often exclusively occupied by a single nationality: this occurs because knowledge about accommodation vacancies is often passed on by word of mouth by an immigrant to his or her fellow-countrymen. Chain migration mechanisms operating via families and friends may lead to immigrants from a single region or village banding together in the same urban neighbourhood; for a good example of this, see Gentileschi's (1979) study of Sardinians in Stuttgart. In Dusseldorf O'Loughlin and Glebe (1981) found that different independent variables had different powers to explain various ethnic distributions. The distributions of Turks, for instance, was best explained by proximity to industrial areas, while the pattern of Greek settlement was more closely correlated to low housing quality. In general, however, the low-status guestworker populations are less segregated from each other than from the host populations, with the oldest-settled groups, such as the Italians, the least segregated.

The movement of migrant households into public housing has been slow almost everywhere, except in Sweden where there has been a moderately effective policy of integration (Hammar, 1985, p. 35). Elsewhere there have been problems over residence qualifications and the length of time it is necessary to spend on waiting-lists. However, slum clearance, for instance in inner Amsterdam and of French *bidonvilles*, has led to some immigrants being rehoused within the public sector. Where inner-city rented tenements have been cleared, public rehousing has often taken place on the city peripheries, leading to a process of marked suburbanisation of the immigrant population. In France this has meant a flow to the peri-urban *grands ensembles*. Decanting inner-city immigrant groups to ex-urban housing estates devoid of services and cultural foci has not generally proved to be a happy solution, however. Nor has the process of resettling *bidonville* residents in *cités de transit* where they were supposed to spend between one and three years being socialised into the behavioural norms associated with 'normal' housing. Unfortunately the minimal-standard *cités* have evolved into a semi-permanent dumping-ground for so-called 'problem families' rather than functioning as a half-way house or training ground (Gokalp and Lamy, 1977). In Britain, where some entry of immigrants into council housing has taken place, a tendency has been noted for those of Afro-Caribbean background to be allocated to the less popular inter-war flatted estates (Peach, 1982).

Penetration of labour migrants into the owner-occupied sector is limited; it is only taking place with the longer-established immigrant groups such as the Italians. The major exceptions to this generalisation are found in Britain where the Asian immigrants have a higher rate of home ownership than the indigenous population. Data from the National Dwelling and Housing Survey of 1977–8 showed that 62 per cent of Asian households lived in owner-occupied dwellings, compared to 45 per cent of whites and only 32 per cent of West Indians (Holmes, 1982, p. 132). The tradition of Asian property ownership plays a major role here, as does the pooling of capital for house purchase by members of extended families.

Socio-cultural issues

The localisation of immigrants in factory and inner-city hostels, poor-quality rented rooms and disjointed peripheral housing estates makes them vulnerable to many kinds of exploitation and deprivation. Castles and Kosack (1985, pp. 240–317) have catalogued some of the worst abuses. Cases of just beds being rented on a shift basis

have not been unknown, one infamous case being the 'merchant of sleep' in Lyons. Access to facilities such as toilets, showers and kitchens is generally poor among immigrants: one study in Brussels found that only 7 per cent of Moroccans and 11 per cent of Turks had access to the use of a bath or shower where they were living (De Lannoy, 1975). Health and welfare-related problems have proliferated in the immigrant communities, further distancing the guestworkers and their families from the host population in terms of social well-being (Knox, 1984, pp. 43–4). Local social services are rarely able to cope with such problems. Those guestworkers who come from non-EC countries have few civil rights, and there is a reaction against according them such rights (Moulier-Boutang, 1986). Educational problems have become acute with the growth of a huge population of immigrant schoolchildren — around 1 million under-16s in both France and Germany, for instance. In many European cities, children of foreign origin or with immigrant parents make up 20 or even 30 per cent of the school population; in some catchment areas they represent the overwhelming majority. The education authorities have so far proved incapable of dealing with the special problems of language and culture posed by immigrant children who have high rates of absenteeism and of examination failure (Beyer, 1980; Bilmen, 1976). The increasing numbers of foreign children are described by Salt (1985) as a social time bomb, for soon these under-qualified young people will be released on to a labour market that has few openings for them beyond the dead-end jobs their parents came to do.

Guestworkers' segregation and marginalisation are also a reflection of, and have a profound influence upon, the overall social formation. Indigenous workers, finding themselves in a superior position because they are no longer required to do the most menial jobs, come to see some justification in a hierarchical social organisation which places the guestworkers at the bottom of the pile. On the other hand, indigenous communities may feel their security threatened by the immigrants' 'invasion' of job markets, housing and welfare provisions. This produces a general anti-immigrant sentiment which has been transmitted into the political arena by the likes of Enoch Powell (in the United Kingdom), James Schwarzenbach (in Switzerland) and Jean-Marie le

Pen (in France), and which occasionally inflates the tension into violence at the local level (Miller, 1982). According to the *Observer* (18 September 1988, p. 52) there were 2,179 'racial incidents' reported in the Metropolitan Police area of London in 1987. In Switzerland in recent years, repeated referenda have been taken on proposals for a reduction in the number of foreigners (Johnston and White, 1977), while the presence of large and increasing numbers of Maghrebins in France has been a major issue in the election campaigns of the 1980s (George, 1986; Wihtol de Wenden, 1987). In Britain the series of riots which began in Bristol in April 1980 and then exploded in 35 other cities across the country in July 1981 produced a massive shock to the popular consciousness of race (Peach, 1985). Although whites formed the majority of those arrested, newspapers resorted to stereotyped images which heightened the apparent racial element in the disturbances (Jackson, 1988). This leads on to the important point about the extent to which immigrants' 'problems' can be traced simplistically to their mere presence in the country, or whether such migrant-related issues are more fundamentally a reflection of the deeper structural ills of the host society at large. More than a quarter of a century ago, Jackson (1963) wrote that social problems were not caused by the immigration of the Irish or West Indians or Pakistanis, rather their presence served to expose the inadequacies of social provision for the community at large. More recently this has been echoed by Peach (1982) who points out that the distribution of Asians and West Indians acts as a kind of barium meal for the X-ray of the problems of the British space economy. The black population of Britain, and particularly the West Indians, have inherited a decay for which part of the social and political system is blaming them. They are the effect of a system for which they are seen as the cause, and the same may be said of the Algerians in France or the Turks in West Germany.

Conclusion: from guestworkers to settlers?

The foreign migrant communities in many countries are now achieving demographic maturity as, for the past 15 years, family reunification has predominated over guestworker migration. As families are reconstituted or formed, rates of natural increase among the immigrant communities rise. Migration is thus losing its significance as the main element of ethnic minority population increase; natural growth has taken over, and it is possible to envisage the future growth of these communities even if all in-migration stopped. The young age-structure of the foreign populations favours demographic growth, and this is further reinforced by fertility levels which, especially in the case of North African and Mediterranean immigrants, are significantly above those of the indigenous populations.

It is arguable whether 'guestworkers' remains an appropriate term to denote Europe's post-war labour migrants. For they appear to be, as summed up in the titles of two recent books, 'guests come to stay' (Rogers, 1985) and 'here for good' (Castles *et al.*, 1984). By taking over certain categories of undesirable employment, migrants have made themselves indispensable. Many have achieved permanent positions, either with private firms or as public employees. Others have stayed on because they have become socialised into the consumer aspirations of the host societies; yet others because they found that, with their aspirations and costs of living constantly redefined upwards, they could not save enough to return home as quickly as they thought; or because economic prospects in their home countries remained bleak. Thus migrants have tended progressively to extend their stays abroad until they have transformed themselves into a more or less fixed structural characteristic of the social and economic system of Western Europe, enduring even the recessions which have occurred since the mid-1970s. The birth of the second and now the third generations indicates an evolving stability of at least some immigrant groups: guestworkers have become settlers. Yet the growth of settled communities of immigrant workers and their families does not mean that they have become satisfactorily integrated, let alone completely assimilated. They still face abundant discrimination and they still cling to their roots and their ethnic origins. The growth of the European motorway and air charter networks means that such communities can become settled without completely sacrificing their links to their Mediterranean homelands and their ancestral villages.

Notes

1. Although it is the unskilled international labour migration of the post-war years that is of greatest significance in conditioning the social and economic geography of contemporary Western Europe, and especially that of its cities and industrial areas, the historical context should not be ignored. Kosiński (1970) has comprehensively surveyed the early international movements and White (1984, p. 99) gives interesting examples of pre-industrial and early industrial cross-border migrations.

2. Despite the numerical importance of international guestworker migration in the Western European context, it is important to realise that not all cross-border movement is of unskilled workers for low-status industrial and tertiary sector jobs. Other types of international migrant flow have been, and still are, of some significance. These include the migration of self-employed entrepreneurs and merchants, the movement of high-level employees of international organisations, and refugee flows. The international migration of skilled manpower within Western Europe has recently been analysed by Salt (1984), while King and Shuttleworth (1988) have examined the particular case of Irish graduate emigrants to the UK and Europe.

3. Data on the emigrant and immigrant flows of OECD countries have been issued every year, since 1973, by the Paris-based organisation Système d'Observation Permanente sur les Migrations (SOPEMI). This source, which unfortunately is only informally available as a mimeographed annual report, is invaluable as an aid to geographers monitoring international migration flows in Western Europe. It needs to be stressed, however, that the SOPEMI annual reports are based on country submissions, and there is very little possibility for standardising data on a strictly comparable format. These considerations should always be borne in mind when handling SOPEMI data (Salt, 1987).

4. The impact of gentrification on immigrant residential patterns has been studied by Cortie and van Engelsdorp Gastelaars (1985) with reference to Amsterdam. In the early 1970s the Mediterranean and Caribbean immigrants were virtually all concentrated in the older, inner parts of the city. Ten years later they had been displaced out of the inner city into newer districts outside the centre, including some public housing. These were still areas with predominantly small and cheap housing, but perhaps more suited than the overcrowded boarding houses, hostels and bedsits of the inner city to family-reunion migration. Meanwhile the inner city has become increasingly the preserve of young Dutch people, either students in bedsits or high-income professionals who are buying and restoring dwellings formerly in multi-occupation.

References

Arena, G. (1982) Lavoratori stranieri in Italia e a Roma, *Bollettino della Società Geografica Italiana*, 11, 57–93.

Beyer, G. (1980) The second generation of migrants in Europe: social and demographic aspects, *European Demographic Information Bulletin*, 11, 49–72.

Bilmen, M.S. (1976) Educational problems encountered by the children of Turkish migrant workers. In N. Abadan-Unat (ed.), *Turkish Workers in Europe 1960--1975*, E.J. Brill, Leiden, pp. 235–52.

Böhning, W.R. (1974) The economic effects of the employment of foreign workers: with special reference to the labour markets of Western Europe's post-industrial countries. In W.R. Böhning and D. Maillat, (eds), *The Effects of the Employment of Foreign Workers*, OECD, Paris, pp. 41–123.

Böhning, W.R. (1979) International labour migration in Western Europe: reflections on the past five years, *International Labour Review*, 118, 401–14.

Carney, J.G. (1976) Capital accumulation and uneven development in Europe: notes on migrant labour, *Antipode*, 8, 30–8.

Castles, S., Booth, H. and Wallace, T. (1984) *Here for Good — Western Europe's New Ethnic Minorities*, Pluto Press, London.

Castles, S. and Kosack, G. (1985) *Immigrant Workers and Class Structure in Western Europe*, Oxford University Press, Oxford.

Clark, J.R. (1977) *Turkish Cologne: the mental maps of migrant workers in a German city*, Michigan Geographical Publications 19, Ann Arbor.

Cortie, C. and van Engelsdorp Gastelaars, R. (1985) Amsterdam: decaying city, gentrifying inner city? In P.E. White and G.A. van der Knaap (eds), *Contemporary Studies of Migration*, Geo Books, Norwich, pp. 129–42.

De Lannoy, W. (1975) Residential segregation of foreigners in Brussels, *Bulletin de la Société Belge d'Études Géographiques*, 44, 215–38.

Dench, G. (1975) *Maltese in London*, Routledge and Kegan Paul, London.

Drewe, P., van der Knaap, G.A., Mik, G. and Rogers, H.M. (1975) Segregation in Rotterdam: an explorative study on theory, data and policy, *Tijdschrift voor Economische en Sociale Geografie*, 66, 204–16.

El Gharbaoui, A. (1971) Les travailleurs maghrébins immigré dans la banlieue nord-ouest de Paris, *Bulletin Economique et Social du Maroc*, 31, 25–49.

Fielding, A.J. (1985) Migration and the new spatial

division of labour. In P.E. White and G.A. van der Knaap (eds), *Contemporary Studies of Migration*, Geo Books, Norwich, pp. 173–80.

Fröbel, F., Heinrichs, J. and Kreye, O. (1980) *The New International Division of Labour*, Cambridge University Press, Cambridge.

Geiger, F. (1975) Zur Konzentration von Gastarbeitern in alten Dorfkernen: Fallstudie aus dem Verdichtungsraum Stuttgart, *Geographische Rundschau*, 27, 61–71.

Gentileschi, M.L. (1978) I lavoratori italiani independenti a Stoccarda, *Studi Emigrazione*, 51, 325–60.

Gentileschi, M.L. (1979) *Sardi a Stoccarda*, Georicerche, Cagliari.

George, P. (1986) Les étrangers en France: étude géographique, *Annales de Géographie*, 95, 273–300.

Gokalp, C. and Lamy, M.L. (1977) L'immigration maghrébine dans une commune industrielle de l'agglomération parisienne: Gennevilliers. In Institut National des Études Démographiques, *Les Immigrés du Maghreb: Études sur l'Adaption en Milieu Urbain*, Presses Universitaires de France, Paris, pp. 327–404.

Griffin, K. (1976) On the emigration of the peasantry, *World Development*, 4, 353–61.

Hammar, T. (1985) *European Immigration Policy: a comparative study*, Cambridge University Press, Cambridge.

Hirschmann, A. (1958) *The Strategy of Economic Development*, Yale University Press, New Haven, CT.

Holmes, C. (1982) The Promised Land? Immigration into Britain 1870–1980. In D.A. Coleman (ed.), *Demography of Minority Groups in the United Kingdom*, Academic Press, London, pp. 1–21.

Hudson, R. and Lewis, J. (1985) Recent economic, social and political change in southern Europe. In R. Hudson and J. Lewis (eds), *Uneven Development in Southern Europe*, Methuen, London, pp. 1–53.

Jackson, J.A. (1963) *The Irish in Britain*, Routledge and Kegan Paul, London.

Jackson, P. (1988) Beneath the headlines: racism and reaction in contemporary Britain, *Geography*, 73, 202–7.

Johnston, R.J. and White, P.E. (1977) Reactions to foreign workers in Switzerland: an essay in electoral geography, *Tijdschrift voor Economische en Sociale Geografie*, 67, 341–54.

Jones, A.M. (1984) Housing and immigrants in Marseille, 1962–1975. In P.E. Ogden (ed.), *Migrants in Modern France*, Occasional Paper 23, Department of Geography and Earth Science, Queen Mary College, University of London, London, pp. 29–41.

Jones, P.N. (1983) Ethnic population succession in a West German city 1974–80: the case of Nuremberg, *Geography*, 68, 121–32.

Kayser, B. (1977) *The Effects of International Migration on the Geographical Distribution of Population in Europe*, Population Studies 2, Council of Europe, Strasbourg.

Kindleberger, C.P. (1965) Emigration and economic growth, *Banca Nazionale del Lavoro Quarterly Review*, 75, 235–54.

King, R.L. (1976a) The evolution of international labour migration movements concerning the EEC, *Tijdschrift voor Economische en Sociale Geografie*, 67, 66–82.

King, R.L. (1976b) Long-range migration patterns within the EEC: an Italian case study. In R. Lee and P.E. Ogden (eds), *Economy and Society in the EEC*, Saxon House, Farnborough, pp. 108–25.

King, R.L. (1979) Return migration: a review of some case studies from southern Europe, *Mediterranean Studies*, 1, 3–30.

King, R.L. (1982) Southern Europe: dependency or development?, *Geography*, 67, 223–32.

King, R.L. (1984) Population mobility: emigration, return migration and internal migration. In A.M. Williams (ed.), *Southern Europe Transformed*, Harper and Row, London, pp. 145–78.

King, R.L. (1985) Italian migration: the clotting of the haemorrhage, *Geography*, 70, 171–5.

King, R.L. (1989) Migration. In H. Clout, M. Blacksell, R. King and D. Pinder, *Western Europe: geographical perspectives*, Longman, London, pp. 40–60.

King, R. and Shuttleworth, I. (1988) Ireland's new wave of emigration in the 1980s, *Irish Geography*, 21, 104–8.

Kinsey, J. (1979) The Algerian movement to Greater Marseille, *Geography*, 64, 338–41.

Kirk, M. (1981) *Demographic and Social Change in Europe 1975–2000*, Liverpool University Press, Liverpool.

Knox, P. (1984) *The Geography of Western Europe: a socio-economic survey*, Croom Helm, London.

Kolodny, E. (1982) *Samothrace sur Neckar: des Migrants Grecs dans l'Agglomeration de Stuttgart*, Institut de Recherches Méditerranéennes, Centre d'Études de Géographie Méditerranéenne, Aix-en-Provence.

Kosiński, L. (1970) *The Population of Europe*, Longman, London.

Linke, W. (1977) *The Demographic Characteristics and the Marriage and Fertility Patterns of Migrant Populations*, Population Studies 1, Council of Europe, Strasbourg.

Lutz, V. (1961) Some structural aspects of the southern problem: the complementarity of emigration and industrialisation, *Banca Nazionale del Lavoro Quarterly Review*, 59, 367–402.

Marshall, A. (1973) *The Import of Labour: the case of the Netherlands*, Rotterdam University Press, Rotterdam.

Mik, G. (1983) Residential segregation in Rotterdam:

background and policy, *Tijdschrift voor Economische en Sociale Geografie*, 74, 74–86.

Miller, M. (1982) The political impact of foreign labour: a re-evaluation of the Western European experience, *International Migration Review*, 16, 27–60.

Moulier-Boutang, Y. (1986) Resistance to the political representation of alien populations: the European paradox, *International Migration Review*, 19, 485–92.

Myrdal, G. (1957) *Economic Theory and Underdeveloped Regions*, Duckworth, London.

O'Loughlin, J. (1980) Distribution and migration of foreigners in German cities, *Geographical Review*, 70, 253–75.

O'Loughlin, J. and Glebe, G. (1981) The location of foreigners in Dusseldorf: a causal analysis in a path analytic framework, *Geographische Zeitschrift*, 69, 81–97.

Paine, S. (1979) Replacement of the West European migrant labour system by investment in the European periphery. In D. Seers, B. Schafer and M.L. Kiljunen (eds), *Underdeveloped Europe*, Harvester Press, Hassocks, pp. 65–96.

Palmer, R. (1984) The rise of the Britalian cultural entrepreneur. In R. Ward and R. Jenkins (eds), *Ethnic Communities in Business*, Cambridge University Press, Cambridge, pp. 89–104.

Peach, C. (1982) The growth and distribution of the Black population in Britain 1945–1980. In D.A. Coleman (ed.), *Demography of Minority Groups in the UK*, Academic Press, London, pp. 23–42.

Peach, C. (1985) Immigrants and the 1981 urban riots in Britain. In P.E. White and G.A. van der Knaap (eds), *Contemporary Studies in Migration*, Geo Books, Norwich, pp. 143–54.

Rist, R.C. (1978) *Guestworkers in Germany: the prospects for pluralism*, Praeger, New York.

Rogers, R. (1985) *Guests Come to Stay*, Westview, London.

Salt, J. (1976) International labour migration: the geographical patterns of demand. In J. Salt and H. Clout (eds), *Migration in Post-war Europe: geographical essays*, Oxford University Press, London, pp. 80–125.

Salt, J. (1981) International labour migration in Western Europe: a geographical review. In M.M. Kritz, C.B. Keely and S.M. Tomasi (eds), *Global Trends in Migration*, Center for Migration Studies, New York, pp. 133–57.

Salt, J. (1984) High level manpower movements in north-western Europe and the role of careers, *International Migration Review*, 17, 633–51.

Salt, J. (1985) Europe's foreign labour migrants in transition, *Geography*, 70, 151–8.

Salt, J. (1987) The SOPEMI experience: genesis, aims and achievements, *International Migration Review*, 21, 1067–73.

Salt, J. and Clout, H. (eds) (1976) *Migration in Post-war Europe: geographical essays*, Oxford University Press, London.

Shanin, T. (1978) The peasants are coming: migrants who labour, peasants who travel and Marxists who write, *Race and Class*, 19, 277–88.

van Amersfoort, H., Muus, P. and Penninx, R. (1984) International migration, the economic crisis and the state: an analysis of Mediterranean migration to Western Europe, *Ethnic and Racial Studies*, 7, 238–68.

Ward, A. (1975a) European capitalism's reserve army, *Monthly Review*, 27(6), 17–32.

Ward, A. (1975b) European migrant labour: a myth of development, *Monthly Review*, 27(7), 24–38.

Wihtol de Winden, C. (1987) France's policy on migration from May 1981 till March 1986: its symbolic dimension, its restrictive aspects and its unintended effects, *International Migration*, 25, 211–19.

White, P.E. (1984) *The West European City: a social geography*, Longman, London.

White, P.E. (1986) International migration in the 1970s: evolution or revolution? In A.M. Findlay and P.E. White (eds), *West European Population Change*, Croom Helm, London, pp. 50–80.

Zelinsky, W. (1983) The impasse in migration theory: a sketch map for potential escapees. In P.A. Momsen (ed.) *Population Movements: their forms and functions in urbanisation and development*, Ordina, Liège, pp. 19–46.

11

Unemployment: regional policy defeated?

M. Wise and B. Chalkley

A tide of deindustrialisation has swept across Western Europe in the past two decades resulting in a dramatic rise in unemployment. The number of jobless in the EC10 rose from 2.33 million in 1971 to 12.5 million in 1985. Unlike the ebb and flow of capitalism's customary economic cycles, this recent recession has had a massive and enduring impact on the numbers without work. Its effects have also been geographically pervasive in that all Western European countries and regions have been affected, although to very differing degrees.

Faced with the unacceptable economic, social and political consequences of high unemployment, national governments have undertaken wide-ranging policy reviews in order to determine how best to meet the challenge of the dole queues. Although many facets of governments' strategies have as a result been revised, few have been changed so substantially as regional policy. Given that a reduction in unemployment has always been a prime objective of regional policies and that regional differences in unemployment rates have widened, one might have expected this field of activity to be enjoying a higher priority and increased funding. At the Community level this has, in fact, occurred, partly as a direct response to unemployment but also in an attempt to smooth progress towards the single European market and, ultimately, monetary union. But at the national scale the reverse has been true. Although the picture varies from country to country, at this scale the general pattern has been one of regional policy in retreat, albeit continuing in a much revised form. Regional policy has clearly not defeated unemployment and has to some degree been defeated by it.

This chapter seeks, therefore, to review the

evolution of regional policies in Western Europe, briefly examining their general strengthening during the 1950s and 1960s and focusing on the important shifts of emphasis made in recent years. National policies are considered first, followed by the European Community's approach and, finally, a brief review of the prospects for regional policy in the 1990s.

The growth of regional policy

In the post-war period the belief became widespread throughout Western Europe that regional policies were needed to reduce spatial disparities in unemployment and standards of living. There emerged a broad consensus, extending across the political spectrum, that national governments should intervene in order to direct capital and jobs from richer regions towards those suffering from lower incomes and higher unemployment. Although notions of egalitarianism and social justice, together with political and electoral expediency, were the primary driving forces, regional policy was also justified on economic grounds. By controlling growth in already overheated areas, plagued by congestion and pollution, and by steering jobs towards the depressed regions, there would be more rational use of each nation's resources. The reservoir of unemployed labour in the problem areas would be put to work, thereby enhancing both output and welfare.

The United Kingdom was a pioneer among Western European countries in developing a regional policy based on these ideas (McCullum, 1979). As early as the recession of the 1930s, the

UK government took a variety of measures to lessen the problems of regions, mainly in the north, which were excessively dependent on contracting industries such as coal-mining, ship-building, heavy engineering and textiles. The Special Areas Act (1934) marks the legislative starting point for UK regional policy. After the war the 1945 Distribution of Industry Act consolidated the position of regional policy which went on to enjoy, notably in the 1960s, a period of considerable expansion. Controls on employment growth in the prosperous South-East England and Midlands were tightened and financial incentives were increased for firms setting up in the assisted regions. There was also a progressive spatial expansion of the areas qualifying for aid, so that by 1975 in geographical terms 60 per cent of the UK (with 40 per cent of its working population) was entitled to some form of help in attracting jobs.

This expansionist picture of UK regional policy was mirrored in many other parts of Western Europe. In Italy the Cassa per il Mezzogiorno programme was introduced and enlarged as the government responded to the sharp contrasts between the advanced industrial cities of the north and the much poorer agricultural areas of the south. In France growing concern over the problem of '*Paris et le désert français*' led to a range of measures designed to control the development of the French capital and to decentralise economic activity to the less-prosperous regions. In Norway and Sweden policies were brought in to deal with the geographically remote northern areas, where rapid changes in agriculture and forestry had created problems of regional imbalance. Even in a small country such as Denmark, measures were introduced to counteract the concentration of economic activity around Copenhagen and to encourage new jobs in western and northern Jutland.

Such is the diversity of Western Europe that in every country the scale and nature of the regional problem have been different, as, in detail, have the policy responses (Clout, 1987). None the less, in general terms regional policies in the 1950s and 1960s had three main characteristics: they were growing both in terms of funding and geographical coverage; for the private sector they could involve both controls in overheated regions[1] and incentives in problem regions ('sticks and carrots'); and for the public sector increased

— though usually still modest — efforts were made to disperse government jobs from the capital cities to the poorer provinces.

This expansion of regional policy reflected the prevailing socio-economic conditions and political ideologies. A period of rapid economic growth across most of Western Europe, associated with Keynsian economic principles, facilitated the implementation of 'welfare state' policies pursuing essentially egalitarian goals, such as the reduction of disparities in living standards between social classes and between regions. Paradoxically, it is easier to pursue notions of 'social justice' and 'equity' through such things as regional policy in times of economic buoyancy (when the problems are actually less severe) than in periods of economic depression (when the difficulties of the least favoured become more acute and widespread). In other words, redistributive policies, including those of a regional nature, have tended to enjoy most general acceptance in prosperous eras of near-full employment, when the wealthier classes and regions have felt that redistribution could be afforded because of growth in the national product.

During the 1970s, as the recession and unemployment penetrated even the most prosperous regions, the political consensus underpinning this approach began to break down across much of Western Europe. National governments entered a period of financial stringency and cuts in public expenditure; the hitherto prosperous regions became increasingly hostile to restraints on their development; and critical appraisals of traditional regional policy raised doubts about its cost-effectiveness and highlighted the need for change. As a result, it is now possible to discern the widespread emergence of a new style of regional policy. This new approach typically involves: a reduction in funding; an effort to target the reduced resources more precisely on the most needy areas; a greater emphasis on indigenous development rather than aid 'parachuted' in from outside; and a shift away from preserving old jobs in traditional large-scale manufacturing industries towards a policy which encourages new jobs in the growing small-business, service and high-technology sectors. One must, of course, be wary of overgeneralisation: the European Community has actually increased spending on its own regional policy, and some national governments have been quicker than others to revise their programmes. None the less, overall, there has been a clear shift from a traditional to a new style of regional policy. In order to understand this shift it is important to review the economic and political setting within which these developments have taken place.

The changing economic and political contexts

In the early 1970s a variety of changes in the global economy converged to extinguish the generally optimistic climate in which regional policy had flourished. In particular, the quadrupling of oil prices by the OPEC cartel after the 1973 Yom Kippur war pushed the Western world firmly towards both recession and inflation. For, as oil prices rose, the prices of many other raw materials, commodities and products rose, too, prompting workers and trade unions to resist cuts in living standards by demanding higher wages. This combination of low or negative growth rates and high inflation was termed 'stagflation'. The response of most Western European governments was to cut public expenditure and introduce anti-inflationary measures, the effect of which was further to increase unemployment.

In these unsettling times it was also becoming clear that many of Western Europe's older manufacturing industries, especially in depressed regions, could no longer compete effectively for world markets. Severe pressure was now being felt from other parts of the industrialised world and from the newly industrialising countries (NICs), particularly around the Pacific rim. For mobile capital, such NICs had — in addition to the advantage of low wage costs — the attractions of weak trade unionism, unrestricted approaches to working conditions, few environmental controls and governments enthusiastic for the expansion of export-based industries. Not surprisingly, therefore, a growing number of multinational companies closed plants in Europe and began locating their manufacturing capacity in such countries as Korea and Taiwan in order to produce goods for world markets at the lowest possible cost (see also Chapter 3; and Dicken, 1986). This global shift was facilitated by transportation improvements such as containerisation and by technological advances which made it possible for sophisticated products to be produced

by largely unskilled cheap labour (Higgot, 1984; Wadley, 1986). Within Europe, too, the application of new technologies often increased unemployment by encouraging the substitution of capital for labour, a process that has been explored in depth by Thwaites and Alderman (Chapter 6). Advances in micro-electronics and computerised systems allowed many more operations to be automated or robotised, while service-sector growth was insufficient to absorb the labour released from manufacturing. Indeed, parts of the service sector were themselves subject to the job-shrinking effects of the new technologies.

The rise of the NICs, the investment strategies of multinational corporations and the impact of the new technologies were therefore key elements in the global environment which surrounded the demise of traditional regional policy. They unleashed economic forces which spatial planning seemed powerless to resist. In the words of Susman and Schutz (1983, p. 175): 'Changes in the world economy have rendered all regions, including old industrial regions, vulnerable to decisions made in remote centres of capital control and subject to a hypermobility of capital that defeats any serious local planning.'

The difficulties posed by these economic and technological changes were compounded by demographic and social considerations. Following earlier periods of high birth rates, there was now a substantial increase in the number of young people entering the labour market, especially in some of the poorer peripheral countries such as Ireland. Furthermore, the proportion of women seeking employment outside the home also increased. Thus, the numbers of people seeking work rose as the number of jobs available fell. In France, for example, the recession of the late 1970s and early 1980s coincided with an increase of some 242,000 new job seekers per annum (Drevet, 1988, p. 71).

These dramatically changing economic and social conditions were paralleled by developments in the political field. Keynsian and socialist ideas advocating substantial central government intervention in the economy were deemed by many to have failed and began to lose their grip on governing groups across Western Europe. A new political orthodoxy arose, extolling the virtues of strict monetary policies ('monetarism') and a free-market economy in which competitive individualism could regenerate economies weighed down by the burdens of collectivism and state controls. Not surprisingly in this new climate, regional policy, sometimes characterised as a subsidy to uncompetitive areas, found it increasingly difficult to flourish. Indeed, its opponents came to see it as one of yesterday's solutions — part of a discredited philosophy.

The degree of political and ideological change has, of course, varied from country to country. In the 'social market' economies such as West Germany, Sweden and Denmark there has been simply a shift of emphasis, whereas in the UK 'Thatcherism' has imposed a much cleaner break with the past. In France and Spain, nominally socialist governments have in practice adopted much of the new free-market thinking. Indeed, in government rhetoric throughout much of Western Europe, notions of social justice have yielded pride of place to the language of efficiency, modernisation and competitiveness. Moreover, as part of this philosophy, tighter controls on the role of the state have ruled out the public sector expenditure programmes which (albeit at risk of further inflation) might have helped to resist the rise in unemployment.

A coalescence, therefore, of the political, economic, technological and demographic circumstances outlined above has provided the setting for what in earlier times would have seemed unthinkable — a combination of sustained high unemployment and a reduced commitment by national governments to regional policy. The paradox is heightened by the fact that, as the next section illustrates, since the recession of the early 1970s regional contrasts in unemployment levels have generally intensified.

Unemployment in the regions

Although during the 1950s and 1960s regional disparities in Western Europe were generally reduced, they quickly reasserted themselves in the years of recession. Within the European Community regional inequalities in unemployment were 2.5 times greater in 1985 than in 1975 (Commission, 1987a). This trend towards inequality was primarily a result of a divergence in national unemployment rates. The unemployment problem intensified more sharply in some countries than others, the worst affected being Spain, where between 1973 and 1985 unemployment

rose by no less than 18 percentage points. Ireland, the Netherlands, Belgium and the UK also suffered serious problems, with increases ranging from 10 to 12 percentage points. By contrast, countries such as Denmark, West Germany and Luxembourg were more successful in restraining unemployment, as were many of the EFTA countries, where increases of under five percentage points were common.

There were also significant regional differences in unemployment trends within states. In the case of Great Britain, for example, in 1974 all ten standard regions had unemployment rates within two percentage points of the national average. However, the fragile economies of the north found it much harder than those of the south to ride out the storms of recession, and now the corresponding figure has quadrupled to eight percentage points (Armstrong and Taylor, 1988).

As a result of the trend towards divergence, both at the international and intranational levels, there is now an enormous contrast between Europe's high and low unemployment regions. Within the European Community (which attempts to produce consistent measures of unemployment rates for all its member states) the range is from 3 per cent or less in Crete, Luxembourg and Stuttgart to over 30 per cent in Andalucia (Eurostat, 1988a). Figure 11.1 shows that the problems are generally most acute in the Mediterranean areas, in the northern and western parts of the British Isles and in parts of north-east France and Belgium. Spain, with 21 per cent unemployment, is especially noteworthy as it contains no less than eight of the EC's 'top ten' unemployment regions.

Figure 11.1 can be interpreted as evidence of a connection between unemployment and peripherality but, before endorsing this notion, it is worth recalling the success of the EFTA countries in combating unemployment. The peak jobless rate recorded in the 1980s for Finland was 6.1 per cent, for Austria it was 4.1 per cent, for Sweden 3.5 per cent, for Norway 3.3 per cent and for Switzerland 0.9 per cent. Such figures are a challenge to those who accept too easily the notion that geographical peripherality, harsh physical environments and remoteness lead inevitably to deprivation.

So far this review has considered unemployment variations only in terms of percentages of the total workforce. However, the geographical disparities go deeper than this. The problem of youth

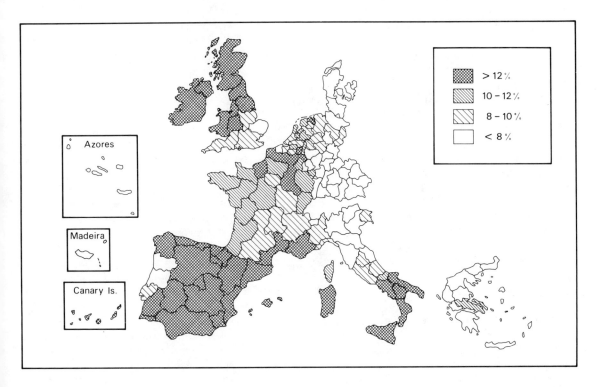

Figure 11.1 *Regional unemployment rates in the European Community, 1987*
Source: Eurostat (1988a).

unemployment, for example, has also grown substantially, both in scale and in its level of geographical inequality (Commission, 1988a). Within the European Community there are now some 5 million under-25s out of work. Even in West Germany, which has an enviable record in youth training, some 21 per cent of this age group are unemployed, and in Spain the figure is 42 per cent. Intranational variations are equally striking: in France, for example, the figures range from 15 per cent in the Ile de France to 30 per cent in the remote region of Languedoc-Roussillon (Eurostat, 1988b). The spatial pattern of the young unemployed, like that of the long-term unemployed and other significant sub-groups, tends broadly to mirror the overall pattern shown in Figure 11.1.

The scale of European unemployment has changed more radically than its geography. The regions with high unemployment in the 1960s were commonly those whose fragile economies suffered the largest job losses in the 1970s and 1980s. Consequently, although the full weight of the recession fell on areas which were commonly

already in receipt of regional assistance, government aid proved quite unable to check the tide of unemployment. Indeed, such was the severity of the recession that the beneficial effects of regional policy in patiently building up employment over many years were often wiped out in a matter of months. In the depressed regions there was no prospect of replacing jobs lost with new ones, since there was often an underrepresentation of growth sectors such as financial services and high technology (Chapters 6 and 7).

In addition to accentuating existing regional inequalities, the recession also hit hard the economy of some previously prosperous areas. For example, the loss of jobs in the UK car industry affected the West Midlands so badly that in November 1984 it was added to the list of assisted areas, despite the general contraction in regional policy. Severe pockets of unemployment were also created in very specific locations, such as the steel towns of Longwy in France and Corby in England, which suffered from the elimination of surplus capacity. The growth of unemployment in some inner-city areas and among the ethnic minorities and Europe's 'guestworkers' (Chapter 10) added further dimensions to the increasingly complex geography of unemployment.

Traditional regional policy could not cope with the dramatic deterioration in the old assisted areas and it was not designed to deal with the new pockets of unemployment. Indeed, with unacceptably high levels of unemployment almost everywhere, many politicians and academics began to ask whether there was still any point in regional policy. Critics on the Right saw it as a wasteful irrelevance, while those on the Left dismissed it as a cosmetic policy designed to conceal the lack of a political will to undertake the fundamental economic reforms necessary to defeat the unemployment problem. The role of traditional regional policy was being challenged from all sides.

Criticisms of traditional regional policy

The growth and persistence of high unemployment levels in many of Europe's assisted areas nurtured the view that regional policies were clearly not achieving their objectives. A review of some 33 studies of regional policies in Western Europe (Nicol, 1980, p. 74) showed that they were generally judged as playing only a very secondary role in determining the geographical pattern of investment. Access to appropriate markets, labour and materials remained the primary determinants of industrial location. In southern Italy, for example, despite heavy investment by the Cassa per il Mezzogiorno since 1950, King (1987) concluded that little progress had been made in catching up with the more prosperous northern areas. Although a considerable number of new jobs had been created in the south, too much of the investment had been in highly 'visible' prestige plants in sectors such as steel, oil and petro-chemicals. These were unconnected with the wider regional economy, and in the recession they faced serious problems of overcapacity. The Cassa, which went into liquidation in 1984, was also accused of being used by the Christian Democratic Party more as a political device than as an instrument of economic development: the expenditure of £500,000 on a soccer stadium in Naples would seem to illustrate the danger of electoral motives overriding genuine considerations of economic planning. Such cases did little to enhance the reputation of regional policy as a serious and effective agent of change.

In Britain, although regional policy has been relatively untainted by political scandals and allegations of corruption, its effectiveness has been very much open to doubt. Estimates of the number of new jobs created by regional policy vary widely. Moore *et al.* (1986) claim that in the 1960s and 1970s it produced some 784,000 jobs in the assisted areas, whereas other studies have suggested figures of around 350,000 (Wadley, 1986, p. 69). But whatever the arithmetic, the underlying north–south gap remained intact and apparently beyond the capacity of regional policy to resolve. Moreover, as the Thatcher government pursued its campaign for cost-effectiveness in public expenditure, automatic regional grants (a key feature of traditional policy) came to be regarded as wasteful because a proportion of the firms setting up in the assisted areas would clearly have done so anyway. For the North Sea petroleum companies, for example, grants for setting up in northern Scotland were simply a windfall. Furthermore, the feeling grew that the firms which really were attracted by regional policy grants sometimes took the money and soon thereafter contracted, closed or relocated. Similar concerns were expressed in other European countries offering automatic grants. In France, survey evidence showed that only 50 per cent of the jobs supposedly created by regional assistance in 1980 were still in existence by 1985 (Drevet, 1988, p. 87). Another common criticism of traditional regional policy focused on the fact that in the UK, France and many other countries, grants were concentrated on the manufacturing sector. It seemed increasingly unrealistic to expect new jobs in the assisted areas to come from a sector which was in relative decline throughout most of Europe, and which often saw the adoption of labour-saving technology as a key to future competitiveness.

As confidence waned in the old-style regional policy, some commentators on the Right even began to see it as part of the problem rather than part of the cure. Years of relying on government 'handouts' had, it was argued, created a culture of dependency which had anaesthetised local economies and prevented the generation of new indigenous businesses. Moreover, by the early 1980s the supply of surplus jobs to be siphoned off from the hitherto prosperous regions had all but dried up. In future the poorer regions would have to look to their own resources and reserves of entrepreneurship.

Confronted with severe economic recession, national governments gave increasing priority to the generation of wealth rather than to its redistribution. The traditionally prosperous cores were therefore to be encouraged rather than restrained. In the words of the French geographer, Kahn (1987, p. 19): 'The capital city is no longer perceived as an obstacle to the development of the regions. Paris no longer carries the responsibility for the economic backwardness of the provinces, but is seen as an asset.' In other words centres such as Paris, London, Barcelona, Milan, Turin, the Randstad and the main West German cities should be allowed to flourish untrammelled by regional policy controls. These centres of strength would maintain the international status of their respective countries in the global economy, provide an example for the more laggardly regions to follow and, hopefully, through spread effects pull these other regions along (Robert, 1987). The answer to high unemployment in the provinces did not lie in restraining employment in the capitals, many of which themselves now contained large pockets of deprivation, particularly in their older inner-city areas.

The standing of regional policy was also undermined by the rise of a battery of non-spatial government programmes designed to deal more directly with the problems of mass unemployment. In the European Community all member states now provide aid to promote the integration of the unemployed into the labour market (Commission, 1988b). In Denmark, Greece, Spain, France and the Netherlands, if employers take on unemployed people they are given sizeable labour-cost allowances as well as tax benefits, and their social security contributions are reduced for a period. Job-creation programmes for the unemployed have been another significant device for dealing directly with the problems of mass unemployment. Such schemes typically involve public works, such as environmental improvements, which would not otherwise be undertaken by governments or private bodies. They are intended primarily for young people, the long-term unemployed and other specific groups targeted by national governments, such as the seasonally unemployed in Portugal or building workers in Greece. These schemes, and especially

those for young people, often include vocational training. This is true, for example, of the *Travaux d'Utilité Collective (TUC)* in France, which also provides young people with work experience. The Belgian *Stages des Jeunes* operate in a similar way. Programmes to help the unemployed become self-employed are also now commonplace. In the UK the Enterprise Allowance Scheme, introduced in 1982, permits young people who have been unemployed for at least eight weeks to receive a grant (currently set at £40 a week for 52 weeks) if they start a new firm.

These kinds of measures have the benefit of being precisely aimed at their client group, the unemployed. By contrast, regional policy is less well targeted. Even in Europe's worst affected regions, two-thirds or more of the labour force are in work. New infrastructure and factories may well enhance the general level of prosperity (a not unwelcome achievement) but there is no guarantee that any new jobs will go to those who need them most, the region's chronically unemployed. The key point here is that in many parts of Europe the last phase of growing unemployment, in the mid-1980s, came through an increase in its average duration. In the European Community two-fifths of the unemployed have now been out of work for over a year, and nearly a quarter for over two years (Commission, 1988b). The long-term unemployed, progressively demoralised and with wasting skills, are regarded by many employers as unemployable; and so in the dole queue it is often 'last in, first out'. Effective personal counselling, retraining and adjustments to the social security and tax systems may well be essential to break down a culture of unemployment (Layard, 1986) and to help such people back into the labour market. The European Community has some 3 million workers who have been without a job for over a year. They are the 'hard core' of European unemployment, but they represent a problem for which traditional regional policy is far from a sufficient answer.

New approaches to regional policy

Although the forces and arguments ranged against regional policy have been formidable, they have not been so powerful as to lead to its total elimination. For electoral reasons, if nothing else, national governments could not be seen to

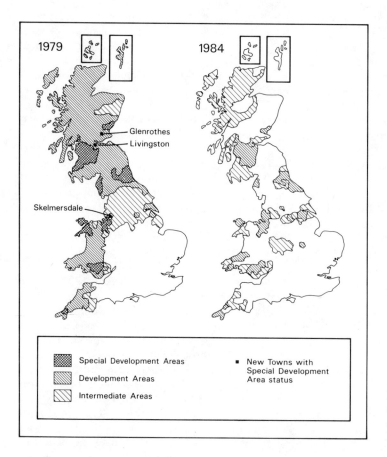

1979

1984

Glenrothes
Livingston

Skelmersdale

Special Development Areas

Development Areas

Intermediate Areas

■ New Towns with
Special Development
Area status

Figure 11.2 *Changes in the UK map of assisted areas, 1979–84*
Source: Department of Trade and Industry.

abandon their depressed regions at a time of high unemployment and deep regional inequalities. Influential lobbying from the assisted areas, and in some cases concern about the political coherence of the state, have also helped to moderate the extent of change. None the less, there have been some very significant revisions and, at the risk of oversimplification, it is possible now to discern two major themes running through the new approaches to regional policy across Western Europe: firstly, there has been a fairly general trend towards reducing the geographical extent of the assisted areas (and hence the cost of the policy); and, secondly, there has been a shift away from geographically redistributive programmes towards the encouragement of indigenous growth.

Despite the fact that more and more regions have experienced serious unemployment problems, the areal extent of the zones receiving assistance has been cut. Changes in the UK map of assisted areas (Figure 11.2) have helped to reduce present expenditure to well under half its mid-1970s level in constant price terms. Table 11.1 shows that similar trends can be found in many other Western European countries. The position has been neatly summarised as follows: 'Whereas ten years ago, assisted area population coverage generally lay between one-third and one-half of the national population, nowadays the "norm" tends to lie between one-fifth and one-third of the national population' (Yuill *et al.*, 1988, p. 22). Even in countries where there has been little or no change in the overall spatial coverage, there is evidence of efforts to target resources more precisely on the most needy. For example, while the Mezzogiorno continues to include 36 per cent of Italian territory, since 1987 the so-called 'sectoral' and 'location' premiums have not been available in its 'relatively advantaged' areas. The latter, which contain 14 per cent

Table 11.1 *Percentage of national population in areas eligible for regional assistance*

	Before cutbacks	1987
Belgium	40	33
Denmark	27	21
France	38	39
Greece	35	42
Italy (Mezzogiorno)	36	36
Netherlands	27	25
Spain	36	41
Sweden	29	14
UK	46	37
West Germany	36	31

Source: Yuill *et al.* (1988).

of the region's population, are therefore deprived of measures which make up two-fifths of the basic regional grant (Yuill *et al.*, 1988, pp. 19–20). Similarly, in France, where the areas qualifying for the *Prime d'amenagement du territoire (PAT)* remain large, the authorities have required that this grant be directed at designated 'poles' of industrial restructuring, research and service development (Yuill *et al.*, 1988, p. 180). The only countries where the total area eligible for assistance has been maintained or extended are those, such as Greece or Spain, where very extensive regions have, compared with Western European norms, low standards of living.

Another element of the effort to restrict and target regional spending has been the move away from the *automatic* award of incentives towards more *discretionary* schemes (Yuill *et al.*, 1988, p. 2). In the UK, for instance, the discretionary Regional Selective Assistance (RSA) award grew in importance relative to the automatic Regional Development Grant (RDG) up to the point where the latter was abolished completely in 1988. Firms now have to persuade government officials that they would proceed with their investment plans in a development area only if a grant were forthcoming. UK experience finds parallels elsewhere. In France, by 1987 the amount of money allocated to the *PAT* scheme (the equivalent of an RDG) had fallen to about 25 per cent of its 1984 level. As a result, the proportion of projects refused a grant rose from 10 per cent to 30 per cent, and even successful firms can longer expect to attain the maximum grant as a matter of course. Similarly, in the Netherlands companies now have

to 'prove' their 'real need' for investment premium (IPR) assistance. Even in Italy, where fixed rates of award and clear eligibility conditions have long dominated an essentially automatic system, since 1986 an element of discretion has been included in the so-called 'planned bargaining' between the Ministry for the South and major industrial groups. A new Regional Incentive System in Portugal also provides more freedom for the central authorities to set different rates of award (Yuill *et al.*, 1988, pp. 8–9). European Community pressure on member states to move away from fiscal incentives has provided another retreat from automatic forms of regional assistance. Whether this shift towards discretionary schemes will achieve the desired objectives remains to be seen. Clearly such procedures, involving complicated applications, time-consuming negotiations with officials and no guarantee of a favourable outcome are less 'transparent' and straightforward than automatic ones. All this could deter would-be investors in areas of high unemployment.

The 'new' regional policy approach has not just been concerned with directing generally more limited amounts of money to smaller areas and more carefully selected projects. There has also been a philosophical change involving a greater emphasis on developing the indigenous potential of the poorer regions and on devising strategies more in tune with local needs and opportunities. One ingredient of this process has been a trend towards decentralisation in the management and implementation of regional policy. Inevitably this generalisation hides significant differences between countries, with some (especially the smaller countries such as Denmark, Ireland and Portugal) still maintaining strongly centralised systems. Decentralisation has proceeded furthest in West Germany (not surprisingly, given its federal structure) and Belgium (with its deep internal divisions between Dutch-speaking Flanders, French-speaking Wallonia and the ever-more independent Brussels region). But other countries, including France, Greece, the Netherlands, Spain and Sweden have also decentralised their regional policies in order to allow more scope for local expertise and initiative.

In parallel with these developments there has been a growing involvement by local authorities in all Western European countries in efforts to promote economic development and reduce unemployment (Commission, 1986a). In the Netherlands, for example, over 30 of the largest towns now have economic affairs departments engaged in efforts to assist their local economies. While in many ways welcome, if local initiatives are pursued in already prosperous areas they may undermine regional policies' efforts to assist the poorest areas. Some controls may therefore be required. In Britain the ability of local authorities to raise money by local taxation and spend it as they will (for example, on local employment initiatives) is contained within strict limits laid down by Whitehall. Another example is Denmark, where only local authorities in regional development areas are permitted to establish incubator workshops for new businesses.

In seeking to encourage indigenous economic development in the assisted areas, the new-style regional policies give much greater attention to the general infrastructure and to the commercial environment. In particular, the stimulation of a wide range of tertiary functions and business services is seen as essential in creating an environment in which new firms can be established and flourish. The manufacturing focus of traditional regional policy is giving way, therefore, to the diversification of the activities which can qualify for aid. In Britain, the 1984 reform of regional policy introduced aid for services such as banking, finance, insurance, tourism, business services and industrial research and development. In 1986 a new Mezzogiorno law extended aid to a similar range of services. Indeed, comparable trends are evident throughout Western Europe. In France, for example, a research and tertiary sector component has been incorporated into the *PAT* scheme. Greece will now provide regional assistance to services associated with high technology, and in Portugal one element of the new 1988 Regional Incentives System is aimed at research and development projects. One tangible expression of this new interest in research and technology is the proliferation of science parks, although this is, admittedly, a field in which the assisted areas are hard pressed to compete against other potentially more attractive areas (Kahn, 1987; Wadley, 1986, pp. 99–106).

The diversification of incentives has also been associated with a shift from supporting massive plant investment by large companies towards stimulating the growth of small and medium-size enterprises (SMEs). Yuill *et al.* (1988) have given

a comprehensive survey of this trend, while elsewhere in this volume an extensive discussion of the rise of small-firms policies is provided by Mason and Harrison (Chapter 5). The reform of UK regional policy was very explicit in this respect. Investment grants of up to 15 per cent are now available for small firms with under 25 employees; when the introduction of any new production method or product is involved, this can rise to as much as 50 per cent of the total cost (Armstrong and Taylor, 1988, p. 30). SMEs in assisted areas qualify for grants of up to two-thirds of the cost of employing management consultants to improve their performance, compared with one-half of the cost elsewhere. Similarly, the new Mezzogiorno law of 1986 discriminates in favour of small firms, especially in terms of award rates. The shortcomings of building large-scale, capital-intensive industrial plants in the Italian south — so-called 'cathedrals in the desert' isolated from the rest of the regional economy — have at last been understood (King, 1987, p. 145). In the Netherlands, Greece and Portugal former restrictions on making grants to SMEs have been relaxed for similar reasons.

The new enthusiasm for small businesses in part illustrates the triumph of the enterprise ideology and the fall from favour of collectivist solutions. On a more practical level, however, there is little prospect of many of Europe's major industries and companies taking on large numbers of new employees; consequently there is perhaps no alternative to a policy of defeating unemployment through the birth and growth of new enterprises. Seen from this perspective, the rise of policies emphasising indigenous regional development is firmly based in pragmatism. Yet it is also necessary to recognise that, in the past at least, Europe's depressed regions have been precisely the areas possessing least dynamism in terms of new business formation and entrepreneurship (Keeble and Wever, 1986). Whether the small-business sector can in practice generate new jobs in the numbers required is therefore very much open to question (Storey and Johnson, 1987). Certainly Pinder (1986), in a survey to which we shall return, has found little evidence that employment goals have been achieved by the European Community's SME programme.

European Community regional policy

Regional policy is not solely the concern of national governments. In the 1950s, even before today's European Community was established, the European Coal and Steel Community (ECSC) was committed to alleviating the economic and social impacts caused by the restructuring of the major industries for which it was responsible (Pinder, 1983). Since 1954 the ECSC has advanced loans and grants totalling 10 billion ECUs to areas hit by the decline of coal and steel. When the European Economic Community was established in 1957, the Treaty of Rome also created the European Investment Bank (EIB), making the reduction of interregional prosperity contrasts, through direct and indirect employment creation, a primary target for this organisation. Unlike national regional policies, which were dominated by industrial grants, the financial principle to which the EIB adhered was that capital could be raised in world money markets where interest rates were low, and could then be disbursed as loans in Community countries where endemically high interest rates were believed to hinder regional development. In its first three decades of operation, the EIB has made loans totalling more that £30 billion in the Community, and two-thirds have been allocated to projects linked with regional development objectives. Recently, wider investment objectives, including energy diversification, the promotion of new technologies and environmental protection, have reduced regional development's share of total EIB activity. Yet, in 1988, 59 per cent of all investment was still associated with regional development objectives, with countries in the Community's southern periphery — Greece, Spain, Portugal and Italy — absorbing two-thirds of the finance available under this heading (Table 11.2).

More detailed accounts of EIB regional strategies have been provided by Licari (1970) and Pinder (1978; 1983; 1986), but two further observations are appropriate at this point. Firstly, the Bank's financing activities illustrate very clearly the view, held by many planners, that regional development efforts should concentrate on infrastructures, as well as on direct finance to industry. In 1987, the most recent year for which data are available, infrastructural loans equalled the value of those made to industry, and the large

Table 11.2 *National allocations of EIB regional development assistance, 1988*

	Regional development loans (million ECU)	Total loans (million ECU)	Regional loans as % of total
Belgium	1.2	4.7	25.5
Denmark	134.4	493.2	27.3
West Germany	154.4	518.8	29.8
Greece	211.9	213.8	99.1
Spain	724.5	1003.1	72.2
France	488.4	756.2	64.6
Ireland	157.1	157.1	100
Italy	1970.1	3396.7	58.0
Luxembourg	–	–	–
Netherlands	203.3	233.2	87.2
Portugal	453.8	488.2	93.0
United Kingdom	413.8	1051.0	39.4
Total	4912.9	8316.0	59.1

Note: 1 million ECU = £0.671 million (average 1988 exchange rate).
Source: European Investment Bank (1989).

majority of infrastructural finance was invested in problem regions. Traditional communications (primarily the road systems) and telecommunications have been major targets for this investment, but recently a quarter of all infrastructure support has been for improved water supplies and waste water treatment.

Secondly, although the EIB is an impressive organisation, there is as yet little evidence that it has made substantial progress in the field of job creation. The evaluation of its efforts is, of course, difficult because activities such as infrastructure investment are intended to have an indirect rather than a direct employment effect. However, it may be significant that, in the Bank's own literature, details relating to employment are few and far between; and it is certainly evident that in the recession-hit 1980s the emphasis in reports changed from 'employment creation' to the 'creation or safeguarding of employment'. Such indicators do not suggest that job generation has been impressive. The one independent study that has attempted to assess effectiveness is also not encouraging (Pinder, 1986). This focused on the EIB's Global Loan Scheme, a programme introduced in 1969, largely to stimulate small businesses in problem regions (Table 11.3). An

Table 11.3 *EIB Global Loans to promote regional development, 1984–88 (million ECU)*

	Infrastructural projects	Industry (allocations to small and medium-sized enterprises) (million ECU)
Belgium	–	1.2
Denmark	5.5	25.0
Greece	93.8	198.8
West Germany	–	8.0
Spain	67.3	195.0
France	383.2	600.7
Ireland	0.7	25.7
Italy	22.7	2265.0
Netherlands	–	8.0
Portugal	–	72.4
United Kingdom	–	75.6
Total	573.2	3475.4

Note: 1 million ECU = £0.671 million (average 1988 exchange rate).
Source: as Table 11.2.

important motivation for this innovation was that job creation per unit of investment was substantially higher in SMEs than in larger capital-intensive enterprises. Yet the study concluded that, at least in terms of direct job creation, the scheme could 'have no more than a marginal effect on the employment prospects of problem regions'.

While the EIB has concentrated on attempting to strengthen regional economies, and thus employment prospects, through a strategy based on loan finance, recently the most publicised form of European Community regional development assistance has been the grant aid offered by the European Regional Development Fund (ERDF). This was introduced in 1975 after considerable political haggling (Wallace, 1977). As with the EIB, and in contrast to national government spending on regional policy, the ERDF has grown steadily in size and will continue to do so into the 1990s (*Bulletin*, 1988, pp. 111–12). Its importance has increased in response to the accession of three of Europe's poorer countries (Greece, Spain and Portugal) and to fears that regional inequalities might be exacerbated by the moves towards a genuine common market by the end of 1992. Although every member state has received some funding from the ERDF, aid is now being increasingly targeted on the poorest countries and regions (Figure 11.3). In 1987, over 87 per cent of total ERDF spending went to the six countries with the most acute regional problems: Italy (26.6 per cent), Spain (18.7 per cent), the UK (17.8 per cent), Portugal (11.0 per cent), Greece (8.5 per cent) and Ireland (4.6 per cent). Aid is confined to the assisted areas as defined by each member state.

The ERDF has enlarged its share of a growing EC budget from 4.8 per cent in 1975 to 8.1 per cent in 1988 (*Bulletin*, 1988, p. 111). None the less, the absolute sums remain unimpressive when placed in the contexts of the Community economy as a whole and that of its individual member states. Even now ERDF expenditure is the equivalent of only about 0.1 per cent of the Community's GDP, though in the case of Greece it represents a more substantial 0.73 per cent of national GDP (the highest member-state figure). Another recent comparison shows that ERDF contributions to private sector industrial projects amount to less than 5 per cent of corresponding national aid (Commission, 1987a, p. 69). Given that the combined total of national government spending on regional aid (which dwarfs that of the Community) has failed to forge any really marked changes in the geography of Western European unemployment or living standards, it is safe to conclude that to date the ERDF will have been still less influential (Wise and Croxford, 1988, pp. 161–82).

This scepticism is intensified when it is realised that ERDF grants often *replace* a proportion of planned national spending rather than *adding* to it. Although in principle Community expenditure should be additional, often in practice it is not. It is widely believed that national governments reduce their own spending in the knowledge that replacement Community funding will be forthcoming. This cannot be definitely proved because of the impossibility of demonstrating what national government expenditure would have been if there were no ERDF. None the less, few would dispute Mény's (1985, pp. 196–8) conclusion that there is an 'infamous unwritten additionality rule, practised by all European governments, according to which European aids are treated as reimbursements of national subsidies'. The limited scope of Community regional policy can best be grasped by analysing its impact on reducing unemployment. Job

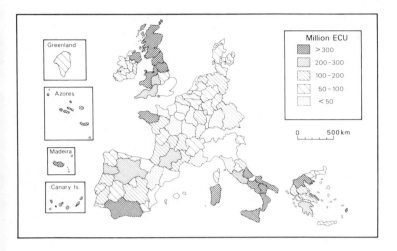

Figure 11.3 *Regional distribution of ERDF assistance, 1975–87*
Source: Commission (1989).

creation or protection is regarded by the European Commission as a fundamental criterion in deciding whether to fund particular project applications (Commission, 1986b, p. 20). Thus every project bidding for aid must predict its likely beneficial effects on employment. It may be safely assumed that such predictions err on the side of optimism, but even such hopeful estimates lead firmly to the conclusion that the ERDF's contribution to easing unemployment has been minimal. For example, it was forecast that the projects receiving ERDF support in 1984 would save or generate some 68,950 jobs. This estimate would have reduced that year's EC10 jobless total of 12.4 million by a minuscule 0.5 per cent. Given that the ERDF contribution to assisted projects in that year averaged around 20 per cent of their total costs, it can be crudely calculated that the Fund's intervention helped a mere 0.1 per cent of the Community's unemployed to obtain a job (Wise and Croxford, 1988, p. 171).

Of course, these macro-level figures conceal significant variations in the ERDF's impact on different countries or regions. For example, during the 1980s, in Greece and Ireland it has been providing up to 20 per cent of public infrastructure expenditure (Commission, 1986b, p. 9). Although the ERDF has failed fundamentally to affect the general scale of interregional inequality, it has clearly formed a significant dimension of public investment in some of Europe's most disadvantaged areas. Moreover, in parallel with the national policies reviewed earlier, the Community is now seeking to encourage such regions to achieve long-term economic progress by improving their business infrastructure and commercial services, thereby assisting indigenous enterprise and the development of small and medium-size companies.

In addition to the ERDF, the Community's other 'structural' funds — namely the European Social Fund (ESF) and the guidance section of the European Agricultural Guidance and Guarantee Fund (EAGGF) — have become increasingly aware of their spatial impact as efforts have grown to integrate them into a wider common regional policy. Thus in 1987 some 69 per cent of EAGGF (Guidance) grants went to the Community's less-favoured regions, as did 44 per cent of the ESF's allocations (Commission, 1989, p. 79).

In 1988, in response to fears that the completion of a genuine single internal market could weaken the poorer peripheral regions still further, the Community adopted measures to double spending on its structural funds (including the ERDF) by 1992 (*Bulletin*, 1988, p. 111). This will also mean that in the new 'Europe without frontiers' some 25 per cent of the EC budget will be devoted to funds with total or partial regional development objectives, as opposed to 17 per cent in 1988. Furthermore, as a result of changes in the Community's allocation guidelines, this enlarged expenditure will be increasingly targeted on the poorest regions in the Mediterranean 'deep south' and on the 'far western' reaches of the Atlantic fringe, where unemployment rates are generally far above the EC average, and incomes far below it.

191

To strengthen the regional impact of this range of budgetary and lending instruments, the Community has been striving to co-ordinate the activities of its various regional development institutions and to avoid the familiar criticism that much assistance is geographically scattered in an ineffective and wasteful manner. Thus several types of integrated regional development programme have been devised; for example, Integrated Mediterranean Programmes, Integrated Development Operations and National Pro-grammes of Community Interest. But progress towards a more coherent approach to Community regional spending has been slow, and like other ideas associated with new regional policy approaches, its long-term effectiveness remains in doubt (Croxford, 1988).

None the less, commitment to the idea of a Community regional policy remains strong, particularly among the poorer countries who stand to gain most. Such member states are also able, within the bargaining arena of Community politics, to extract concessions from their richer partners by making agreement on moves towards a more genuine 'common market' dependent upon a strengthening of common efforts to reduce regional disparities. But these efforts have some way to go before the criticism that they are essentially 'cosmetic' in character can be dismissed with conviction.

Future prospects

It is simply too early at this stage to make confident long-term predictions about the effects of the recent significant changes in EC and national government strategies. Indeed, researchers cannot even agree on the impact of the traditional policies of the 1960s and 1970s. This final section will therefore attempt no more than some brief speculations on the prospects for regional policy and its recipient areas in the 1990s.

The outlook for Europe's depressed areas has, of course, never been wholly or even largely dependent on regional policies. The state of the European and national economies, the health of each region's main sectors and companies, the behaviour of multinational corporations and the effects of factors such as technological innovation and changing locational preferences have all been and will continue to be as important as, or more

important than, regional policy. The regional impact of other government spending programmes (such as transport, education, defence, and health) will also continue to exert a major influence on regional economies. Moreover, despite increased EC expenditure, generally reduced national budgets may mean that the role of regional policy becomes still more marginal. This is not to deny, however, that the new approaches will prove more cost-effective: the money spent per job created is indeed likely to fall. The emphasis on discretionary payments and labour, rather than on capital-intensive schemes, is bound to increase financial efficiency. But efficiency is not the same thing as effectiveness, and it is very doubtful if the new style of regional policy can produce enough jobs to make a really significant dent in the unemployment figures.

The key issue here is the job-generation potential of indigenous growth and small businesses. Keeble and Wever (1986) have shown that the stock of new small firms is now growing in virtually every Western European country. However, most small firms stay small, they suffer from a very high death rate, and very few go on to provide significant numbers of new jobs. Moreover, the lowest rates of new firm formation are to be found in Europe's old specialised urban-industrial regions dependent on coal, steel, ship-building, heavy engineering and textiles. These are precisely the areas worst hit by deindustrialisation and very much in need of new firms and jobs. In communities dominated by a few major employers, such as coal-mines or iron and steel plants, there has been no small-business tradition, and measures to sow the seeds of entrepreneurship may take some time to yield a harvest. Moreover, it is often in such areas that a combination of large-scale redundancies and the near-absence of growth sectors has produced a pool of long-term unemployed for whom the culture of unemployment will not be easily replaced by the culture of enterprise. Although the new emphasis on indigenous development may indeed be ultimately the right kind of regional policy, it is not a short-term answer. Moreover, it will need to be integrated closely with the whole range of other, non-spatial, policies currently aimed at promoting the small-business sector (Chapter 5).

This theme of co-ordination is also becoming increasingly important because the near-monopoly which national governments used to enjoy in the regional policy field has been broken. At one end of the spectrum many local authorities have become heavily involved in economic planning, job creation and measures to help the unemployed. At the other end, the European Community is enlarging its role. In addition, many regions also have their own development and promotion agencies. This proliferation of bodies runs the risk of duplication, competition, inflated bureaucracies and inefficiency. If the new-style policies are to be effective, they must be delivered by a system which is clear, streamlined and carefully orchestrated.

Despite the large number of development agencies now working at the local and regional levels, the future of regional policy will still depend essentially on the willingness of the European Community and national governments to allocate funds. In the past the EC's richer nations, such as West Germany, have been somewhat reluctant to foot the bill for regional policy. In future, with moves towards an open Community-wide labour market, and with prosperity returning to many 'core' regions, an increase in regional imbalance might lead once again to substantial international migration (Chapter 10) and to a concomitant range of cultural, social and political problems for the recipient countries and areas. Faced with this prospect, the richer member states may therefore be more willing to fund projects designed to absorb surplus labour in countries such as Spain and Portugal. At the national level, too, the future of regional policy will depend on political as well as economic considerations. Indeed, in shaping such policy, the lengthening or shortening of the dole queues will no doubt be secondary to the ideologies of governments, the pressure of regional lobbies and crude electoral calculations. Harsh economic trends may force some governments to leave their ailing regions to the mercy of market forces, but for countries containing regions with separatist tendencies (such as Belgium, Spain, France and the UK) concerns about the coherence of the state will demand the survival of some form of regional strategy.

Finally, aside from these political-geographical considerations, governments cannot forget the continuing reality of mass unemployment. Because industrial production and real wages for those in work have risen, it is tempting to imagine that things are going well again. In fact, in Western Europe as a whole unemployment has fallen only very slightly: in the European Community, for example, it fell from 10.8 per cent to 10.6 per cent between April 1986 and April 1987. Moreover, it is in the least prosperous areas that economies are slowest to revive. Europe may have grown accustomed to living with regions of mass unemployment, but the problems have not been resolved. And just as traditional regional policy had little success in resisting the rise of unemployment, so it is difficult to see how the new, generally leaner, regional policy can have much more success in effecting its reduction.

Note

1. In practice, regional restraint measures were not widely employed. Only in France and the UK were they a major policy feature for a significant period.

References

Armstrong, H. and Taylor, J. (1988) *Regional Policy and the North–South Divide*, Employment Institute, London.

Bulletin of the European Communities (1988) 21, no. 6, 111–12, Official Publications of the European Communities, Luxembourg.

Clout, H.D. (ed.) (1987) *Regional Development in Western Europe*, 3rd edn, David Fulton, London.

Commission of the European Communities (1986a) *Programmes for Research and Action on the Development of the Labour Market: the role of local authorities in promoting local employment initiatives*, Official Publications of the European Communities, Luxembourg.

Commission of the European Communities (1986b) *ERDF: Eleventh Annual Report (1985)* COM (86) 545 final, 20 October, Brussels.

Commission of the European Communities (1987a) *The Regions of the Enlarged Community: third periodic report on the social and economic situation and development of the regions of the Community*, Official Publications of the European Communities, Luxembourg.

Commission of the European Communities (1987b) European regional policy, *European File*, 14, Official Publications of the European Communities, Luxembourg.

Commission of the European Communities (1988a) *Programmes for Research and Action on the Development of the Labour Market: youth unemployment policies*, Official Publications of the European Communities, Luxembourg.

Commission of the European Communities (1988b) *Programmes for Research and Action on the Development of the Labour Market: main report*, Official Publications of the European Communities, Luxembourg.

Commission of the European Communities (1989) *ERDF: Thirteenth Annual Report (1987)* COM (88) 728 final, 10 January, Brussels.

Croxford, G. (1988) The implementation of European Community regional policy. Unpublished doctoral thesis, Department of Geographical Sciences, Plymouth Polytechnic, Plymouth.

Dicken, P. (1986) *Global Shift: industrial change in a turbulent world*, Harper and Row, London.

Drevet, J. (1988) *1992-2000: les régions françaises entre l'Europe et le déclin*, Souffles, Paris.

European Investment Bank (1989) *Annual Report, 1988*, European Investment Bank, Luxembourg.

Eurostat (1988a) *Rapid Reports: regions*, Official Publications of the European Communities, Luxembourg.

Eurostat (1988b) *Basic Statistics of the Community*, 25th edn, Official Publications of the European Communities, Luxembourg.

Higgot, R. (1984) Export-oriented industrialisation, the new international division of labour and the corporate state in the Third World: an exploratory essay on conceptual linkage, *Australian Geographical Studies*, 22, 58–71.

Kahn, R. (1987) La nouvelle politique régionale des pays européens, *Problèmes Economiques*, 2049, 17–24.

Keeble, D. and Wever, E. (1986) *New Firms and Regional Development in Europe*, Croom Helm, London.

King, R. (1987) Italy. In H. Clout (ed.), *Regional Development in Western Europe*, 3rd edn, David Fulton, London, pp. 129–63.

Layard, R. (1986) *How to Beat Unemployment*, Oxford University Press, Oxford.

Licari, J. (1970) The European Investment Bank, *Journal of Common Market Studies*, 8, 192–215.

McCullum, J.D. (1979) The development of British regional policy. In D. Maclennan and J.B. Parr (eds), *Regional Policy: past experience and new directions*, Martin Robertson, London, pp. 3–41.

Mény, Y. (1985) French regions in the European Community. In M. Keating and B. Jones (eds), *Regions in the European Community*, Clarendon Press, Oxford, pp. 191–203.

Moore, B., Rhodes, J. and Tyler, P. (1986) *The Effects of Government Regional Economic Policy*, HMSO, London.

Nicol, W.R. (1980) *An Appreciation of Regional Policy Evaluation Studies: a comparative study, final report*, Wissenschaftszentrum, Berlin.

Pinder, D.A. (1978) Guiding economic development in the EEC: the role of the European Investment Bank, *Geography*, 63, 88–97.

Pinder, D.A. (1983) *Regional Economic Development and Policy: theory and practice in the European Community*, George Allen & Unwin for the University Association of Contemporary European Studies, London.

Pinder, D.A. (1986) Small firms, regional development and the European Investment Bank, *Journal of Common Market Studies*, 24, 171–86.

Robert, J. (1987) Paris et le désert français? Pour en finir avec un mythe!, *Problèmes Economiques*, 2046.

Storey, D.J. and Johnson, S. (1987) *Are Small Firms the Answer to Unemployment?*, Employment Institute, London.

Susman, P. and Schutz, E. (1983) Monopoly and competitive firm relations and regional development in global capitalism, *Economic Geography*, 59, 161–77.

Wadley, D. (1986) *Restructuring the Regions*, OECD, Paris.

Wallace, H. (1977) The establishment of the Regional Development Fund: common policy or pork barrel? In H. Wallace, W. Wallace and C. Webb (eds), *Policy-making in the European Communities*, Wiley, London, pp. 137–63.

Wise, M. and Croxford, G. (1988) The European Regional Development Fund: Community ideals and national realities, *Political Geography Quarterly*, 7, 161–82.

Yuill, D., Allen, K., Bachtler, J. and Wishlade, F. (1988) *European Regional Incentives*, EPRC, University of Strathclyde, Glasgow.

12

Supranational environmental policy and pollution control

Richard Williams

This chapter will address the twin questions of environmental policy-making and policy implementation at the European scale. It is therefore concerned with the development and implementation of environmental planning and pollution-control measures, rather than with the manifestation of pollution as such. Environment and pollution problems have mainly urban origins, but clearly transcend national boundaries: obvious examples include the problems of acid rain, pollution of the Rhine and the North Sea. Thus environment policy is, as much as any issue addressed in this book, a question on which there is a role for international agreement and supranational action.

Key themes in this chapter will be the place of environment policy within the framework of spatial policies designed to operate at a European scale; and the institutional and governmental structures which enable policy to be implemented Europe-wide. The chapter will concentrate on European-scale policies, programmes and institutions rather than individual countries; and it will consider not only environment policy and pollution-control measures as such, but also the question of how environmental planning at this scale interrelates with the other spatial issues and policy sectors discussed in this book.

The broad structure for the chapter is as follows: firstly, it considers the nature of pollution problems in Europe; secondly, it examines the steps being taken to overcome them; and thirdly it discusses what this adds up to in terms of an effective response to these problems. Within this structure, substantial emphasis is given to the framework necessary for effective policy-making and implementation at the European level, and to those supranational institutions which have powers to act in this regard. The prime focus therefore is on the European Community (EC), which has legal authority to legislate and enforce policies directed at environmental problems, although some discussion of the role of the Council of Europe (CoE) and the Organisation for Economic Co-operation and Development (OECD) is also appropriate.

The nature of the problem

Two quite different perspectives can be taken on the nature of international environmental pollution problems in Europe. In the first place, the problems may be expressed in terms of the substantive pollution issues which require a concerted international, rather than national, response if they are to be tackled effectively. Secondly, given the focus of this chapter on policy-making and implementation, the problems may also be expressed in terms of the need to establish the legal and institutional framework to formulate, implement and enforce policy at the supranational level.

A number of substantive pollution issues of international or even global significance are very much in the public eye: acid rain, the greenhouse effect, climatic change resulting *inter alia* from deforestation in South America, diversion of river waters in the Soviet Union, contamination of the Mediterranean and disasters such as Bhopal are just a few examples. Within Western Europe some of the worst excesses have been avoided, but there have been very great problems associated with industrial dereliction; emissions from industry into water courses and the atmosphere; asbestos pollution; discharges of titanium oxide causing the marine 'red mud' phenomenon; cadmium in soil; the biological death of certain rivers including the Rhine; and disasters such as those at the chemical works of Seveso in Italy in 1976 and Flixborough in the UK in 1974. There are several reports available detailing such issues and responses to them (see, for example, OECD, 1985; Commission, 1987a; DocTer International, 1987 — the last two published to coincide with the European Year of the Environment, 1987–8).

Identification of manifestations of environmental pollution which require an international response if they are to be tackled effectively is only part of the problem. To turn to the second perspective, effective action requires international agreement, co-ordinated action, and the legal and institutional framework for formulating, implementing and enforcing policy on a supranational basis. As has been indicated, the main thrust of this chapter is concerned with this second perspective and, in particular, with the establishment of the necessary institutional infrastructure.

The traditional method of achieving international agreement on environmental action has been by international conventions and treaties. A number of these that are applicable to Western Europe are discussed in Haigh (1987). Major examples include the Rhine Convention, agreed in Bonn in 1976 by the riparian states including Switzerland; the Geneva Convention on long-range transboundary air pollution, drawn up under the auspices of the United Nations Economic Commission for Europe in 1979; and others concerning water, the atmosphere, and wildlife and the countryside. In these and other cases, the European Commission was either a party to the agreements at the outset or has become associated with them at a later date.

In post-war Western Europe there has been a succession of supranational bodies (OEEC, UNECE, OECD, CoE, ECSC and EEC, now EC) capable of promoting concerted action and, in the case of the EC, of legislating on a supranational basis. The problem, as far as environmental policy and pollution control are concerned, has been one of gaining acceptance of the idea that this is a legitimate area of intervention, within their respective legal competence and consistent with their respective overall objectives. Given the economic orientation of most of these organisations, it is scarcely surprising that these objectives

have typically been expressed in economic development terms.

Which Europe? The legal and institutional basis for supranational policy

Three organisations — the EC, the CoE and the OECD — now play a significant role in promoting awareness of the international dimension of environmental and pollution-control issues, and in encouraging the development of an international response. It is therefore appropriate that discussion should focus on these organisations. Before embarking on this, however, it is important to pause and consider which definition of Europe is most appropriate. This is a question of institutional, as well as geographical, definition.

The EC and the CoE do not constitute the only possible definitions of Europe, but they are the most obvious candidates for attention: the CoE because it includes all the countries associated with Western Europe; the EC because its legal competence is unparalleled by any other supranational body and is increasingly analogous with the federal level in the United States, reflecting the vision of Monnet and Schuman of an embryonic United States of Europe.

THE EUROPEAN COMMUNITY

The EC, currently with 12 member states, has from the environmental policy viewpoint some important gaps in its territorial extent, notably Scandinavia, Switzerland and Austria. However further enlargement remains a possibility: Turkey is at present a candidate for membership; Malta, Austria and possibly Cyprus intend to apply for membership; and Sweden and Norway (again) are seriously considering the possibility of an application. The twelve members of the EC have made it clear that no further enlargement will be contemplated before 1992, but it is likely to occur later in the 1990s, to include some of these countries though probably not all of them.

In spite of these remaining territorial gaps, which are partially overcome in any case by bilateral agreements, the EC is a potentially powerful influence on environmental policy because it has legislative powers which can be used to achieve supranational regulation on a legally enforceable basis. The Community has had an environment policy since 1973, although in the earlier years it was seen as a relatively peripheral part of the Commission's activities. Most recently, the Community's power has been greatly strengthened by the passage of the Single European Act (SEA), which came into legal force on 1 July 1987. This has had the effect of amending the Treaty of Rome, *inter alia*, by adding an Environment Title (or section) to the Treaty.

In the 1980s, one of the fundamental components of the EC's environment policy, and one of the first building blocks of an overall strategy aimed at pollution control and the harmonisation of regulatory provisions, has been the Directive on the Environmental Assessment of Projects. This measure, commonly referred to as the EA Directive, was adopted in 1985 (Commission, 1985). Prior to adoption, it was the subject of some of the most extensive negotiations, consultations with member states' governments and interest groups, and redrafting ever required by a Commission proposal. The discussion returns to this Directive, and to the SEA, at length below (pp. 203–5).

THE COUNCIL OF EUROPE

The CoE has the larger territory, with 23 member states. Membership is open to all plural democracies: the Council comprises all Western Europe and the Mediterranean, including island states such as Cyprus and Iceland and neutrals such as Switzerland. Its principal territorial limitation, as far as environment policy is concerned, is the exclusion of the communist states of Eastern Europe. Unlike the EC, it does not have any supranational legislative powers, and it must therefore operate by persuasion, power of example, dissemination of ideas and experiences, and the adoption and promotion of protocols, agreements, charters, etc. Activities such as the European Architectural Heritage Year in 1975, the European Campaign for Urban Renaissance in 1981, and the European Year of the Environment in 1987–8 (the latter jointly with the EC) are examples of CoE initiatives.

Discussion of CoE bodies responsible for the promotion of spatial and environmental policies is

given at greater length in Williams (1988b). In general, CoE action is guided by the European Regional/Spatial Planning Charter which was adopted in 1983 by the CoE Conference of Ministers responsible for regional planning. (This conference is known by its French acronym CEMAT.)

The 'fundamental objectives' of the Charter, which were agreed by the sixth CEMAT, held in Torremolinos, Spain, in 1983 are: balanced socio-economic development of the regions; improvement of the quality of life; responsible management of natural resources and protection of the environment; rational use of land (CoE, 1984, pp. 7–8). As well as identifying goals, the Charter goes on to discuss implementation. Above all, it stresses the need to promote European international co-operation; to co-ordinate the various policy sectors relevant to the above objectives, including environmental protection; and to ensure that decision-making takes place at the appropriate level, whether local, regional, national or European. Clearly, in the minds of the ministers and their representatives who drew up the CEMAT Charter, it made sense to consider the European scale when formulating spatial and environment policy. However, it is far from clear that this way of thinking has penetrated the planning system generally.

Subsequently, in April 1987, a European conference of regional officials responsible for regional planning and development was held in Valencia, Spain, with the intention of giving further momentum to the spatial planning process at the European scale, and to the promotion of the Torremolinos Charter. The programme of the Valencia conference, and the *Declaration and Conclusions* adopted there (CoE, 1987), made explicit reference to the regional and sectoral policies of the EC, including environment policy.

THE ORGANISATION FOR ECONOMIC CO-OPERATION AND DEVELOPMENT

The OECD was founded in 1961, in accordance with an international convention signed in Paris in 1960. It has 24 member countries world-wide, including all the main economic powers. Nineteen non-communist European countries are members, and Yugoslavia is associated with the OECD for certain purposes, including environmental matters. Like the CoE, therefore, the OECD's definition of Western Europe is very wide. Structurally, it is an intergovernmental organisation engaging in research and policy evaluation, and its typical approach does not include international declarations or agreements. Its prime objective is the promotion of economic growth, but it is concerned with qualitative as well as quantitative aspects of expansion. Since 1979, periodic reports on the state of the environment have been prepared by the OECD Environment Committee. The most recent, in 1985, is a comprehensive review of problems and progress on all the main environmental issues, and is therefore particularly valuable (OECD, 1985).

Against this background, the difference between these bodies may be readily summarised. The CoE has the fullest territorial extent of any Western European supranational organisation, but its powers are limited to persuasion, awareness campaigns and the promotion of specific intergovernmental agreements. The OECD covers most of Western Europe, its powers are similarly limited and it can only address itself formally to national governments. On the other hand, as a world body capable of taking a wider view, it can be influential in making governments aware of the need for concerted international action on environmental issues. The EC, meanwhile, has the power of legislation and its supranational jurisdiction is unique. The key question is the extent to which it uses its legislative competence to put in place and enforce an effective environmental policy and pollution-control regime.

Although the CoE and the OECD are of considerable importance, the main priority here must be to consider the EC because of the political and legislative power to act on the environment which it alone possesses.

Political factors affecting progress

A theme which underlies any discussion of environment policy and environmental action, at the European level as much as at any national level, is the question of the speed with which progress is made towards effective political action. Several factors serve to impede progress, while others act as an imperative to increase the pace.

Reasons for slow progress fall into two main categories. One is the wide range of interests potentially affected by an environmental protection measure. Remedial action (the main thrust of earlier measures) has frequently been directed at specific substances or industrial processes, and therefore at specific industries or sectors of the economy. With this approach the targets have known who they were, and have often strongly resisted environmental controls. The greater emphasis more recently on preventative measures has not reduced resistance from industrial interests: preventative action potentially has an impact on a whole range of economic interests which have tended to lobby collectively for a minimal approach, fearing the possible catch-all nature of proposals such as the EA Directive.

The second set of reasons falls into the category of political support. Fundamentally, environmental and pollution-control measures depend for their effectiveness on political will. In the face of strong industrial and agricultural lobbies, political will has not always been sufficiently strong to satisfy the demands of the environment lobby. Today, however, it is important to recognise the change in the balance of power between economic and environmental interests that has occurred in recent years. This has been linked with the strengthening of the political power of the Green parties, in particular through their representation in several provincial and national legislatures, including those of West Germany, Belgium, France and, most recently, Sweden. Also, since 1984 there has been a Green Party group in the European Parliament. This trend is, of course, very much related to the growth of public awareness and concern about environmental questions, not only in these countries but in the UK and several others. The balance of political pressures is now much more supportive of environmental action, and governments have felt they must be seen to respond.

Recent progress has also been helped by increased acceptance that environmental protection and pollution-control measures do not necessarily have a negative impact on economic interests. Indeed, there are a number of ways, discussed below, in which a stronger environment strategy can have a positive effect on the economy. It was necessary to achieve acceptance for this point of view, not only to overcome opposition to stricter environmental controls, but also to justify an environment policy within the framework of an 'economic' Community.

Lastly, circumstances contributing to improved control have come about because the institutional and legal frameworks for implementing environment policy at the European level have been greatly enhanced in recent years. As will be made plain later, particular landmarks have been the establishment of a Directorate-General in 1981 and, above all, the addition of an 'Environment Title' to the Treaty of Rome by the adoption of the Single European Act in 1987.

The final set of political factors which must be taken into account is that relating to the EC's fundamental objective of creating a true common market in Europe after 1992. National pollution-control measures, and the variations between them, have to be regarded as non-tariff barriers, impeding harmonisation and allowing the possibility of pollution havens. Therefore the EC's 1992 programme creates a strong imperative for action to eliminate, as far as possible, disparities in environmental standards in order to remove any distorting effect on economic activities and development.

European environment policy: basic principles

The place of the 'environment' as a sectoral concern for an 'economic' Community has always needed justification, although the Commission has consistently taken the position that environmental problems are intrinsically transnational and therefore are an appropriate issue to be addressed by a supranational body such as the EC. This justification was elaborated by the Commission in successive *State of the Environment* reports (Commission, 1979, pp. 9–14; 1987a, pp. 351–8). Official reasoning has been developed on the following lines.

It is argued that a clean environment is essential in order to create the conditions necessary for several important sectors of the economy. These obviously include fisheries, agriculture and the tourism industry, which is a vital component of the economy of many peripheral regions of the EC (see Chapters 13 and 15). But several modern sectors of the economy also need clean atmospheric conditions or, if these are not available, they require them to be created by air

filtering or scrubbing. This is true, for example, of many high-technology and computing activities. In addition, many other industries are capable of much greater operational efficiency with the benefit of cleaner air and water, or the absence of acid depositions, etc. This highlights another linkage between the economy and environmental protection: emission control to achieve higher environmental standards of air and water quality creates the potential demand for enterprises manufacturing filtration equipment. This is a well-developed industry in West Germany, but one in which the UK is lagging far behind.

It is undoubtedly the case that the positive employment effects of a successful environment policy can be considerable, when both direct and indirect benefits are considered, although any figures must be treated with caution as this is a notoriously difficult phenomenon to measure accurately. Direct effects include employment in environmental recovery programmes, derelict land reclamation and the environmental technology industry; indirect effects include associated raw material production, training, administration, more opportunities for tourist development, and the general benefits of a more efficient and buoyant economy.

Accurate figures cannot easily be put to these effects. The Commission has published some examples of claimed gross aggregate employment benefits attributable to both direct and indirect effects of a positive environment programme. It claims, for example, 70,000 jobs in the Netherlands produced in 1982 on the basis of environmental initiatives, and 380,000 in West Germany in 1980 through expenditure on environmental protection (Commission, 1987a, p. 356).

Finally, it is increasingly acknowledged that quality-of-life criteria affect people's decisions concerning where they wish to live, and where they may wish to locate and invest in an industrial or service sector development. In particular, those industries which need to recruit high-quality and scarce expertise find that they are at a severe competitive disadvantage if they are located in an unattractive location which cannot offer a good quality of life. Clearly environmental quality is a major factor in the standard of the residential environment offered in different locations.

In formulating the basis on which its environment policy should operate, certain general and largely self-explanatory principles have been established by the Commission. These are: that action should be taken at the appropriate level of government (whether by local, regional or national authorities) and not necessarily by the Commission itself; that the polluter pays; and that prevention is better than cure because it is more economical in the long run. In the case of the first, there is both a practical and a political basis. The practical reason is that the bureaucracy in the 22 Directorates of the Commission in Brussels is quite small (smaller than that of a medium-sized metropolitan district council in the UK) and it would be quite impossible for them to administer all their programmes themselves. Secondly, in order to integrate procedures into domestic policy and ensure sensitivity to local conditions and local politics, devolution to the appropriate level of government is necessary. The polluter-pays principle is a simple idea, but sometimes poses great difficulties in identifying the polluter and enforcing some form of payment. Later discussion of the EA Directive will illustrate how these principles are put into practice.

The link between, on the one hand, environmental and pollution controls and, on the other, the drive to harmonise regulations and eliminate non-tariff barriers to trade must not be underestimated. Williams (1986) has argued that, with the creation of a common market, a hypothetical multinational facing quite large emission control costs to minimise potentially substantial pollution would, *ceteris paribus*, seek to locate where the cost of meeting local standards would be least. Such areas could be in danger of becoming pollution havens. Conversely, enterprises for which a clean environment is essential for either technical or recruitment (quality-of-life) reasons might, *ceteris paribus*, choose to locate in an area with higher than normal pollution-control standards. For sectors of the economy in either of these categories, disparities of environmental regulation lead to distortions of competition and spatial disparities. EC environment policy has sought to overcome these disparities and to establish a 'level surface' in the form of common standards of regulation. In fact, these are closer to being common minimum standards, since it is politically impossible (as well as environmentally undesirable) to reduce higher national pollution-control standards where these already exist. The EA Directive is the most ambitious attempt so far to establish this 'level surface'.

The elimination of non-tariff barriers to trade is a major element of the 1992 programme, and the implications of the existence of a common market after 1992 affect environmental regulation as much as other sectors of policy. The argument referred to above, about the need for common standards in order to avoid pollution havens and distortions of competition, will apply with more force after 1992.

The European Community: powers and policies

Powers and administration: 1973 to the Single European Act

Unlike the normal position with governments, the EC has no legal basis for intervention in any policy sphere unless there is a specific article in the Treaty of Rome calling for such action and defining the necessary powers and duties of the Commission. When the EC instituted its environment policy in 1973 this was problematical, since at that time there was no Treaty reference to environmental protection measures. Therefore, it was legally necessary to base the environment policy on two non-specific articles in the Treaty of Rome: Article 100 and Article 235.

Article 100 confers general powers to take action to achieve harmonisation:

The Council shall, acting unanimously upon a proposal from the Commission, issue directives for the approximation of such provisions laid down by law, regulations or administrative action in Member States as directly affect the establishment or functioning of the Common Market.

The other power sometimes invoked was under Article 235, which confers a general power to initiate policies not otherwise provided for in the Treaty. Since this was not a strong legal basis, it always required unanimity from the Council of Ministers. Much depended on political pressure or the will to achieve effective environmental measures, and acceptance of their place in an 'economic' community. Thus the passage in 1987 of the SEA, which provides a comprehensive legal basis for an environment policy, was an exceptionally important step. For further discussion, see Lodge (1986), and also Kromarek (1986) and Klatte (1986).

From the environmental viewpoint, the main provisions of the SEA are as follows. Article 25 adds a new Title, VII Environment, to Part III of the Treaty of Rome, imposing on the Commission and Council the duty to ensure that action relating to the environment is taken by the Community. Among the new articles which the SEA adds to the Treaty of Rome under this Title, Articles 130B and 130R are particularly important to note here.

Article 130B stresses the need for economic and social cohesion in the Community, and makes clear that, in preparing action on the environment, the Commission is to take into account environmental conditions in the various regions of the Community, together with the need for balanced development of the regions.

Article 130R requires action on the environment to have the following objectives: to preserve, protect and improve the quality of the environment; to contribute towards protecting human health; and to ensure a prudent and rational utilization of natural resources. It also states that environmental protection 'shall be a component of the Community's other policies' (Article 130R (2)) and that the Community shall 'take action . . . to the extent to which the objectives . . . can be attained better at Community level' (Article 130R (4)).

It is therefore evident that the duties imposed on the Commission by the new Environment Title of the Treaty establish a close link between regional or spatial policy and the environment. They also provide a basis for the consolidation of environment policy within other policy sectors, and for seeking to ensure that environmental protection is built into the programme for the co-ordination of the structural funds. These links have been strengthened by the interpretation placed on the Environment Title by the Commission's Fourth Action Programme on the Environment (see p. 203).

One final point relating to the SEA must be emphasised. Article 18 amends the Treaty of Rome by adding new Article 100A, which provides for qualified majority voting in the Council of Ministers in respect of many areas of EC policy, including the environment. It also provides that the Commission, in preparing environment proposals, 'will take as a base a high level of protection' (Article 100A, para. 3). Today, therefore, unanimity is not essential for the introduction of binding regulations, and there

is no ambiguity concerning the importance accorded to environmental protection.

In the administrative sphere, to date two significant phases can be identified. Initially, the Commission established an 'Environment and Consumer Protection Service' rather than the more normal form of a Directorate-General (DG), such as those responsible for other areas of EC policy. The rationale was that environmental issues impinged on all sectors of EC policy and that, as a service, it could serve all the DGs — just as the Translation Service obviously does. This was seen by some as giving environment policies second division status since, unlike the Translation Service, its contribution to other policy areas was not seen to be indispensable. Moreover, interaction often did not in fact take place with other areas. In 1981, however, environment policy achieved a considerably higher profile through the formation of a Directorate-General for Environment, Consumer Protection and Nuclear Safety (DG XI). Since then it has had its own Commissioner (since January 1989 the Commissioner has been Mr Ripa di Meana, during 1985–8 it was Mr Stanley Clinton Davies). Since the creation of DG XI, the extent of integration between environmental and other policy areas has in fact increased greatly, to the undoubted benefit of environment policy as a whole.

Environmental measures and Action Programmes

Space does not allow a comprehensive listing and analysis of environmental measures already enacted. For this purpose, Haigh (1987) and the *1986 State of the Environment Report* (Commission, 1987a) are recommended. However, two basic trends must be noted.

Firstly, there has been a move away from specific remedial measures designed to tackle problems caused by individual pollutants or industrial processes. The pendulum has swung instead towards initiatives designed to put the principle 'prevention is better than cure' into operation. Secondly, despite the unsatisfactory legislative framework which existed before 1987, the number of environmental protection measures introduced by the Commission was far from negligible. If all measures introduced by the Regulations, Decisions and Directives (the three

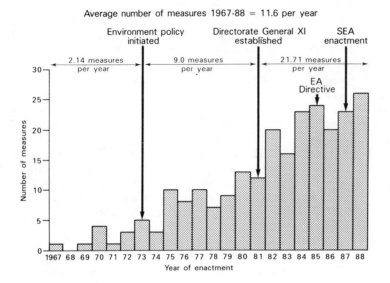

Average number of measures 1967-88 = 11.6 per year

Figure 12.1 *EC Directives, Regulations and Decisions concerning environment policy and pollution control.* *Source:* Haigh (1987), Appendix 1 and private communication.

forms of Community legislation) are aggregated, no less than 243 protectionist steps can be identified over the period 1967–88 (Figure 12.1). A few of the early examples pre-dated the official introduction of the Community's environmental policy in 1973, and no less than 74 per cent were introduced between that date and the enactment of the SEA. The steady growth that has characterised the EC's environmental legislation is clearly evident.

So far as the targets of these measures are concerned, those intended to overcome water pollution and improve water quality form a major category (Guruswamy *et al.*, 1983; Taylor *et al.*, 1986). Within this group, several are concerned with the discharge into watercourses and the marine environment of detergents; oil and chemicals; and waste metals or related chemicals, such as mercury, cadmium and titanium dioxide. (The latter has been responsible for the notorious red mud in the Mediterranean and North Seas.) A number of others relate to water for human consumption, or to water standards affecting the food chain in other ways. These include the standards desirable in areas producing freshwater fish for human consumption, as well as around shellfish fishing grounds. In addition, the water quality of bathing beaches has received widespread attention as a result of regulations introduced under this heading.

Other headings have rather fewer measures, yet are none the less significant. With respect to solid waste, steps have been taken to introduce site licensing of toxic waste disposal and control its transfrontier shipment. Similarly there are regulations to control oil and sewage sludge. In the field of air quality, the aims have been emission control, particularly of pollutants such as sulphur dioxide and nitrogen dioxide. In this context progress towards the reduction of lead in petrol has also been pioneered, while an important Regulation has introduced a programme running from 1987 to 1992 to monitor the relationship between acid rain and forest damage (Commission, 1986). Among a number of initiatives concerned with chemicals in the atmosphere, one landmark has been a 1980 Decision on chloroflourocarbons, with the aim of protecting the ozone layer (Commission, 1980). This problem was far from being in the public eye at that time, although of course there is now great public concern about this issue. The so-called Sevesco Directive of 1982 must also be considered outstanding (Commission, 1982). This is a major preventive measure requiring safety procedures wherever hazardous industrial processes are undertaken. There have also been a number of wildlife-protection and noise-control initiatives.

While it is important to recognise that the expanding volumes of Community environmental

Table 12.1 *The European Community's Action Programmes on the Environment*

Action Programme	Period covered	Date approved	Official Journal reference
I	1973–76	22-11-73	C112, 20-12-73
II	1977–81	17-05-77	C139, 13-06-77
III	1982–86	7-02-83	C46, 17-02-83
IV	1987–92	19-10-87	C328, 7-12-87

Source: Official Journals of the EC, as above.

legislation has been targeted at a number of priority problems, it is also necessary to appreciate that the majority of measures have been introduced in the context of a series of Action Programmes on the Environment. Once again, space does not permit detailed consideration of all of these, but the sequence is outlined in Table 12.1, which also provides official reference details. Discussions of the earlier Programmes are to be found in Commission publications (1979; 1987a; 1987b).

Particular mention must be made of the fourth Action Programme (1987–92). This was published to coincide with the European Year of the Environment (EYE), and is underpinned by the third *State of the Environment Report* (Commission, 1987a) and the powers which became available under the SEA. The fourth Programme outlines the objectives and principles behind the Community's environment policy, reviewing substance-oriented and source-oriented controls, and concluding that a flexible approach is necessary to accommodate different technical factors and national traditions. It examines the specific sectors of atmospheric pollution, fresh- and seawater, chemicals, biotechnology, noise and nuclear safety.

It also outlines policies for the management of environmental resources, for natural resource conservation, for soil protection and for waste management. In addition, policies for urban, coastal and mountain areas are proposed, while attention is also paid to research requirements and the need for action at the international level.

This approach is based on the legal authority provided by the Environment Title in the SEA, and on the knowledge that decisions concerning the adoption of environmental measures by the Council of Ministers will be by majority voting, rather than the unanimous decisions which have

hitherto been necessary. However, perhaps the prime importance of the fourth Action Programme is that it goes much further than previous Programmes in proposing procedures to ensure that environmental considerations are given full weight in other EC policy areas and that policy integration does, in fact, occur.

Reform of the Community's structural funds was begun on 1 January 1989. The funds involved are the Guidance section of the European Agricultural Guidance and Guarantee Fund (EAGGF) — which is part of the Common Agricultural Policy (CAP) — the European Regional Development Fund (ERDF) and the European Social Fund (ESF); see Chapters 11 and 13. Although there is no structural fund attached to Community environment policy, and therefore the Regulations governing the reform of these funds do not directly refer to environment policy, the fourth Action Programme makes the linkage clear. Particularly explicit are the statements that co-ordination with the structural funds 'will clearly assist in taking full account of environmental needs' and that 'the Commission is already working on effective internal procedures to ensure that environmental requirements are built into processes of assessing and approving proposals for all development to be financed from such funds' (Commission, 1987c). Other financial instruments such as the European Investment Bank (EIB) are also subject to this general strategy of policy co-ordination. Loans from the EIB have always strongly reflected regional policy objectives, but there has recently been a distinct pattern of growth in loans for projects with an environmental dimension, especially in relation to water- and air-pollution problems, as Figure 12.2 illustrates. The need for policy co-ordination on the above lines has been felt for some time, and is in part a response to calls from the European Parliament (1983) and the Council of Europe (1984; 1988) for a co-ordinated spatial planning strategy at the European scale.

Environmental assessment

The EA Directive, or, to give it its full title, 'Council Directive on the assessment of the effects of certain public and private projects on the environment' (Commission, 1985) was notified in 1985 and came into legal force in July 1988. It has

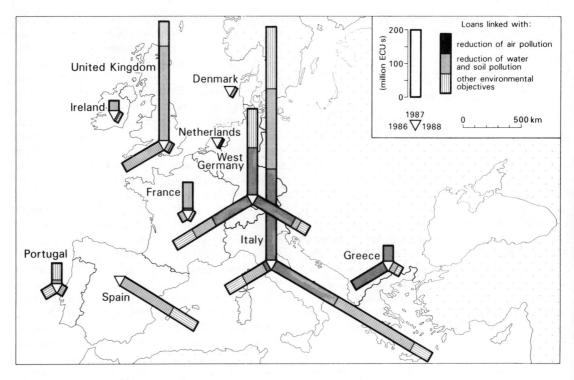

Figure 12.2 *European Investment Bank loans for environmental protection, 1986–88*
Source: European Investment Bank, *Annual Reports.*

attracted a great deal of attention, and in business circles many have opposed the measure, fearing the impact of its catch-all nature on their commercial interests. A considerable literature has grown around it, discussing its evolution and potential significance (see, for example, Tomlinson, 1986; Williams, 1983; 1986; 1988a; Wood, 1988). Although its introduction is too recent to allow firm conclusions to be drawn, it can be argued that this is the most important single measure that has yet taken effect during the period of the fourth Action Programme. As such, it requires specific consideration.

The concept is derived from the US experience of Environmental Impact Statements (EIS), which are required under the US National Environment Protection Act of 1969, but it is very much looser in its formal demands. Under the EA Directive, would-be developers of a whole range of projects having potential to damage the environment must, when they apply for authorisation to develop, submit an environmental assessment indicating their project's likely effects. They must also specify action to be taken to overcome adverse environmental impacts.

Two lists define the projects covered by the Directive. The first of these, detailed in Annex I of the Directive, is mandatory. It includes oil production and processing, nuclear and major thermal power stations, radioactive and toxic waste disposal installations, integrated iron and steel plants, asbestos extraction and processing plants, integrated chemical installations, and transport infrastructure such as motorways, major airports, waterways, sea and inland ports.

There is also a longer list of activities for which assessments are required whenever significant environmental effects can be anticipated. Activities appearing on this list include a variety of agricultural operations; all forms of oil, gas, coal, peat and mineral extraction or deep drilling; a wide range of nuclear and non-nuclear energy installations; metal processing; glass manufacture; chemical and pharmaceutical processes not included under Annex I; the food processing, textile, leather, wood, paper and rubber industries; infrastructure projects including industrial estates, urban redevelopment, pipelines and tourist facilities such as ski lifts, marinas and holiday villages; dams; light rail systems; and scrap yards, waste-disposal plants and a number of other necessary but unpleasant features of urban existence.

A significant linkage has been made by the Commission between this Directive and the operation of the European Regional Development Fund (Chapter 11), illustrating the strategy of integration between policy sectors discussed above. Before an ERDF grant can be authorised, it is now necessary to demonstrate that any EA Directive requirements which may arise have been fulfilled. Although this stance is recognised by the fourth Action Programme to be only a first step, it is already indicative of how environmental considerations may become more integrated within EC policy as a whole.

The EA Directive is seen by the Commission as a major element in its strategy of constructing a framework of preventive measures operating throughout the Community. As has been indicated, with 1992 in mind it is critical that there is no great variation in pollution control regimes which might allow pollution havens to develop and distort competition by influencing industrial location. This directive was always intended to be the first of several designed to use the land-use authorisation or planning procedures which already exist in all member states to identify potential polluters and ensure, when deciding whether to give authorisation, that adequate environmental safeguards are built into the development. Initially, it was intended that this Directive, requiring the assessment of projects, should be followed by Directives requiring environmental assessment of statutory plans and programmes of development. This approach has now been replaced by a strategy of closer integration with other policy areas such as regional policy, referred to above, and with the development of an environment programme within the programme funding strategy being developed by the Regional Policy Directorate (DG XVI).

Referring back to the discussion above of the principles on which Community environment policy is based, the EA Directive is totally founded on the principle that prevention is better than cure. It attempts to ensure that the polluter pays by requiring, firstly, identification of adverse environmental effects and, secondly, action to overcome these in advance of authorisation of the development. It is argued that it is in the developer's interest to do this, since remedial action for which the developer is responsible is likely to cost more than it would to take environmental factors into account at the initial planning stage. This is even true of project redesign when construction is at an advanced stage.

The concept of EA is only workable and administratively practical if it is integrated into existing national systems of land-use planning, most of which are administered at subnational levels of government, so the principle of taking action at the appropriate level of government is also embodied. Additionally, questions of local policy and politics inevitably enter into the authorisation of any major development of the type requiring an EA, and these can be taken into account since EA is integrated with the planning system.

Conclusion

It is not easy to draw concrete conclusions to a story which is continually unfolding, especially when the main thrust of action has been a very large number of specific measures, a smaller number of more generally applicable ones, but above all the establishment of a legal and institutional framework for legislating and enforcing environment policy. It is not possible within the scope of this work to assess the numerous specific measures. Indeed, in the literature as a whole such assessments are as yet rare, the work of Guruswamy *et al.* (1983) and Taylor *et al.* (1986) being exceptional in this respect. Perhaps the fairest summary to date is provided by Stanley Clinton Davis: 'Notwithstanding the substantial successes of Community environmental policies, however, much remains to be done' (Commission, 1987a, Preface). One problem is, of course, that new technologies and research are regularly identifying new environmental problems, and there is a strong sense, therefore, of running in order merely to stand still.

However, we are now in a situation in which the legal and political frameworks are in place for enforceable international action against pollution and for positive environmental improvements. This is a major achievement, and Western Europe is uniquely well served in this respect. The Environmental Assessment Directive, for example, is probably unrivalled as a supranational land-use planning instrument. Although it has many limitations, not least because it will be interpreted in very different ways by different sub-national or national authorities, its very existence is of major significance.

While the principle of such supranational measures, and of supranational legislation and enforcement of environmental protection, is now accepted, their full and effective exploitation remains to be demonstrated. Ultimately, effective action, as with so many issues, depends on political will. With the legal framework now in place, there is no institutional impediment to political will being expressed as action. Western Europe has a great asset when facing the challenges and changes of the 1990s.

Acknowledgements

The author is grateful to Nigel Haigh of the Institute for European Environmental Policy, London, for permission to use data from his book *EEC Environmental Policy and Britain*, 2nd edition, 1987, and also for further information, supplied privately, in order to construct Figure 12.1.

References

Commission of the European Communities (1979) *State of the Environment*, second report, Brussels.

Commission of the European Communities (1980) Decision 80/732/EEC, *Official Journal* L90, 03-04-80.

Commission of the European Communities (1982) Directive EEC/82/501, *Official Journal* L230, 05-08-82.

Commission of the European Communities (1985) Directive EEC/85/337, *Official Journal* L175, 05-07-85.

Commission of the European Communities (1986) Regulation 3528/86, *Official Journal* L326, 21-11-86.

Commission of the European Communities (1987a) *The State of the Environment in the European Community 1986*, Brussels.

Commission of the European Communities (1987b) The European Community and Environmental Protection, *European File*, 5/87, 3-11.

Commission of the European Communities (1987c) Fourth Action Programme on the Environment, *Official Journal* C328, 07-12-87.

Council of Europe (1984) *European Regional/Spatial Planning Charter*, Strasbourg.

Council of Europe (1987) European conference of regional representatives responsible for regional planning and development, *Valencia Declaration and Conclusions*, Strasbourg.

Council of Europe (1988) *Draft European regional planning strategy*, CEMAT, Strasbourg.

DocTer International (1987) *European Environmental Yearbook, 1987*, Milan.

European Parliament (1983) *Report on a European Regional Planning Scheme*, Document 1-1026/83, 21 November.

Fairclough, A.J. (1983) The Community's environmental policy. In R. McRory (ed.), *Britain, Europe and the Environment*, Imperial College, London, pp. 19-34.

Guruswamy, L.D., Papps, I. and Storey, D.J. (1983) The development and impact of an EEC Directive: the control of discharges of mercury into the acquatic environment, *Journal of Common Market Studies*, 22, 71-100.

Haigh, N. (1987) *EEC Environmental Policy and Britain*, 2nd edn., Longman, Harlow.

Klatte, E. (1986) The past and future of European environmental policy, *European Environmental Review*, 1, 32-4.

Kromarek, P. (1986) The Single European Act and the environment, *European Environmental Review*, 1, 10-12.

Lodge, J. (1986) The Single European Act: towards a new Euro-dynamism?, *Journal of Common Market Studies*, 24, 203-23.

OECD (1985) *The State of the Environment, 1985*, Paris.

Taylor, D., Diprose, G. and Duffy, M. (1986) EC environment policy and the control of water pollution: the implementation of Directive 76/464, *Journal of Common Market Studies*, 24, 225-46.

Tomlinson, P. (1986) Environmental assessment in the UK: implementation of the EEC Directive, *Town Planning Review*, 57, 458-86.

Williams, R.H. (1983) Land use policy, pollution control and environmental assessment in EC environment policy, *Planning Outlook*, 26, 54-9.

Williams, R.H. (1986) EC environment policy, land use planning and pollution control, *Policy and Politics*, 14, 93-106.

Williams, R.H. (1988a) The Environmental Impact Assessment Directive of the European Communities. In M. Clark and J. Herington (eds), *The Role of Environmental Impact Assessment in the Planning Process*, Mansell, London, pp. 74-87.

Williams, R.H. (1986) European spatial planning

strategies and environmental planning. In G.J. Ashworth and P.T. Kivell (eds), *Land, Water and Sky: European Environmental Planning*, Geopers, Groningen, Netherlands, pp. 9–17.

Wood, C.M. (1988) The genesis and implementation of environmental assessment in Europe. In M. Clark and J. Herington (eds), *The Role of Environmental Impact Assessment in the Planning Process*, Mansell, London, pp. 88–102.

Part III
Challenge and change in rural Europe

13

The challenge of land redundancy

Brian W. Ilbery

Agricultural modernisation in the post-war period

Agriculture in Western Europe, and especially the European Community (EC), has experienced profound change in the post-war period. This reorientation is commonly referred to as the second agricultural revolution (Healey and Ilbery, 1985), and in many areas agriculture has been transformed from little more than a peasant venture into a major business. Farms have become larger, fewer in number and more specialised. Farming has been modernised and, in some cases, industrialised as links with food processing and agricultural supply industries (for chemicals, fertilizers, machinery and feedstuffs) have been strengthened. Capital inputs into agriculture have increased substantially, as farmers have striven to obtain economies of scale to counteract declining farm incomes. There has been a large reduction in the agricultural labour force, especially of farm labourers but also of farmers themselves, with notable consequences for rural population densities and employment (Figure 13.1 and Chapter 14).

Bowler (1985a; 1986) identified three dimensions to the modernisation process in Western European agriculture: *intensification*, as a consequence of which output per hectare of land has increased, aided by mechanisation, capitalisation and developments in biotechnology; *concentration*, which has led output and resources to be confined to fewer but larger farm businesses, regions and countries; and *specialisation*, leading to less-profitable activities being eliminated, as land, labour and capital have concentrated on fewer products. At the regional scale, the pattern of agricultural specialisation is fairly complex, although it appears to have increased in those regions where a product was already an important element of the agricultural system (Figure 13.2). The overall pattern has been shown to be a function of the location of food processing plants, poor environmental quality, large farm sizes, and regions being within one of the first six countries to join the EC (Bowler, 1987). Two-thirds of the administrative regions of the EC10 have become more agriculturally specialised since 1960 (Bowler, 1986).

Agricultural modernisation has been 'driven' by technological developments and favourable government policies. Throughout the 1960s and 1970s, the Common Agricultural Policy (CAP) attempted to solve the social problem of low farm incomes with structural measures and productionist price-support policies (see Hill, 1984; Bowler, 1985b; and Ilbery, 1986). Over-generous levels of price support (guaranteed prices) have provided farmers with a relatively risk-free economic environment within which to operate and increase output faster and further than warranted by market demand. Such a policy has not only raised the cost of food to the consumer; it has also favoured 'northern' products such as beef, cereals and milk, and larger farms in the 'northern' regions of the EC. As price support levels also vary from country to country, due to differences between 'green' and market rates of exchange,[1] and because of the operation of Monetary Compensatory Allowances (MCAs)[2] the CAP has had a spatially variable impact (Ritson and Tangerman, 1979; Bowler, 1985b). North–south contrasts have been exaggerated, just as favoured 'core' regions such as East Anglia,

211

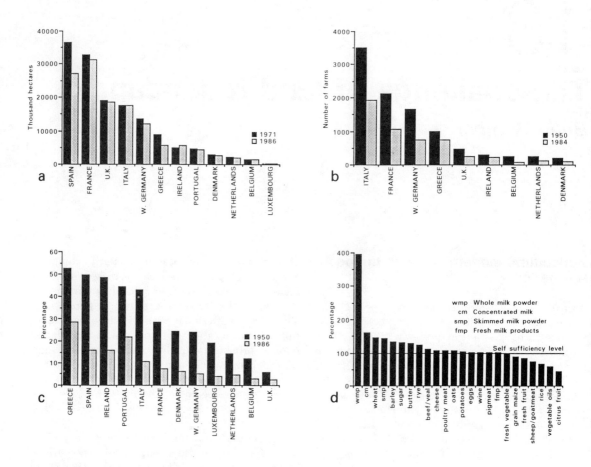

Figure 13.1 *Indicators of agricultural change in European Community countries*
(a) Utilised agricultural area, 1971–86
(b) Decline of farms > 1.0 ha.
(c) Working population in agriculture, percentage change, 1950–86
(d) Self-sufficiency levels for selected agricultural products in the Community of Ten, 1984

northern Belgium, north-east France, Schleswig-Holstein and the Netherlands have benefited more than 'peripheral' regions in western, southern and northern Europe.

One major consequence of the CAP has been the huge 'mountains' and 'lakes' of supported products that have developed, even though the utilised agricultural area (UAA) has declined in all EC countries except Ireland (Figure 13.1). Other consequences include higher consumer prices, widening farm income differences and environmental degradation from a more intensive and industrialised system of farming (see Chapter 12 for details of pollution control). The enormous cost of storing and disposing of the burgeoning

surpluses has threatened the financial viability of the CAP. For example, cereal production in the EC has increased by 40 per cent since 1975 (Smith, 1987), and in 1987 27 per cent of the cereal crop was surplus to requirements (Stanford, 1988). As storage costs are £114 (168 ECU)[3] per tonne for cereals, £224 (360 ECU) for butter and £1832 (2695 ECU) for beef (Schopen, 1987), reform of the CAP has become a top priority. The EC simply cannot afford such vast sums of money as £5 million (7.35 million ECU) per week to store surplus butter and skimmed milk powder.

The logical and most economic solution to the problem of surpluses is to lower the guaranteed prices of supported products by as much as 20–30 per cent (Bowler, 1987). However, this is politically unacceptable to the European Commission, not least because many farmers would become bankrupt overnight and many politicians would be in danger of losing their jobs. (The farm lobby remains remarkably strong on the European continent.) A less painful solution is to reduce production, either by withdrawing vast areas from agriculture, or by reassigning them to alternative uses. The concept of land redundancy is quite complex and involves compensating farmers for taking agricultural land out of production. Up to 10 million hectares of agricultural land in the EC are surplus to requirements, demonstrating the scale of the problem facing policy-makers. These issues are examined in detail in later sections. Land retirement and/or diversion is indeed part of a series of production control measures designed to solve the ills of the CAP. However, at the time of writing, attitudes towards specific policy measures are divided, with major contrasts being apparent between 'northern' and 'southern' countries and between such individual countries as France, West Germany and Denmark. Little empirical evidence is yet available and so the effect of different schemes must, like the rest of this chapter, remain speculative. As the CAP dominates Western European agriculture, discussion will concentrate upon the EC and the EC10 in particular. The recent entry of Spain and Portugal into the EC (1986) means that they are exempt from many CAP measures until the early 1990s. Nevertheless, they have already added to the problem of food surpluses and so will be grappling with some of the issues raised in this chapter during the 1990s.

Structural policies and land retirement/diversion programmes

A series of structural measures was introduced into EC agriculture during the 1970s and early 1980s to complement the price-support system of the CAP (Hill, 1984; Bowler, 1985b). One of the most geographically sensitive was the Less Favoured Areas (LFAs) Directive (75/268), designed to maintain rural population densities, conserve and manage the landscape, and safeguard agricultural activities in regions with physical handicaps to production (Winchester and Ilbery, 1988). In such physically marginal areas,

an annual compensatory allowance is available to cover increased costs of production; it is paid in terms of a grant per hectare or head of livestock on a farm. The Directive also provides favourable rates of aid for structural modernisation and investment in such non-agricultural activities as tourist and craft facilities. Additional support for farmers in problem southern regions came in the form of the 'Mediterranean package' (1978), integrated rural development programmes (1979) and the integrated plan for the Mediterranean regions (1983). At no time were production controls imposed on farmers in the more favourable agricultural regions, who thus continued to benefit most from price-support measures.

Changing agricultural policy, 1984–88

Since the early 1980s it has been increasingly recognised that a system of agriculture which does not produce for the market has no sound long-term prospects. This was echoed in a relatively recent statement from the Commission of the EC (1986, p. 8): 'European agriculture has to accept economic realities and learn to produce for the market, to adapt to commercial demands and to continue to modernise.' A series of decisions in 1984 concerned with a restrictive price policy represented the beginning of a change in agricultural policy. These included revisions of the guaranteed thresholds for cereals, sunflower seeds, processed fruit and raisins (revisions which now allow guaranteed prices to fall after a certain volume of production is exceeded); co-responsibility levies (which require producers to pay part of the cost of storage or disposal through a tax on production); and, most dramatically, compulsory production quotas in the milk sector. Further signs of change were evident in the Commission's 'Green Paper' of 1985 (Commission, 1985). This recommended the withdrawal of land along environmentally strategic buffer zones, within ecological corridors along field boundaries, and around water bodies (Burnham *et al.*, 1986). For the first time, environmental problems were recognised and conservation of the farmed landscape stressed. Environmentally Sensitive Areas (ESAs) were recommended, within which farmers would receive an annual premium per hectare to introduce or maintain farming practices that were

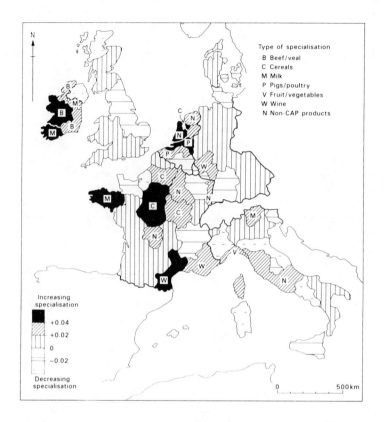

Figure 13.2 *Changing regional specialisation in agriculture in the EC9, 1964–77*
Source: Bowler (1986), 21.

compatible with the protection of the environment and of natural resources. The Green Paper also suggested income aids for full-time farmers in financial difficulties from the restrictive price policy, social aids to very poor farmers, pre-pensions for farmers over 55 years of age and the abandonment of agricultural land.

Few of these ideas were incorporated immediately into the new structures Regulation of 1985 (797/85), in which structural improvements and modernisation were again emphasised above environmental issues and solutions to the problem of mounting food surpluses. However, ESAs were introduced, as was an EC farm woodland scheme and an increase in the investment ceiling for diversification into tourism and crafts in the LFAs. Moreover, in the following year the European Commission forwarded proposals to amend Regulation 797/85 in order to help farmers, especially in marginal areas, adjust to new market conditions created by restrictive price policies (Revell, 1987). The centrepiece of the proposals was a pre-pension scheme to encourage farmers to

leave farming and either fallow their land or use it for non-agricultural purposes such as forestry and recreation. The 1986 package also included proposals to reduce production (through extensification and conversion to non-surplus products), to provide Community funding for ESAs, and to introduce measures on training, advisory services, forestry, marketing, and research and development.

While a pre-pension scheme was eventually introduced in 1988 (see below), extensification was the first of these proposals to be brought into force. In 1987 Regulation 1760/87 established a scheme to compensate farmers for reducing their output of cereals, beef or wine by at least 20 per cent. The Regulation, which was compulsory for member states but voluntary for individual farmers, caused considerable discussion (Jenkins and Bell, 1987). But the Regulation was short-lived as well as controversial, largely because of movement towards a form of extensification known as 'set-aside', under which farmers are compensated for taking land out of production altogether. Thus, in 1988 Regulation 1094/88 introduced an arable set-aside scheme and, in doing so, simultaneously amended Regulation 797/85 and replaced Regulation 1760/87.

Agricultural policy in 1988

The year 1988 could prove to have been an important turning point for the CAP. Although structural inefficiencies still exist in European agriculture, measures have at last been introduced to deal with overproduction and storage costs. Two important Regulations, with different objectives but both dealing with the retirement of agricultural land, are now in force. It is too early to measure their effects, but the details of the schemes, and some of their likely geographical consequences, can be examined.

REGULATION 1096/88: CESSATION OF FARMING

Although it aims to reduce production potential and improve farm structures, the main objective of this Regulation is to help farmers over the age of 55 who are no longer able to adjust to the 'new situation'. Motivation for the measure is, therefore, primarily social. Farmers practising farming as their main occupation will receive compensation provided that they satisfy one of two conditions:

1. Agricultural production on the holding can be halted throughout the whole period, from the time of cessation to the time when the farmer reaches normal retirement age (as fixed by social security arrangements in member countries). This period must be not less than five years. The vacated land can be left fallow, afforested or converted to non-agricultural uses, but the quality of the environment and the countryside must be preserved.
2. The agricultural area of the holding can be used to enlarge that of one or more other agricultural holdings, provided the farmers on the enlarged holdings practise farming as their main occupation and undertake not to increase the production of surplus products on the new total area.

In either case, compensation takes the form of an annual allowance and a supplementary annual

Table 13.1 *Regions eligible for EC support under the cessation of farming Regulation (1096/88)*

1. Regions receiving 50% of eligible expenditure:	
Italy	Abbruzzi, Sardegna
United Kingdom	Northern Ireland
Spain	País Vasco, Rioja, Balearics
2. Regions receiving 25% of eligible expenditure:	
Italy	Campania, Molise, Puglia, Basilicata, Calabria, Sicily
Greece	Whole territory
Spain	Whole territory, except regions listed under 1 above
Portugal	Whole territory
Ireland	Whole territory

Source: Official Journal of the European Community, 20.8.88.

premium per hectare of land withdrawn from production. An annual allowance or a lump-sum payment is also available for permanent paid agricultural workers and permanent family helpers. The maximum allowance per holding is 3000 ECU (£2,041) per year, for a maximum duration of ten years, plus an annual supplementary premium of 250 ECU (£170) per hectare. If the redundant land is afforested, the supplementary premium is extended up to a maximum of 20 years. The EC will contribute towards the cost of the Regulation in two groups of regions (Table 13.1). Where a 'composite indicator'[4] is over 75 per cent but less than 85 per cent of the Community average, the EC will contribute 50 per cent of eligible expenditure. This ceiling falls to 25 per cent in regions where the composite indicator is less than 75 per cent of the Community average. Member states can opt not to apply the scheme in mountain and hill areas and in regions where production should not be abandoned because of natural conditions or the dangers of depopulation. As Table 13.1 indicates, the scheme is aimed firmly at southern regions which suffer from weak farm structures and unviable units. Dessylas (1987) has estimated that up to 55,000 farmers will cease production over the first five years of the scheme, taking 0.5 million hectares out of production. It would, however, be wrong to present this scheme as one likely to make a major contribution to the reduction of agricultural surpluses, since Dessylas also estimates that up to 80 per cent of the land may be under forest. The effect on production will be minimal, therefore, but this is not surprising as the Regulation has social rather than economic or environmental objectives.

REGULATION 1094/88: SET-ASIDE OF ARABLE LAND AND THE EXTENSIFICATION AND CONVERSION OF PRODUCTION

As was indicated above, this Regulation introduces a set-aside scheme for arable land and modifies the arrangements for extensification and conversion as outlined in Regulation 1760/87. Although involved with improving the efficiency of agricultural structures, the main aim of the Regulation is to help re-establish the balance between output and the market's capacity to absorb it. Member states must introduce the Regulation, even though it is voluntary for individual farmers. However, on receipt of a reasoned application, the Commission can authorise member states not to apply Regulation 1094/88 in those areas or regions where, because of natural conditions or the threat of depopulation, it should not be applied. As is indicated below, there are three main schemes, and farmers may receive aid from just one of these schemes for any particular piece of land.

1. *Set-aside of arable land.* At least 20 per cent of a farmer's arable land must be set aside for a minimum of five years, with the possibility of termination after three years. Producers who set aside at least 30 per cent of their arable land will not pay the co-responsibility levy on the first 20 tonnes of their crops. Land withdrawn from cultivation should be left fallow (with rotational possibilities), wooded or used for non-agricultural purposes. Participants must ensure that retired land is kept in good agricultural condition, with a view to protecting the environment and natural resources. Member states can also authorise, within some or all of their territory, the use of set-aside land for either extensive livestock grazing or the production of chickpeas, lentils and vetches.

Compensation for set-aside may be granted for all arable land, irrespective of the crops grown, providing that they are covered by the price-support system of the CAP and the land has been in cultivation for a certain reference period. The level of compensation is determined by member states on the basis of 'loss of income'. However,

minimum and maximum annual figures of 100 and 600 ECU (£68 and £408) per hectare have been fixed, although in exceptional cases the maximum can rise to 700 ECU (£476). The EC contribution is 50 per cent for the first 200 ECU of the premium, 25 per cent for 200–400 ECU and 15 per cent for 400–600 ECU. Rates of aid can vary according to the use to which retired land is put. National rules for set-aside have to be cleared by the European Commission and the scheme has been effective since autumn 1988.

2. *Extensification of production.* Extensification is defined as a 20 per cent reduction, over a period of at least five years, in the output of the product concerned — without any increase in other surplus production capacity. Countries have had to introduce the scheme, restricted until December 1989 to beef, veal and wine. In the case of livestock, the number of animals must be reduced by 20 per cent, whereas for wine the yield per hectare must be cut by that amount. As with the set-aside scheme, member states determine the amount of aid to be paid for reducing output, the conditions for granting aid, and the reference period for the purposes of calculating the reduction. The EC's contribution towards the cost of the scheme is 25 per cent of eligible expenditure.

3. *Conversion of production.* Conversion is defined as a switch in production to non-surplus products. The Council of the EC, acting on a proposal from the Commission, will adopt a list of products towards which production may be converted, and will also establish the conditions and procedures for the granting of aid. Detailed rules for the conversion scheme are still to be adopted, although the EC will reimburse member states 25 per cent of their eligible expenditure.

Land-retirement problems

Land-retirement and conversion programmes are not new in the Western world and have long been in existence in the USA, where they have met with limited success (Buckwell, 1986a; Potter, 1986; Ervin, 1987; 1988). Major problems with such schemes include slippage and a selectivity effect. The former occurs when the fall in production is proportionally lower than the amount of land set aside, causing the overall impact to be less than expected. Slippage of up to 50 per cent can be caused by farmers retiring their least-productive land first; intensifying production on their remaining land; using retired land for other crops; or bringing more land into production, outside the scheme. The selectivity effect involves farmers participating in schemes to subsidise farm changes that would have occurred anyway. This is especially the case with farmer retirement, but has also been shown to be applicable to farmers interested in farm woodland and conservation projects (Potter and Gasson, 1987).

Other problems surrounding land abandonment include the dangers of rural depopulation; the increased risk of soil erosion on bare land; pollution from the spraying of weeds with pesticides; and poor financial incentives. If rates of compensation are low, set-aside will only attract farmers in marginal cereal areas and not the major producers in 'core' regions. This is especially likely with fixed rates of aid. Indeed, some commentators have advocated a national bidding system by the farmers themselves, where bids are accepted in order of relative cheapness to the EC, up to a pre-set level (Potter and Gasson, 1987; House of Lords Select Committee, 1988). With careful planning, some of the listed problems are preventable. However, American experience of acreage-reduction programmes is not encouraging, and Regulations 1094/88 and 1096/88 could have significantly less effect in cutting agricultural production than envisaged.

Locational aspects of land redundancy

Most agricultural policy measures in the EC have had a spatially variable impact (Bowler, 1985b; Winchester and Ilbery, 1988). Land-retirement and conversion schemes are also likely to vary in their effects, at national, regional and local scales. American experience suggests that the lower returns obtained by farmers in southern and other marginal agricultural areas of the EC will make compensation more attractive than in the high-return areas of northern Europe. Much depends, however, upon whether southern countries attempt to exempt their more marginal areas from the schemes. In fact, the idea of retiring agricultural land has received little support in southern states, and even in northern countries the degree of enthusiasm has varied. In Denmark, for example, it is believed that, as long as agricultural production is reduced by a set amount, each

country should introduce its own measures. In contrast, it is felt in France that set-aside and conversion programmes will destroy the country's position as the world's second-largest exporter of agricultural products (Delormé, 1987). Reducing output is not a priority, and French policy is to keep marginal areas in production by encouraging large livestock-grazing farms. The United Kingdom and West Germany are stronger advocates of production controls and set-aside in particular. In the former this is seen, together with price constraints, as a way of controlling cereal output, while in the latter it is seen more as a means of helping elderly farmers to retire.

In 1986, West Germany adopted a fourfold approach to restoring the balance between agricultural supply and demand (*Marktentlastungsprogramm*). This advocated the promotion of production alternatives (for deficit products), new uses for agricultural commodities, compulsory production quotas, and a voluntary scheme to take out of production entire farms, or parts of farms, operated by elderly farmers. Schopen (1987) examined the likely impact of the last scheme among 2,000 farmers and found that part-timers were most interested, possibly reflecting a selectivity effect. However, the scheme was of interest to operators with farms of all sizes in all areas, with the crucial influencing factor being the rate of compensation. West Germany is advocating a figure of around £400 (588 ECU) per hectare to set aside arable land, whereas the United Kingdom has agreed on rates of between £150 (220 ECU) and £200 (294 ECU) per hectare (depending on the use to which the retired land is put and whether or not it is in an LFA). Such wide discrepancies between member states could lead to marked national differences in the uptake of land-retirement schemes.

The distribution of abandoned farmland will depend on the answers to three related questions: how much land? which land? and which farmers?

HOW MUCH LAND?

Estimates of the amount of land that is surplus to requirements in the EC10 range from 7 to 12 million hectares (Buckwell, 1986b; Conrad, 1987; Harvey, 1987). Employing data on the balance between supply and utilisation for five major surplus crops in the EC, Buckwell calculated that

Table 13.2 *Land expected to be set-aside from wheat, barley and sugar-beet, EC10*

Country	Wheat	Barley	Sugar-beet	Total		% of UAA	Cereals self-sufficiency
		('000 ha)			(%)		(%)
West Germany	471	380	67	918	17.4	7.6	95
France	1,411	449	84	1,944	36.9	6.2	192
Italy	484	66	37	587	11.1	3.3	82
Netherlands	38	8	19	65	1.2	3.1	30
Belgium	54	25	19	98	1.9	6.9	55
Luxembourg	0	4	0	4	0.1	3.1	55
UK	493	417	32	942	17.9	5.0	118
Ireland	17	63	6	86	1.6	1.5	89
Denmark	55	277	12	344	6.5	12.1	117
Greece	221	60	6	287	5.4	3.1	105
Total	3,244	1,749	282	5,275	100.0	5.2	116

Source: Buckwell (1986b, p. 10) and Bowler (1987, p. 4).

8.75 million hectares (or 8.5 per cent of the UAA of the EC10) need to be retired to eliminate recent levels of overproduction. As the author remarks, this is equal to the combined UAA of Denmark and Ireland, or 43 per cent of the UAA of the United Kingdom. The 8.75 million hectares consist of 5.2 million from wheat and barley, 2.32 million from milk, 0.94 million from beef and veal, and 0.28 million from sugar-beet.

Arguing that set-aside is not a practical instrument for milk, beef and veal, Buckwell (1986b) estimated the amount of land that needs to be set aside from wheat, barley and sugar-beet in each of the ten EC countries (Table 13.2). Of particular interest are the final three columns in this table. At first sight it appears that France, accounting for nearly 37 per cent of the total land needing to be set aside in the EC10, should bear the largest burden. West Germany and the United Kingdom make the next largest contributions (at 17.4 and 17.9 per cent, respectively), followed by Italy (11.1 per cent). However, in terms of the percentage of UAA, Denmark is asked to make the largest sacrifice (12.1 per cent), followed by West Germany (7.6 per cent), Belgium (6.9 per cent) and France (6.2 per cent). The figure for the United Kingdom is very close to the EC average of 5.2 per cent, although this is based on a conservative estimate of 0.94 million hectares of land being made redundant. If the assessment is based on overproduction, as measured by self-sufficiency

levels in cereals, yet a different combination of countries emerges. France again leads the way (192 per cent self-sufficient in cereals), but this time followed by the United Kingdom (118 per cent) and Denmark (117 per cent). However, it is not a foregone conclusion that countries overproducing will be the ones to make the necessary adjustments. As Bowler (1987) suggests, these countries may take the view that, because they hold an economic advantage in the production of cereals, adjustments should be made in countries such as the Netherlands, Belgium and Ireland, which account for a small proportion of EC agricultural production.

WHICH LAND?

Although it has already been suggested that land-retirement schemes will be most successful in marginal agricultural areas, the situation is not that straightforward. At the outset, it is important to distinguish between agricultural and environmental set-aside (Potter, 1987). The former, which has few environmental benefits, aims to control agricultural production through acreage-reduction programmes and is favoured by EC ministers. It is arable-specific and of a voluntary and possibly temporary nature; a random pattern of uptake is likely, which coincidentally might result in an 'optimal' distribution (Buckwell,

1986b). Environmental objectives become apparent only in the use(s) to which abandoned land can be put. Many authors (Buckwell, 1986b; Potter, 1987; Burnham *et al.*, 1986; Baldock and Conder, 1987) believe this to be a missed opportunity, as agricultural set-aside could also have been presented as an environmental policy.

Environmental set-aside attempts to link land retirement to explicit environmental objectives. To be successful, it would need to be permanent and geographically sensitive, with strict controls on alternative uses of land. Specific locations could be 'targeted', on the basis of criteria such as conservation and recreational potential, environmental vulnerability and land suitability (where there is a mismatch between land use and land quality). A scheme of this nature has been outlined for the United Kingdom by Green and Potter (1987), who suggested that environmental set-aside could operate at three scales — the field, headland (a zone of certain width around the edge of fields), and parts of farms or whole farms. Clearly, the two types of set-aside have different objectives — economic and ecological — and neither satisfies both, which is disappointing when farming and the countryside should be economically and ecologically inseparable (Baldock and Conder, 1987).

Concentrating upon agricultural set-aside, a key variable likely to affect which land is involved is the level of compensation. Member states are able to vary the rate of aid offered in different parts of their territory. The United Kingdom, in the first formal set-aside scheme to be submitted to the EC, has offered farmers in the LFAs £20 per hectare less to retire land than farmers elsewhere. Despite this, the scheme is unlikely to be very popular in productive areas because the rate of compensation is quite low: £200 per hectare for permanent fallow or woodland, £180 for rotational fallow, or £150 for non-agricultural uses. Indeed, the initial uptake was very disappointing, with only 2,000 farmers (offering just 61,000 hectares) participating in the scheme in 1989 (Collins, 1988). With certain regions likely to be excluded from the programmes altogether, land retirement appears to be aimed at the marginal cereal areas (where the rate of compensation is effectively worth more than in the prosperous agricultural areas). In a study of the effects of changes in the CAP on British agriculture, Harvey *et al.* (1986) indicated that up to 1 million

hectares of 'low gross margin' land were at risk. Such land was especially concentrated in the lowland areas of the East and West Midlands, and South-West England. In contrast, the more marginal agricultural areas showed signs of demonstrating remarkable stability.

The pattern of land retirement is also likely to reflect the position of different farm enterprises within the larger agricultural system of each region in the EC. For example, a study of changes in the number of different types of farm in England between 1976 and 1985 identified counties with a weak competitive ability in beef and cereals (Bowler and Ilbery, 1989). Compared with other areas, these counties were more likely to be involved in extensification and set-aside schemes (Table 13.3). But it was also found that such developments were unlikely to be associated exclusively with weak competitive ability. In particular, counties with a strong 'competitive component' in mixed and part-time farming (and thus with viable alternative land uses) were shown to be potentially strong candidates for land-retirement and diversion schemes. Potter and Gasson (1987), in a survey of reactions to land abandonment, similarly found that farmers with mixed farms were likely to set aside larger areas than those operating purely arable or livestock farms, although this was partly a function of the larger size of mixed farms in the sample. Three regional groupings of counties emerge from Table 13.3: South-West, South-East, and East-Central England. Unfortunately, these do not coincide with areas that would have been 'targeted' in a policy of environmental set-aside based on the criteria previously listed. Similar analyses could usefully be undertaken for other regions in the EC and at sub-regional levels where, for example, the influence of large cities on potential rates of land retirement could be examined.

WHICH FARMERS?

Spatial diffusion theory has identified certain regularities in the spread of agricultural innovations (Brown, 1981; Ilbery, 1985). These include a neighbourhood effect, where a clustering of adopters occurs around an innovator or diffusion agency, and a hierarchical effect, whereby larger farms adopt an innovation before smaller ones. Research has also indicated that innovators tend

Table 13.3 *Counties in England with (A) a low and (B) a high competitive ability, by farm type*

| Type of farm: | | | | |
A: Beef	A: Beef with sheep	A: Cereals	B: Mixed	B: Part-time
Kent	Suffolk	Cornwall	Avon	West Sussex
Durham	Cambridgeshire	Devon	Cheshire	Surrey
Northumberland	Lincolnshire	Cheshire	East Sussex	East Sussex
Humberside	Bedfordshire	Dorset	Warwickshire	Dorset
Shropshire	Humberside	Somerset	Somerset	Hampshire
Hertfordshire	Northamptonshire	East Sussex		Berkshire
		West Yorkshire		Bedfordshire

Source: Bowler (1987, p. 10).

to be young and share characteristics such as a high level of education and social status. They are also associated with large and often specialised operations (Jones, 1975). Clearly, different farm and farmer characteristics will affect the pattern of land abandonment in the EC.

Work by Munton *et al.* (1987) and Potter and Gasson (1987) has indicated that 'accumulators' or 'enabled' farmers (with larger, well-established owner-occupied farms) show less resistance to change than 'survivors' or 'constrained' farmers (with small farms, limited choice of enterprise and high vulnerability to policy changes). In Potter and Gasson's survey in England, the amount of land offered for retirement, and the level of financial compensation demanded, varied spatially. For example, farmers in Suffolk (a prosperous agricultural area) wanted more money to enrol fewer hectares than farmers on the South Downs (an ESA). There was little significant difference according to the age and education of the farmers, but stage in the family life cycle was important in influencing both the level of the bid and the amount of land offered. Farmers with sons and daughters of working age and living at home were prepared to set aside more land than other farmers; this again reflected a farm-size effect. Enabled farmers were shown to offer more land than either moderately or severely constrained farmers. As Potter and Gasson (1987, p. 25) remark, 'this throws some doubt on the ability of voluntary schemes to initiate land use changes on farms which lack a history of conservation or woodland management, for instance'. Close monitoring of the type of farm and farmer registering for set-aside and extensification

programmes will be needed if the locational aspects of land abandonment are to be fully understood.

Alternative uses for abandoned land

Land scarcity is no longer a problem for agriculture in the EC. Instead, decisions regarding the use(s) to which retired farmland should be put have to be made. A wide range of alternative crops and livestock, from borage and teasels to lamoids and snails, have been discussed (Carruthers, 1986; Conrad, 1987). Most are not economically viable at present, and certainly would not favour the smaller-scale farmer in marginal agricultural areas. Others require specific climatic conditions and are possible only in particular parts of the EC. For example, lentils, chickpeas and vetches, allowable on land set aside for an experimental period of three years, are suited mainly to southern climates.

Realistically, there is a limited number of alternative uses to which set-aside land can be put. The choice lies between some kind of fallowing (grazing, rotational or green) and various types of 'farm diversification', such as afforestation, tourism, recreation and conservation (Figure 13.3). Conrad (1987) maintains that fallowing could serve important nature-conservation purposes in areas with marginal soils. Bare fallowing should not be encouraged because the risk of soil erosion is increased and there are few wildlife benefits. Rotational fallowing, where a farmer is able to set aside different parcels of land each year as part of a traditional arable rotation, is favoured

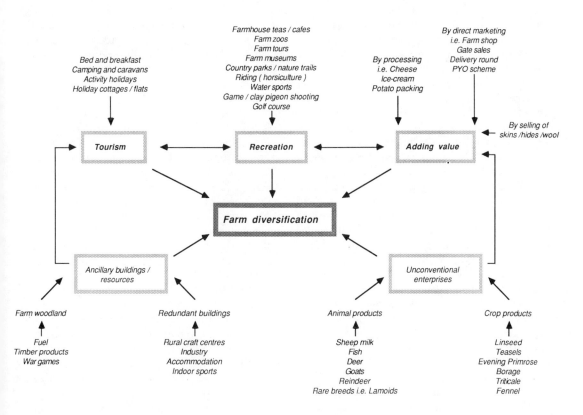

Figure 13.3 *Schematic flow diagram of the major types of farm diversification*
Source: Ilbery (1988), p. 36.

by farming bodies and could be attractive in areas with higher-grade soils. Indeed, the set-aside scheme in the United Kingdom distinguishes between permanent fallow, where the same parcel(s) of land are set aside for the full five-year period, and rotational fallow, which attracts £20 per hectare less in compensation. With either option, the land must be kept in good agricultural condition, either by sowing a green cover crop, or by allowing the naturally occurring vegetation to regenerate. The use of fertilizers and pesticides is, as a general rule, prohibited.

The use of a ground-cover crop is commonly referred to as 'green fallowing', a system favoured by West Germany. A trial scheme, introduced in Lower Saxony in 1986, involved paying farmers between £360 and £430 (530 and 632 ECU) per hectare to transform arable land into green fallow on a permanent or rotational basis. Full- and part-time farmers were eligible, with a minimum of one hectare and a maximum of 20 hectares to be

fallowed. The 'green' plant cover could not be fertilized or sprayed with pesticides, marketed or used as fodder. What must be noted is that — at least initially — the results of this scheme were not encouraging. Less than 7,000 farmers (working 32,000 hectares) took part in the first year, and uptake was concentrated mainly in areas of marginal (heath) land, suggesting that the rate of compensation was too low for farmers in the more productive regions. Conversely, however, a substantial increase in aid in 1987 resulted in too many applicants, proving that financial incentive is a crucial factor in land-retirement schemes.

Grazed fallow permits retired land to be used for extensive livestock farming. Although it is not permitted under the UK's set-aside scheme, elsewhere it has been allowed for a three-year trial period, on the basis of French pressure. Farmers will receive a lower rate of compensation than for other types of fallow and must ensure that stocking levels are not increased. But grazed fallow is a controversial use for surplus land because, while it can be environmentally beneficial, there are major problems over monitoring and knock-on effects on the livestock (beef and sheep) sector.

Forestry and non-agricultural uses are the other alternatives for retired land. The EC imports much of its timber, and farm woodland is a permitted option under the set-aside scheme. Indeed, a farm woodland scheme was introduced in the EC in 1986 to encourage the long-term planting of trees. In much of the EC this policy can build on a substantial farm woodland tradition, but in the United Kingdom this tradition and associated management expertise are largely absent. The UK is, however, attempting to expand farm woodland, especially in lowland areas, through both a woodland option in the set-aside scheme and a Farm Woodland Scheme. In the latter, planting grants are available from the Forestry Commission, with higher rates for broadleaved trees than for conifers. In addition, farmers will receive an annual payment of £190 per hectare (280 ECU) for up to 40 years and a maximum of 40 hectares. (These rates fall to £150 or 220 ECU in disadvantaged areas, and £100 or 147 ECU in severely disadvantaged areas.) The scheme began in October 1988 and is limited to 36,000 hectares over the first three years. Yet, despite these efforts, the signs are that in practice forestry has difficulty in competing with more profitable land uses.

Retired land in the EC could profitably be used for alternative activities such as golf courses, riding schools, livery stables, nature reserves, caravan and camping sites, football pitches and tourist facilities (Figure 13.3). Such 'diversification' has received considerable publicity and academic interest in recent years (Slee, 1987; Ilbery, 1988), and tourist and craft activities have been encouraged in the LFAs since 1975. The United Kingdom is again one of the leaders in this field. British farmers can receive £150 per hectare (220 ECU) through the set-aside scheme to develop non-agricultural alternatives. In LFAs, the grant falls to £130 (190 ECU). In addition, capital and marketing grants are available through the national Farm Diversification Scheme introduced in 1988.

Interest in rural recreation and tourism has increased in recent years (see Chapter 15). Farm-based activities would seem to be a more viable option in physically marginal (tourist) areas, and

on the fringes of major metropolitan areas, than in the more prosperous agricultural lowlands, where land can be either fallowed or afforested. A spatial mismatch therefore appears to exist between supply and demand: the demand for outdoor recreation is highest in some of the most densely populated lowland areas of the EC, whereas the greatest supply of, for example, national parks and nature reserves is in the more marginal uplands. Davies (1983) estimated that over 4,000 farms in the LFAs of England and Wales (20 per cent of the total) offer tourist and recreational facilities, although only 30 per cent of them earn more than 5 per cent of their total income in this manner. This relatively low figure reflects the fact that farm tourism is usually supplementary to, and not fully integrated into, the farm business.

Other factors likely to affect the distribution of non-agricultural uses include farm size and the stage reached by the family in its life cycle. Most types of farm diversification tend to be introduced by farmers with larger holdings (Gasson, 1987; Robson, 1987) and/or older children (Frater, 1982; Davies, 1983). In contrast, off-farm employment and home businesses are more typical of smaller farms, leaving little time for alternative uses on any retired land. The United Kingdom is lagging behind her EC partners in terms of farmers with other gainful activities (OGAs) on and off the farm; less than 25 per cent have an OGA, compared with 43 per cent in West Germany and 40 per cent in Greece.

Interest in 'alternative' land uses is likely to vary spatially, within the EC generally and in individual countries in particular. Notable contrasts between urban fringes, marginal fringes and prosperous agricultural lowlands will be complicated by the influence of different farm and farmer characteristics. Various 'resistance' factors — ranging from farmers' attitudes, location and tenancy restrictions, to rural planning controls and a lack of finance and marketing skills — will also complicate the picture (Ilbery, 1988).

Conclusions

The CAP has been, and still is, dominated by a system of high support prices. This has distorted the normal economic relationship between supply and demand by increasing the former at a much greater rate than the latter. Faced with the mounting costs of food storage and disposal, agriculture ministers have at last recognised the need to control agricultural production. A package of measures, ranging from milk quotas and stabilisers to land retirement, conversion and extensification, is now on the statute books. Significantly, there have been no cereal quotas or substantial cuts in support prices.

American experience of acreage-reduction programmes has not been encouraging (Potter, 1986; Ervin, 1988) as problems associated with slippage and the selectivity effect have reduced their potential. The EC is able to learn from such weaknesses, although its own set-aside scheme appears problematical and does not address the root cause of overproduction. Regulation 1094/88 controlling set-aside is essentially an agricultural policy with few environmental objectives; it is a means of 'buying time over the problem of cereal surpluses' and 'a temporary device of market management' (Bowers, 1987, p. 8). Rates of compensation are to be fixed by member states, and the limited evidence so far available suggests that these will be too low to attract the more productive farmers in the prosperous agricultural lowlands of northern Europe. Certain authors have advocated a national system of sealed bids by the farmers themselves (Potter and Gasson, 1987; House of Lords Select Committee, 1988). Not only would this be more economic, but regional controls on the amount of land to be set aside could be imposed, to limit the impact of the more competitive (cheaper) bids from farmers in marginal areas. Whatever the system, inadequate money appears to have been allocated by the EC to land-retirement schemes.

The real solution to agricultural overproduction in the EC does not lie in set-aside schemes. Cereal quotas would be more effective in controlling cereal supplies, as has been demonstrated by the effect of milk quotas on the butter 'mountain' (Erlichman, 1988). However, most agricultural economists are agreed that the only answer to the problem lies in major cuts in CAP prices (possibly linked to set-aside schemes), increases in returns for non-agricultural uses, and incentives to encourage environmentally sensitive farming (Buckwell, 1986b; Bowers, 1987; Harvey, 1987). Such a solution is, of course, unacceptable to most farmers and ministers and, consequently, the

problem will remain unresolved. Agricultural dualism — under which intensive, market-oriented production in favoured areas contrasts with marginal, locally oriented production offering few realistic employment opportunities in less-favoured areas — will continue. Much more attention needs to be devoted to the spatial diversity of agriculture in the EC, and to the regional and local impacts of alternative uses of land. Such enormous variations necessitate spatially explicit measures rather than uniform policy approaches.

Government policies have interfered with the natural process of agricultural development, and government policies will have to solve the problems of the CAP. This does indeed represent a major challenge and one which cannot be solved by land-retirement schemes alone. As Harvey (1987, p. 18) aptly remarks, 'schemes to encourage the transfer of land out of production are an attempt to push the agricultural system in an unnatural direction'.

Notes

1. 'Green' in this context refers to CAP financial transactions, not to environmental issues. Green rates were originally the same as the market rates of exchange for each national currency. However, many member states floated their currencies in the late 1960s and early 1970s, but refused to alter their green rates of exchange accordingly. This reflected fears as to the likely adverse effect adjustments would have on farm incomes.
2. MCAs were introduced to take account of the divergence between national 'green' and market rates of exchange in the context of agricultural trade (see Bowler, 1985b, for further details).
3. The ECU is the European unit of account. In 1988, one ECU was equivalent to 68p.
4. The composite indicator is weighted in the ratio of 75 per cent from the GDP per person, and 25 per cent from the share of agriculture in total employment.

References

Baldock, D. and Conder, D. (eds) (1987) *Removing Land from Agriculture*, Council for the Protection of Rural England and Institute for European Environmental Policy, London.

Bowers, J. (1987) Set-aside and other stories. In D. Baldock and D. Conder (eds), *Removing Land from*

Agriculture, Council for the Protection of Rural England and Institute for European Environmental Policy, London, pp. 5–18.

Bowler, I.R. (1985a) Some consequences of the industrialization of agriculture in the European Community. In M.J. Healey and B.W. Ilbery (eds), *The Industrialization of the Countryside*, GeoBooks, Norwich, pp. 75–98.

Bowler, I.R. (1985b) *Agriculture Under the CAP: a geography*, Manchester University Press, Manchester.

Bowler, I.R. (1986) Intensification, concentration and specialisation in agriculture: the case of the European Community, *Geography*, 71, 14–24.

Bowler, I.R. (1987) Locational aspects of Council Regulation (EEC) No. 1760/87. In N.R. Jenkins and M. Bell (eds) *Farm Extensification: implications of EC Regulation 1760/87*, Research and Development Paper 112, Institute of Terrestrial Ecology, Grange-over-Sands.

Bowler, I.R. and Ilbery, B.W. (1989) The spatial restructuring of agriculture in the English counties, 1976–1985, *Tijdschrift voor Economische en Sociale Geografie*, 80, (forthcoming).

Brown, L.A. (1981) *Innovation Diffusion: a new perspective*, Methuen, London.

Buckwell, A. (1986a) *Controlling Cereal Surpluses by Area Reduction Programmes*, Discussion Paper in Agricultural Policy 86/2, Department of Agricultural Economics, Wye College, London.

Buckwell, A. (1986b) What is a set-aside policy?, *Ecos*, 7, 6–11.

Burnham, P., Green, B. and Potter, C. (1986) *A Set-aside policy for the United Kingdom*, Set-aside Working Paper 3, Wye College, London.

Carruthers, S.P. (ed) (1986) *Alternative Enterprises for Agriculture in the United Kingdom*, Centre for Agricultural Strategy, Reading University, Reading.

Collins, R. (1988) 2,000 take up cereal cash, *The Guardian*, 26 October, p. 12.

Commission of the European Communities (1985) *Perspectives for the Common Agricultural Policy*, Commission of the European Communities, Brussels.

Commission of the European Communities (1986) *Europe's Common Agricultural Policy*, Commission of the European Communities, Brussels.

Conrad, J. (1987) Alternative land-use options in the European Community, *Land Use Policy*, 4, 229–42.

Davies, E.T. (1983) *The Role of Farm Tourism in the Less Favoured Areas of England and Wales*, Report 218, Department of Agricultural Economics, Exeter University, Exeter.

Delormé, H. (1987) An outline of French views on land conversion programmes. In D. Baldock and D. Conder (eds), *Removing Land from Agriculture*, Council for the Protection of Rural England and Institute for European Environmental Policy, London, pp. 40–2.

Dessylas, D. (1987) The Commission's proposals adapting agriculture to new market conditions — preservation of the countryside. In D. Baldock and D. Conder (eds), *Removing Land from Agriculture*, Council for the Protection of Rural England and Institute for European Environmental Policy, London, pp. 31–4.

Erlichman, J. (1988) Butter mountain melts but gloom spreads to cheese, *The Guardian*, 9 August, 3.

Ervin, D.E. (1987) Cropland diversion in the US: are there lessons for the EEC set-aside discussion? In D. Baldock and D. Conder (eds), *Removing Land from Agriculture*, Council for the Protection of Rural England and Institute for European Environmental Policy, London, pp. 53–63.

Ervin, D.E. (1988) Set-aside programmes: using US experience to evaluate UK proposals, *Journal of Rural Studies*, 4, 181–91.

Frater, J. (1982) *Farm tourism in England and overseas*, Memorandum 93, Centre for Urban and Regional Studies, University of Birmingham, Birmingham.

Gasson, R. (1987) The nature and extent of part-time farming in England and Wales, *Journal of Agricultural Economics*, 38, 175–82.

Green, B. and Potter, C. (1987) Environmental opportunities offered by surplus production. In D. Baldock and D. Conder (eds), *Removing Land from Agriculture*, Council for the Protection of Rural England and Institute for European Environmental Policy, London, pp. 64–71.

Harvey, D.R. (1987) Extensification schemes and agricultural economics: who will take them up? In N.R. Jenkins and M. Bell (eds) *Farm Extensification: implications of EC Regulation 1760/87*, Research and Development Paper 112, Institute of Terrestrial Ecology, Grange-over-Sands.

Harvey, D.R. *et al.* (1986) *Countryside Implications for England and Wales of Possible Changes in the CAP*, Centre for Agricultural Strategy, University of Reading, Reading.

Healey, M.J. and Ilbery, B.W. (eds) (1985) *The Industrialization of the Countryside*, GeoBooks, Norwich.

Hill, B.E. (1984) *The Common Agricultural Policy: past, present and future*, Methuen, London.

House of Lords Select Committee on the European Communities (1988) *Set-aside of Agricultural Land, 10th Report, Session 1987–88*, HL65, HMSO, London.

Ilbery, B.W. (1985) *Agricultural Geography: a social and economic analysis*, Oxford University Press, Oxford.

Ilbery, B.W. (1986) *Western Europe: a systematic human geography*, Oxford University Press, Oxford.

Ilbery, B.W. (1988) Farm diversification and the restructuring of agriculture, *Outlook on Agriculture*, 17, 35–9.

Jenkins, N.R. and Bell, M. (eds) (1987) *Farm Extensification: implications of EC Regulation 1760/87*,

Research and Development Paper 112, Institute of Terrestrial Ecology, Grange-over-Sands.

Jones, G.E. (1975) *Innovation and Farmer Decision-making*, D203 Agriculture, The Open University, Milton Keynes.

Munton, R.J., Eldon, J. and Marsden, T. (1987) Farmers' responses to an uncertain policy future. In D. Baldock and D. Conder (eds), *Removing Land from Agriculture*, Council for the Protection of Rural England and Institute for European Environmental Policy, London, pp. 19–30.

Potter, C. (1986) *Environmental Protection and Agricultural Adjustment: lessons from the American experience*, Set-aside Working Paper 1, Wye College, London.

Potter, C. (1987) Set-aside; friend or foe?, *Ecos*, 8, 36–9.

Potter, C. and Gasson, R. (1987) *Set-aside and Land Diversion: the view from the farm*, Set-aside Working Paper 6, Wye College, London.

Revel, B.J. (1987) The extensification scheme and structural policy: the lessons of previous experience. In N.R. Jenkins and M. Bell (eds), *Farm Extensification: implications of EC Regulation 1760/87*, Research and Development Paper 112, Institute of Terrestrial

Ecology, Grange-over-Sands.

Ritson, C. and Tangerman, S. (1979) The economics and politics of monetary compensatory amounts, *European Review of Agricultural Economics*, 6, 119–64.

Robson, N. (1987) The changing role of part-time farming in the structure of agriculture, *Journal of Agricultural Economics*, 38, 168–75.

Schopen, W. (1987) Removing land from agriculture – the German view. In D. Baldock and D. Conder (eds), *Removing Land from Agriculture*, Council for the Protection of Rural England and Institute for European Environmental Policy, London, pp. 35–9.

Slee, R.W. (1987) *Alternative Farm Enterprises*, Farming Press, Ipswich.

Smith, E.J.G. (1987) Initiatives on alternative land use. In D. Baldock and D. Conder (eds), *Removing Land from Agriculture*, Council for the Protection of Rural England and Institute for European Environmental Policy, London, pp. 43–8.

Stanford, T. (1988) Voluntary set-aside: which way to turn, *Farmland Market*, February, 13–15.

Winchester, H. and Ilbery, B.W. (1988) *Agricultural Change: France and the EEC*, John Murray, London.

14

Counterurbanisation: threat or blessing?

A.J. Fielding

Introduction

This chapter reflects on the nature and implications of a number of fundamental changes that have occurred in rural Western Europe during the post-war period. It begins by contrasting the rural landscapes of 1950 with those of the late 1980s and moves on to examine related trends in population redistribution. Particular emphasis is given to the way in which a greater concentration of population (urbanisation) came to be replaced, sometime around 1970, by the deconcentration of population or counterurbanisation. Causes of this reversal are considered, and a model is used to link the rise of counterurbanisation with changes in the geography of production. Likely demographic, economic and social effects of counterurbanisation are then explored, and the chapter ends with a discussion of the social costs and benefits which the process has brought to many rural regions of Western Europe.

The transformation of rural landscapes 1950–90

How can one characterise the rural landscapes of Western Europe in 1950? The answer to this question is 'with some difficulty' since the rural landscapes of that time varied greatly from one region to another. This variety arose in part because of the uses to which the land was put; separate, distinctive landscapes were produced by each of the main branches of agricultural activity — viticulture, cereal and beet production, horticulture, dairying and other forms of livestock farming and arboriculture. But in addition, and equally important, there were variations in the social relations of production, many of which were related to differences in land tenure. Thus there were distinctive peasant farming landscapes, such as those of north-west France and northern Portugal; landscapes dominated by large estates, such as parts of southern Italy and south-west Spain; and landscapes of 'modern' commercial farming, examples being much of eastern England, Denmark and the Netherlands. And yet, despite these differences, most rural landscapes revealed the combined and multiple effects of two overriding circumstances: first, the very large or even total dependence of rural areas on agriculture and, second, a state of crisis which permeated the agricultural sector.

Heavy dependence upon agriculture meant that farm and forest land-uses predominated in all rural areas, other than those in which, usually for physical environmental reasons, large tracts of land were left uncultivated. It also meant that the settlements in rural areas remained agriculturally based. Their non-agricultural populations were, for the most part, linked to agriculture; they processed the products of agriculture, traded in agriculture goods, and provided services to agricultural households. The crisis in agriculture meant that rural landscapes contained the manifestations of what was, in many areas of Western Europe, a desperate rural poverty. Marginal land was going out of cultivation, levels of farm mechanisation were low, rural roads were in a very poor condition, farm buildings were often dilapidated, and houses were typically poorly equipped and often structurally unsound. There were, in addition, many signs that poverty was causing rapid rural depopulation. Small holdings, and sometimes even whole villages, were being abandoned. Village schools and shops were being closed, and the small towns providing local services were declining. Most of the rural areas of Western Europe at this time were 'no hope' areas. To get on, you had to get out. And because so many young people and young adults had decided to do just that, the human figures in our rural landscape were increasingly elderly as well as poor.

The rural landscapes of Western Europe today (and especially those in north-western Europe) are remarkably different! While rural depopulation continues in some remote areas, and few would be so foolish as to claim that rural poverty had disappeared (farm labourer incomes remain at, or near to, the bottom of the earnings table), it remains clear that the prospects for rural areas have dramatically improved. First, agriculture has almost universally become a business and, for some farmers, a highly profitable business at that. In this way the diversity of rural social relations typical of the 1950s has been much reduced, to be largely replaced by the relationships of a capital-intensive, market-sensitive and technologically sophisticated commercial agriculture. This high-productivity sector now employs far fewer workers than it did in the 1950s; consequently, even in the most rural regions of Western Europe, employment in agriculture is typically smaller than that in the services and is often smaller than in manufacturing industry. But second, and even more important, there is the fact that this manufacturing and service employment is now no longer based on agriculture — it uses non-agricultural inputs to produce goods and services for (very predominantly) non-agricultural consumers. Thus, although much of the landscape is still dominated by farming (albeit in much of north-western Europe a rationalised, machine-modified landscape), few of the buildings are farm, or farm-related, structures. They are occupied by people who live in the countryside, but whose jobs and lifestyles are fully 'urban'. Their jobs are urban in the sense that they conform to that sophisticated division of labour and interdependence of tasks which economic historians regard as both a prerequisite for, and a product of, urbanisation (Lampard, 1955). Their lifestyles are urban in that, however strongly motivated they are to realise that 'village in the mind', they are in fact members of a much wider

society or *Gesellschaft*, rather than of the local rural community or *Gemeinschaft* (Mellor, 1977).

The origins of this urbanisation of the countryside go back further than 40 years, and it was almost 30 years ago that Ray Pahl (1965) studied the social transformation of villages in Hertfordshire, England, in his justly celebrated monograph *Urbs in Rure*. At that time the *embourgeoisement* of the rural areas was very largely confined to the zones within commuting distance of the centres of the largest metropolitan cities, but after the mid-1960s it was not just these areas which experienced the transformation. It was rural areas in general, with new manufacturing and service jobs being located in free-standing cities, and in small and medium-sized towns in rural regions (Fothergill and Gudgin, 1982). All this can be read in the landscape, not just in lowland Britain but over much of Western Europe. Farm houses and farm workers' cottages have been gutted and modernised throughout, and now stand in manicured garden plots, fronted by expensive cars parked on gravel driveways. Former farm buildings such as barns and haylofts, and former schools, shops and village halls have been converted into comfortable middle-class residences. Small estates of new houses have appeared in villages and small towns where virtually no new buildings had been seen for a century or more. The local towns are usually prosperous, and now have shopping centres, modern public buildings and decent roads. In addition to all this, there is almost invariably a new industrial estate, tucked away around the 'back' of the town. This landscape reflects a degree of rural prosperity undreamt of in the 1950s. But, in particular, it reflects the population rejuvenation and net migration gains which almost inevitably accompany such prosperity. So great, and yet so unheralded, has been this transformation of rural areas in Western Europe, that we can correctly describe it as the 'quiet revolution' (Ambrose, 1974).

Population redistribution in Western Europe since the 1950s

Many processes contributed to this quiet revolution. Some were very general in character, notably the overall improvements in living standards in Western European countries during the long post-

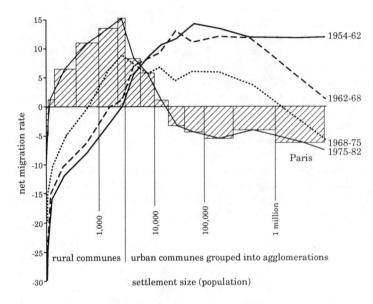

net migration rate

1,000　10,000　100,000　1 million

rural communes | urban communes grouped into agglomerations

settlement size (population)

1954-62

1962-68

1968-75
1975-82

Paris

Figure 14.1 *France: annual net migration rates, 1954–82, per thousand population by settlement size category*
Note: The shaded areas are proportional to the absolute gains and losses in each size category.
Source: Censuses

war economic boom (Chapter 1). Others were specific to rural areas, for example the restructuring of agricultural production. But central to this transformation was the drastic reduction of rural depopulation, and its replacement in many rural areas by population growth. This population 'turnround' in rural areas was not restricted to Western Europe; it was a general feature of Western industrial societies in the 1960s and 1970s, and it is now often depicted as a replacement of urbanisation by counterurbanisation (Fielding, 1986a).

Although there continues to be some debate about the term 'counterurbanisation', there is no longer much doubt about the changes to which it refers, and there is less and less uncertainty about what caused it. Counterurbanisation refers to a pattern of population redistribution in which the larger settlements decline (or stagnate) in their populations due to net migration losses, while the smaller ones, such as villages and small or medium-sized 'free-standing' towns, increase their populations through net migration gains. It implies, therefore, a population deconcentration or decentralisation. However, counterurbanisation is not in every respect a simple reversal of what went before. The period of urbanisation saw massive shifts of people out of agriculture, and into manufacturing and service employment.

Counterurbanisation does not imply an end to this trend and, partly because of this, it is sometimes called 'decentralised urbanisation'.

The changeover from urbanisation to counterurbanisation after the Second World War can be effectively demonstrated by recording the statistical relationship between net migration rates and settlement size, where settlements are defined as labour-market areas and size is measured in terms of resident population. Figure 14.1 shows how this relationship has changed in France over the period 1954 to 1982.[1]

In the inter-censal period 1954–62, the relationship was clearly positive (upward-sloping); the larger the place, the more likely it was that the settlement was experiencing population growth through net migration gains. Smaller (rural) places were experiencing major net migration losses at this time. As one moves forward in time the relationship changes until, in the 1975–82 period, it is the smaller places which are gaining by net migration, and the larger ones which are losing. Similar measurements have been made for each of the 14 major countries of Western Europe (that is, those lying west of a line drawn from the Baltic to the Adriatic), often at several spatial scales. The detailed results of this research are reported elsewhere (Fielding, 1982; 1986a) but the broad features can be briefly summarised.

In the 1950s the relationship between net migration rates and settlement size (measured through population density[2]) was positive in every country of Western Europe; urbanisation was the dominant trend in population redistribution during this period.

In the 1960s this positive relationship began to break down in a number of countries. In particular, the 'core' countries of north-western Europe began to show signs of a shift towards a more dispersed pattern of population growth. Suburbanisation was very important at this time, but what is being referred to here is much more than suburbanisation. By the end of the 1960s it was clear that decentralisation beyond the commuting fields of the major cities was becoming an important element of the population redistribution patterns of a number of countries.

The early and mid-1970s saw the relationship between net migration and settlement size become quite clearly and unequivocally negative in a majority of countries in Western Europe. Countries (and regions) which were outside the 'core' area, and which still retained sizeable proportions of their total workforces in agriculture, saw their positive relationships between net migration rates and settlement size continue through the 1960s and sometimes into the 1970s. But if one takes the 1970s as a whole, it can be said that in only one country, Spain, did urbanisation dominate. In six other countries (Austria, Ireland, Italy, Norway, Portugal and Switzerland) the urbanisation relationship had come to an end (sometimes because of the tourist development of sparsely populated regions), to be replaced by a situation in which no clear relationship between net migration and settlement size was discernible. In the remaining seven countries (Belgium, Denmark, France, the Netherlands, Sweden, the United Kingdom and West Germany) the relationship was unequivocally negative; in these countries counterurbanisation had replaced urbanisation as the dominant form of population redistribution.

Since the early 1980s the situation has changed again. Four countries (Austria, France, Portugal and Spain) can provide us with virtually no information on population redistribution in this period, but the main feature of the remaining ten is that the counterurbanisation trends of the 1970s have tended to diminish. In Norway and Sweden this movement away from counterurbanisation has proceeded so far that a significantly positive

relationship between net migration and settlement size has become re-established. In contrast to this, in Italy and West Germany it seems that counterurbanisation continued at least into the early 1980s. But, for the majority of countries, no significant association between net migration and population density can be discerned.[3] This does no imply, however, that there is an absence of spatial pattern in the net migration figures. In almost every country, the 1980s witnessed the replacement of an urban–rural redistribution pattern by one possessing a broader regional character. The chief feature of this was that a majority of places (irrespective of settlement size) in one region were experiencing net migration gains, while a majority of places in another region were making net migration losses (Fielding, 1989a). This is nicely demonstrated in the migration patterns for West Germany and Italy (Figure 14.2).

The implications of these redistributions for rural areas in Western Europe are, at one level, fairly obvious. In the 1950s and early 1960s, with the exception of a small number of tourist areas, only the rural areas in close contact with the largest cities experienced net migration gains, and then only through suburbanisation. In the late 1960s and the 1970s, however, there was a general (though not universal) bias towards net migration gains in rural areas, and in the free-standing small and medium-sized towns found in rural and peripheral regions. Subsequently, this general bias towards rural areas has been replaced by net gains in the rural areas of high-growth regions only. In place of the relatively unselective counterurbanisation of the early 1970s, we see a pattern of rural net migration gains which is highly region-specific. Because these gains are not limited to settlements which have become part of the commuting zones of the largest cities, we might be tempted to speak of a continuation of counterurbanisation but, if so, it is clear that it is a different kind of counterurbanisation and that the former spatial generality of this process has been lost.

It is one thing to know the facts about population redistribution trends, but quite another to recognise the social content of the redistribution. So, before we can say anything sensible about the social consequences of counterurbanisation, we need to develop an understanding of the processes which brought it about, and which might indicate

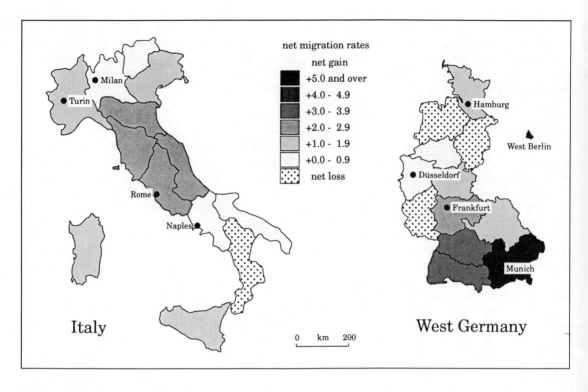

Figure 14.2 *Italy and West Germany: annual net migration rates, 1985/6, per thousand population*
Source: Registration data from Government Statistical Offices.

the kinds of people who were involved, and the circumstances of their migration. Such an understanding can only emerge from a discussion of counterurbanisation theories.

Counterurbanisation theories

During the 1970s, as it came to be realised that urbanisation was giving way to counterurbanisation, a lively debate developed on the causes of this turnround (Fielding, 1986b). An early frontrunner was the argument that the reversal of the trend was not real, but was the product of the well-known and widely acknowledged problem of the statistical 'underbounding' of urban areas. As cities expanded they spilled over their administrative boundaries into the contiguous areas of the surrounding rural districts; the building of a single housing estate on the edge of a rural district might be more than sufficient to cancel out the net migration losses of the rest of that district. It was

soon realised, however, that the scale of the changes was much greater than this; distant regions such as western Ireland, Brittany and southern Italy were now experiencing net migration gains after a century or more of losses.

The spotlight then turned on the migrants themselves. It was argued that they were the actors in this drama, and that one must examine what was making them choose the smaller and more rural places in preference to the larger metropolitan cities. The answer seemed to lie in a widely shared anti-urbanism. This characterised cities as the sites of stress and conflict, and the countryside as the realm of harmony and sociability. A pertinent fact used to support this argument was that, at this time, some of those who were free from the constraints of work — the prosperous retired — were tending to migrate to villages and small towns in pleasant rural environments. Most commentators eventually came to accept, however, that these individual preferences, though clearly significant, were not sufficient to explain the turnround.

An alternative explanation that was advanced was that the turnround was due to the ways in which Western European governments were influencing the investment decisions of firms through their urban and regional development policies. The trouble was, if receipt of state aid had been the key factor in bringing about an altered pattern of population redistribution, why was it that the rural areas in receipt of aid experienced the turnround from net loss to net gain, when the urban industrial areas in receipt of aid did not? Today, the sensible stance to take on this issue seems to be, not that state activity was irrelevant to the emergence of counterurbanisation, but that it was probably more effective in facilitating decentralisation through mechanisms other than that of overt spatial redistribution policy. In particular, Western European governments tended to standardise the 'general conditions of production' over their territories, for example through investments in transport and communications and public utilities, and through the provision of standardised services such as those in education, health and social security. In this way they facilitated a decentralisation of economic activity (Fielding, 1982).

The view that economic activity was fundamental to the turnround was then strengthened by the emergence of a new conceptual scheme. As a result of this, the balance of opinion is now that the key decisions which affect the distribution of population are not taken by ordinary people but by large organisations, notably private sector companies, and that these organisations do not shift their geographical patterns of investment unless they expect to gain greater profitability or efficiency from doing so.

The scheme in question comes from the literature on regional industrial restructuring (Massey, 1979). While not denying the importance of changes in transport and communications, or the problems of industrial site development in major built-up areas, this political economy approach lays its greatest emphasis on changes in the organisation of production (for example, the growth of multi-plant firms), and in particular on management–labour relations.

The starting point for this approach is that, until the 1960s, the geography of production was characterised by regional sectoral specialisation (RSS). This means that each city and region tended to specialise in the production of particular goods or services. Many regions were almost entirely agricultural, some specialised in textiles, others in engineering products, others in tourism, and so on. Thus the spatial division of labour coincided with the social division of labour; each of the main branches of the economy was to be found highly concentrated in specific regions of the country.

RSS is, of course, still present in the countries of Western Europe today, but during the post-war period, and especially during the 1960s and 1970s, it came to be overlain and partially replaced by a rather different geography of production. In this hierarchical or new spatial division of labour (NSDL), the differences between regions were not to be found in the goods or services they produced, but in the functions that their workforces performed in the overall production process. Thus a small number of metropolitan regions came to be the centres for corporate decision-making and control. Semi-rural 'prestige environments' near to these major metropolitan cities often became the areas specialising in research and development activities. Existing industrial cities and conurbations became the locations of those activities which still required traditional male manual skills; and routine branch-plant production activities, often employing a high proportion of female labour, were dispersed among old industrial conurbations and the remainder of the national territory, including rural and peripheral regions. Although applying most forcefully to manufacturing industry, this separation of tasks has also been observable in services, where the 'back offices' of metropolitan city-based firms and government departments have often decentralised routine activities to the provinces (see also Chapter 7).

The social consequences of counterurbanisation

Armed with this RSS to NSDL transition model, and with the other theories of counterurbanisation discussed above, we can begin to speculate about the possible effects of counterurbanisation upon those who live in rural areas. This can be done in three stages. In the first, we can identify the expectations, arising from the theories of counterurbanisation, about the volume, direction and

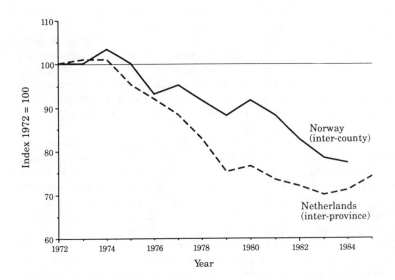

Figure 14.3 *Norway and the Netherlands: interregional migration rates, 1972–85, per thousand population*
Source: Registration data.

social content of migration flows into rural areas. In the second, we can check those expectations against data for England and Wales covering the period 1971–81. These data (uniquely) record the social class compositions of the urban–rural and rural–urban migrations, together with the social-class changes which accompanied those flows. Finally, we can list the possible effects of such migrations under two headings: those which enhance the life chances of the people who live in rural areas (the blessings); and those which diminish those life chances (the threats).

Our theoretical discussion suggests that the following migration developments affecting rural areas might be expected to accompany counter-urbanisation. First, the overall volume of migration between labour market areas might be expected to decrease. The reason for this is that, under RSS, the growth or decline in specific branches of the economy would be automatically reflected in movements of people from the regions of declining sectors to the regions of growing sectors; the depopulation of agricultural regions in the 1950s and early 1960s is a case in point. Under NSDL, however, this migration is unnecessary, not only because organisations are seeking out reserves of labour, but because regional sectoral diversification permits sectoral shifts in employment to take place without interregional migration. In particular, the migration of lower-level manual and white-collar workers would be expected to be less. Evidence from many of the

major countries of Western Europe shows that the volume of migration between labour-market areas did indeed decline in the 1970–85 period (Figure 14.3).

Second, the major migration flows should now favour free-standing cities, and small and medium-sized towns in rural and peripheral regions, rather than the metropolitan cities and industrial conurbations. This is because industrial and service restructuring usually involves a disinvestment in routine production in the large cities and an expansion of investment in smaller places (often on 'greenfield' sites). Almost all countries in Western Europe saw an urban–rural shift in manufacturing employment in the 1970s (Klaassen and Molle, 1983), and the decentralisation of service employment was also widespread. However, it is conformable with the explanation of counterurbanisation developed above that the very largest cities should, after a period of decline due to deindustrialisation, see their employment levels rise again as a result of the further accretion of company headquarters and the associated growth of producer services (see Chapter 7). This is precisely what seems to have happened in such cases as London and Paris.

Finally, the social composition of the migration flows to and from the rural areas would also be expected to change. Flows from the rural area would contain fewer young working-class men and women, but more professional, technical and managerial employees (that is, members of the 'service class'), since they are likely to be moved on by their employers to new jobs in other regions as their careers develop within the internal labour markets of large organisations. Flows to the rural areas would contain many people at or close to retirement age. This would reflect the preference frequently held by prosperous people for village and small-town environments, and the wish of many former rural out-migrants to return to their regions of origin. It would not mean, however, that these flows into rural areas would normally be dominated by the elderly; the main thrust of the counterurbanisation literature is that the turnround was related primarily to changes in the labour-market situations of rural and peripheral regions. Most of the in-migrants would, therefore, be expected to be working-age adults and their children. Thus the main components of the migration flows to the rural areas would be members of the service class relocated to these areas by their employers, and others moving in to take advantage of the new job opportunities in factories, on building sites, and in shops, offices, schools, hotels, and so on.

Given the importance of a knowledge of the social composition of migration flows for assessing the social consequences of counterurbanisation, it would be useful to have comparable data on this subject for each of the major countries of Western Europe. Unfortunately, this is not possible. Those countries which have registration-based data (Belgium, Denmark, Italy, the Netherlands, Norway, Sweden, Switzerland and West Germany) provide remarkably detailed information on the locations of the origins and destinations of the migrations, but they usually provide little or nothing on their social composition. Only from local case studies, such as those on return migration to southern Italy carried out by King and others (see King, 1986), can one discover what kinds of people moved and how their circumstances changed as a result of their migration. Rather more information about the social and economic characteristics of migrants sometimes comes from those countries which have migration questions in their censuses (France and the United Kingdom), but it is still exceptional to have information on the migrants both before and after their migration. This means that the Longitudinal Studies (LS) of England and Wales, and of Scotland, are uniquely valuable sources of information because, as large-sample national surveys, they permit detailed analyses of the links between social and geographical mobility (Fielding, 1989b).

The LS links the information available on about half a million individuals from the 1981 census with information on the same individuals from the 1971 census. This allows one to study not only the social class compositions of the out- and in-migration flows, but also the changes in status (for example, in occupation or housing tenure) which occurred during that decade. Thus the LS data can be used to test some of the generalisations about the social content of counterurban migration flows listed above. To show this we present information on the class-specific net migration rates for East Anglia and the South-East (London) Region (Table 14.1).

The size of a social class in a region can be seen to increase or decrease as a result of four changes, each of which is considered in Table 14.1. The first derives from migration into and out of the region by those who were in the labour market in both 1971 and 1981; the second also involves migrants, but comes about as a result of entries to, and exits from, the labour market (entries are typically from education or child-rearing, exits are typically to retirement or child-rearing); the third change is produced by the movements between social classes of those who were in the labour market at both dates and who did not migrate; and the final change results from the entries to, and exits from, the labour market of non-migrants.

The data show, first, that East Anglia was a major gainer by net migration between 1971 and 1981. Second, a high proportion of this region's in-migrants were members of the service class of professional, technical and managerial workers and, third, the area also attracted an appreciable number of self-employed people. Fourth, the metropolitan South-East Region lost heavily through the migration flows of those in the labour market at both dates, whereas East Anglia benefited significantly from this source. Fifth, while East Anglia showed net gains in all social classes, the rates of growth were higher in the two 'middle-class' categories than in the two working-class ones. This contrasted with the South-East, which showed net losses in all four categories. Finally, the East Anglian labour market also gained from the migration flows of those who were entering or leaving the labour market, and at

Table 14.1 *East Anglia and South-East England: components of social change, 1971–81*

All values are percentages of the numbers estimated to be in each class in 1976.

(i) *East Anglia*

| | Migrants | | Non-migrants | |
	in labour market	entries and exits	in labour market	entries and exits
Service class	+13.31	+11.02	+12.24	+1.00
Petite bourgeoisie	+8.99	+2.29	+15.24	−17.84
White-collar workers	+4.32	+4.59	−6.51	+6.40
Blue-collar workers	+3.16	+2.79	−7.20	−5.36
Unemployed	+13.76	+10.81	+21.13	+24.08
Total	+6.38	+5.13	0.00	−0.79

(ii) *South-East England*

| | Migrants | | Non-migrants | |
	in labour market	entries and exits	in labour market	entries and exits
Service class	−1.29	+3.92	+13.14	+5.47
Petite bourgeoisie	−1.28	−1.32	+13.57	−12.35
White-collar workers	−2.00	−0.49	−7.61	+4.32
Blue-collar workers	−1.78	−0.51	−8.18	−8.35
Unemployed	+0.67	+3.70	+30.16	+36.06
Total	−1.61	+0.66	0.00	+0.38

Note: Service class are professional, technical and managerial workers; SEGs 1, 2.2, 3, 4, and 5.1.
Petite bourgeoisie are self-employed and owners of small businesses; SEGs 2.1, 12, 13, and 14.
White-collar workers are those in low-level non-manual occupations; SEGs 5.2, 6, and 7.
Blue-collar workers are those in manual occupations; SEGs 8, 9, 10, 11, 15 and 17.
Source: Unpublished tables from the Office of Population Censuses and Surveys' Longitudinal Study. Crown Copyright reserved.

a rate which was higher than that for the South East.

These results confirm in almost every detail the picture that was emerging from the more theoretical discussions, but they also add something new — the importance of the shift into self-employment among the counterurban migrations. Of course, one cannot assume that the situation in East Anglia is typical of rural areas in other parts of Western Europe. Indeed, in few other regions has agricultural restructuring resulted in such a capital-intensive form of agriculture, or productive decentralisation resulted in such a rapid growth of manufacturing and

service employment. But having evidence on the social content of counterurbanisation in this instance provides a basis for broader speculations on the social consequences of the process.

The 'blessings' of counterurbanisation

Counterurbanisation was without doubt a blessing for some. It meant that people with poorly developed skills and lacking formal qualifications (many of them new entrants to the labour market), could now hope to find work locally, where previously they would have had to leave for the nearest metropolitan city. Two groups were particularly favoured by this arrival of new jobs in factories, offices and shops — women and farm workers. There had previously been very few jobs in rural areas for women. They had supplied the domestic labour, worked as family labour on the smaller farms and had sometimes earned wages as casual labour (for example, at harvest time). Now there were manual jobs in manufacturing, and both blue- and white-collar jobs in the expanding services. Almost every region in Western Europe experienced an increase in female activity rates during the post-war period, but the increases were particularly significant in the rural areas and, in so far as this meant enhanced household incomes, it improved rural living standards. Male agricultural employees, meanwhile, could now expect to find manual jobs in factories, in transport and building work and in the public sector. Thus, for ordinary working people in rural areas, counterurbanisation brought with it new jobs, a greater diversity of job opportunities, higher wages, shorter working hours, better working conditions, less casual labour and less un- or underemployment.

New opportunities were provided for those who owned land and property in rural areas. Farmers and small landowners were particularly affected, since many enhanced their wealth, through the sale of small pieces of land for housing purposes, or through the renovation and sale of old farm buildings. Numerous small fortunes were made from such developments. The effects of this varied; some used their windfall gains to shore up a farm business which was threatened by low profitability or debt encumbrance. For others, such deals permitted an exodus from the cares, responsibilities and sheer hard labour of farming. Those who were more substantial owners of land and property often became agents in the development of local towns through their investments in new shopping precincts, industrial estates, and so on.

In-migrant working-class households also benefited from counterurbanisation. True, they may have had to accept lower wages than their city cousins, but their costs of living were lower, due largely to much lower housing costs. Also, unlike the situation in the past, they no longer had to forgo the material advantages of urban living (such as decent schools and medical services, gas, electricity, running water and mains drainage) on

moving to the small town or rural area. Furthermore, they usually now worked in establishments which were smaller, friendlier and less prone to industrial conflict.

An additional significant development was that the fortunes of many self-employed people, and of the owners of small and medium-sized firms in rural areas, were made or restored. The arrival of new customers, many of them fairly affluent, meant good business for retailers, and for those offering personal services (such as garage owners and restaurateurs). Of course, most of these firms were located in the small and medium-sized towns in the rural areas. The increased personal mobility of their customers ensured that village-level services would continue to decline in all but the most remote or touristically most attractive areas. In-migrant business people also did well, often by putting their savings into bars and shops, guest houses and riding stables. While weekend and summer visitors (some of them with second homes in the vicinity) were a part of the market for these services, the main target was the service class of professional, technical and managerial workers, who were now beginning to form a significant proportion of the rural population.

These professional groups gained considerable advantages by migrating to rural areas. They were often on rates of pay that were nationally negotiated and reflected the costs of living in the main metropolitan cities. They were now able to avoid the heavy costs in time and money of long-distance commuting, and they could live in pleasant rural and semi-rural surroundings without that loss of distinction associated with suburban life in a large city. Many of the early in-migrants made considerable gains through the purchase of housing and land, and the later arrivals ensured the creation of a community of well-educated, 'progressive' people who shared the more cosmopolitan lifestyles (foreign holidays, modern furnishings, exotic cuisine) of the 'new middle class'. As this community of interests developed, so did the political and social confidence of its members. Men and women who had been treated as outsiders, even intruders, by locals used to very long-established patterns of social leadership and deference, now began to run the rural areas into which they had moved, even though many of them would be moving on within a few years of their arrival.

Among the in-migrant service class and business

community there were many who were at or close to retirement age. They benefited from counter-urbanisation because, while the drawbacks of rural living (such as the lack of modern facilities) had largely disappeared during their lifetimes, the benefits (such as a slower pace of life, lower crime rates, less noise and pollution) largely remained. And with the television as the main form of social communication, what did it matter if you did not know, or did not like, your rural neighbours?

Finally, the benefits of counterurbanisation accrued in full measure to those whose investment-location decisions had been so important in bringing it about. Those who invested in the small and medium-sized towns in the rural and peripheral regions found many of the things they had hoped for: reliable, manageable workforces untainted by the work practices of the large factory or office, prepared to work hard and co-operatively for wages which were well below what would have had to be paid in a metropolitan region; land for modern buildings in spacious surroundings, providing decent car parking space and room for expansion; and helpful and interested local authorities, keen to attract and retain new businesses. In short, for a time at least, counterurbanisation was the 'spatial fix' which allowed major firms to achieve profitability and growth in the face of sharpening international competition.

This discussion has related the benefits of counterurbanisation to individuals, but it is also helpful to adopt more generalised perspectives. The population turnround brought about a number of social, economic and environmental changes which many would regard as progressive. First, it quickened the erosion of some of those formerly powerful bases of authority in Western European society, notably the Church, the family and the aristocracy. It did so because it was in the rural areas that these bases of authority had held out most successfully against the modern social and political relations of the city. These changes brought more individual freedom to those who lived in rural areas, fewer strictures on those who failed to conform, and a destruction of the monopoly of influence held by the members of the old elites, notably the landowners and the priests.

Second, it transformed rural economies. It brought in new income in the form of more and higher wages, and it changed occupational structures away from declining sectors and towards

high-growth ones. This led to a virtuous circle of development assisted by the local multiplier effects of the new investments. Finally, the newcomers often improved the environment by renovating dilapidated farm buildings and old town houses, and also brought with them a knowledge of environmental issues and an enthusiasm for conservation. These qualities, combined with a familiarity with the principles and practices of large bureaucracies, meant that the in-migrants could use planning law and local authorities' policies to effect actions favouring environmental protection. Sometimes this resulted in powerful anti-growth coalitions which fought against further population increases in areas already strongly affected by counterurbanisation. But it also resulted in campaigns to protect endangered habitats, preserve historic buildings, encourage organic farming methods, and so on.

Counterurbanisation as a 'threat'

With such a list of 'blessings' how could one imagine that counterurbanisation might be viewed by some as a 'threat'? Well, first, there were many people in rural areas whose interests were damaged by the population turnround. Those on low incomes, for example, often saw house prices and rents increasing to levels which placed local housing outside their reach. While this was not a problem, perhaps, for those who already owned their houses, it was a serious obstacle for others, often including their own sons and daughters, who wished to remain in the rural areas. On their low wages they could not compete with the in-migrating professionals and managers, who could always outbid them for properties, even when their intention was to use those properties solely as second homes. Paradoxically, at just that moment when jobs were more available than they had been for a century or more, young people were being forced out of the villages and small towns towards the cities, because it was there that large stocks of poor but affordable housing were to be found. The increasing general prosperity of rural areas had other detrimental effects on the poor. The high levels of car ownership of the newcomers, and their ability to use privatised forms of health and education services, left those who were not so prosperous dependent upon services which were locally becoming less available, such as village shops, public transport, state schools and public health services.

Many of these problems, some of them conceptualised as 'mobility deprivation', were experienced in particular by working-class women. For them, the arrival of the service class implied higher prices in the local shops, a less-frequent bus service to the local town, and an altered, more difficult environment within which to raise a family. Those women who now worked found that their wages did not constitute 'pin-money', but formed an essential part of household income, needed to meet housing costs or other essentials. Thus the transition to wage employment was not that liberating; it represented the substitution of one type of subordination for that of another.

A further drawback was that new jobs for both men and women were not always as secure as they might have seemed. Branch-plant employment suffers from the basically temporary nature of the branch plant itself and, particularly after the onset of recession in the 1970s, external decision-making undertaken at corporate headquarters led to the closure of many rural branch plants. Often, the service-class employees of such plants would be redeployed but, typically, the other employees would be less fortunate.

Owners of small businesses also found that the counterurban migration flows were a mixed blessing. The increase in business activity was fine, but the arrival of many new competitors for that business was not so welcome. In many rural areas of Western Europe, a serious oversupply of certain kinds of services occurred, leading to low incomes and business failure (King, 1986). More generally, the small and medium-sized firms that developed in rural regions often found that the circumstances which pertained at the time of their establishment, changed to their disadvantage in succeeding years. For example, some discovered that the threshold sizes needed to remain profitable were being raised to levels which excluded small-scale businesses located in rural regions.

It was stressed earlier that the turnround was particularly associated with the migration into rural areas of members of the rapidly expanding service class. Surely, these professional people, managers and technicians had nothing to lose by this migration? Not so! In the first place, their employment in a rural area made them highly

vulnerable through their dependence upon their present employer. Being highly specialised in their qualifications or experience, their alternative sources of employment were typically located elsewhere, often in the major cities. Furthermore, to a considerable degree they were socially isolated, cut off from the attainment of that 'cultural capital' which is so important for further upward social mobility. Perhaps this isolation was felt most acutely by the adolescents and young adult children of service-class couples. Not only did they have to leave the area for higher education and training, and for jobs which offered the prospect of social promotion, but until this happened they were restricted by the social facilities of rural areas, which could not compete with the variety and nature of those in the city.

Those who migrated on, or close to, retirement usually spent several years as fully active members of the village or small-town communities which they joined. But after a time, their situations changed. They became older, slower, less healthy, less independent, more liable to be living on their own, less mobile. Suddenly, even if they remained fairly wealthy, they began to experience the mobility deprivation of the poor, and the distance from their family and former friends often came to mean more to them than the attractiveness of their rural environment. Rural retirement migration can sometimes end in a fearful loneliness.

A common experience for the in-migrant has been to witness his rural utopia being scarred by the built-form manifestations of the very same process of economic development which brought him into the area. As villages and small towns grew, they tended to lose their traditional character and became increasingly dominated by the new developments. They lost those very qualities which attracted the service class in-migrant — their quiet streets, their lack of 'time-economy', their balance of the 'human' and the 'natural'.

Finally, for many of the big investors in rural areas, the 'spatial fix' turned out to be only a temporary thrill. Workers did not remain compliant for long, and the arrival of competitors in small-town labour markets quickly forced up the level of wages. Furthermore, it became increasingly difficult to get service-class employees to move on at the request of their employers; they now lived in two-career households, and their household situations often connected them in complicated ways with their communities. 'Spiralists' within large organisations became harder to find.

As with the blessings, the disbenefits of counterurbanisation were not just experienced at the level of individuals in different class and gender positions. There were important negative impacts on rural social relations. The old leadership had been self-seeking and overtly class-conscious, but its paternalism had helped to cement a certain kind of solidarity. Rural social relations, and the 'moral order' that accompanied them, were quickly displaced by the influx of new people. With this change came the loss of distinctive localisms in culture, dialect, work traditions and so on. Rural places became banal; they increasingly conformed to popular conceptions of what a village or small market town was like. Rural regions had been the main repository for much of the variety in Western European culture; now they saw their cultural differences rubbed out by the counterurban movement, which carried with it the messages, rules and practices of middle-class taste, mass society and popular culture.

The economic effects of counterurbanisation were also equivocal. While attracting new activities to rural areas, it also placed these activities at the end of the queue for new technologies such as those developing rapidly in the fields of information technology, telecommunications and broadcasting services. But one economic fact stands out above all others as a condemnation of counterurbanisation. Changes since the early 1960s have resulted in the external control of Western Europe's rural and peripheral regions. The inhabitants of these areas now have even less command of the economic circumstances of their lives than their predecessors in the 1950s.

Finally, despite the environmental awareness of many in-migrants, the turnround had a sinister significance for the landscapes and environments of Western Europe's rural regions. It turned rural areas into 'urbanised countryside', with all that this implies for atmospheric and water pollution, the destruction of natural habitats and general environmental degradation. But it also brought about a transformation of the pre-existing built environment, notably the destruction of vernacular architecture, or its inappropriate modification through renovation and modernisation. As in so many other respects, the rural landscapes came to reflect the relationships of the general and the global, rather than those of the local and the specific. Issues raised by these trends are explored further in Chapters 12 and 16.

Conclusion

Counterurbanisation, though its precise form has varied, has been a major agent of change for several decades. It has produced employment growth in both the manufacturing and service sectors in rural and peripheral regions — regions which previously experienced a long period of economic decline. It has brought new investment by firms and governments and, associated with those investments, a new service class has significantly magnified and modified the nature of demand. The prosperous retired have also made an important contribution to changing the scale and quality of demand in rural economies. Valuable though this economic revolution has been, however, the disadvantages of counterurbanisation should not be ignored. The traditional social relations of village life have been eroded. Economic dependency on outside agencies (governments and firms) has been created. New social groups and work relationships have been inserted into rural and small-town environments. The economic problems of the poorer elements in rural communities, who find economic competition with prosperous newcomers increasingly difficult, have been exacerbated. And, not least, physical pressure on the rural environment has grown significantly. Faced with these contrasting pointers, the primary conclusion to be drawn is clear. Counterurbanisation must be seen as both a threat and a blessing, a double-edged sword that has brought — and will continue to bring — far-reaching change to rural Western Europe.

Notes

1. See Fielding (1986b) for a discussion of the strengths and weaknesses of this French material.
2. The reasons for expressing the relationship in this way are explained in Fielding (1982).
3. The following results detail the relationships between net migration rates and population density for the major countries of Western Europe between 1981 and 1987.

Austria: no data.

Belgium: data for 9 provinces, 1981–3
$Y = 0.28 - 0.00107X$; $r = -0.407$; $T = -1.18$
i.e. no relationship (weak counterurbanisation).

Denmark: data for 11 regions (counties + capital region), 1981–6
$Y = 0.88 - 0.000423X$; $r = -0.041$; $T = -0.12$
i.e. no relationship.

France: no data.

Ireland: data for 9 planning regions, 1981–6
$Y = -3.72 - 0.00735X$; $r = -0.348$; $T = -0.98$
i.e. no relationship (weak counterurbanisation).

Italy: data for 13 regions (*regioni*, sometimes grouped), 1981–6
$Y = 2.03 - 0.00236X$; $r = -0.165$; $T = -0.56$
i.e. no relationship (weak counterurbanisation);
data for 69 regions (provinces, sometimes grouped), 1981–3
$Y = 8.20 - 0.00768X$; $r = -0.246$; $T = -2.07$
i.e. counterurbanisation.

Netherlands: data for 11 provinces (polders excluded), 1981–6
$Y = -0.86 + 0.00211X$; $r = +0.391$; $T = +1.27$
i.e. no relationship (weak urbanisation).

Norway: data for 8 regions (groups of counties), 1981–6
$Y = -3.36 + 0.166X$; $r = +0.690$; $T = +2.34$
i.e. urbanisation.

Portugal: no data.

Spain: data of doubtful reliability.

Sweden: data for 12 regions (counties, sometimes grouped), 1981–7
$Y = -1.93 + 0.0575X$; $r = +0.775$; $T = +3.88$
i.e. urbanisation.

Switzerland: data for 11 regions (groups of districts), 1981–3
$Y = 27.9 - 0.0738X$; $r = -0.506$; $T = -1.76$
i.e. no relationship (weak counterurbanisation);
data for same regions, 1986–7
$Y = 4.14 - 0.00132X$; $r = -0.061$; $T = -0.18$
i.e. no relationship.

United Kingdom: no reliable data (National Health Service Central Register data possible source, but not used).

West Germany: data for 12 regions (groups of districts) 1981–3
$Y = 5.15 - 0.0181X$; $r = -0.633$; $T = -2.59$
i.e. counterurbanisation;
data for 50 regions (groups of districts), 1981–3
$Y = 5.97 - 0.0169X$; $r = -0.449$; $T = -3.48$
i.e. counterurbanisation;
data for 12 regions, 1985–6
$Y = 2.03 - 0.000917X$; $r = -0.077$; $T = -0.24$
i.e. no relationship.

References

Ambrose, P.J. (1974) *The Quiet Revolution: social change in a Sussex village 1871–1971*, Chatto, London.

Fielding, A.J. (1982) Counterurbanisation in Western Europe, *Progress in Planning*, 17, 1–52.

Fielding, A.J. (1986a) Counterurbanisation in Western Europe. In A. Findlay and P. White (eds), *West European Population Change*, Croom Helm, London, pp. 35–49.

Fielding, A.J. (1986b) Counterurbanisation. In M. Pacione (ed.), *Population Geography: progress and prospect*, Croom Helm, London, pp. 224–56.

Fielding, A.J. (1989a) Migration and urbanization in Western Europe since 1950, *Geographical Journal*, 155, 60–9 and 80.

Fielding, A.J. (1989b) Interregional migration and social change: a study of South East England, *Transactions of the Institute of British Geographers*, 14, 24–36.

Fothergill, S. and Gudgin, G. (1982) *Unequal Growth: urban and regional employment change in the United Kingdom*, Heinemann, London.

King, R. (ed. (1986) *Return Migration and Regional Economic Problems*, Croom Helm, London.

Klaassen, L.H. and Molle, W. (eds) (1983) *Industrial Mobility and Migration in the European Community*, Gower, Aldershot.

Lampard, E.E. (1955) The history of cities in the economically advanced areas, *Economic Development and Cultural Change*, 3(2), 81–102.

Massey, D. (1979) In what sense a regional problem?, *Regional Studies*, 13, 233–44.

Mellor, J.R. (1977) *Urban Sociology in an Urbanized Society*, Routledge, London.

Pahl, R.E. (1965) *Urbs in Rure: the metropolitan fringe in Hertfordshire*, London School of Economics, Geographical Paper 2.

15

Tourism and development

Gareth Shaw and Allan M. Williams

Introduction

The tourist industry is one of the most important growth sectors within Western Europe, with the region dominating the international tourism market. From 1965 to 1986 the number of international tourist arrivals more than doubled, whilst tourist movements within Western Europe increased at a similar rate (World Tourism Organisation, 1987). Within the European Community, expenditure and revenue from international tourism increased more than sixfold between 1970 and 1984; and by 1984 it accounted for 5 per cent of all credits within the Community's balance of payments (Pearce, 1988). Much of the tourist activity originates within the region, and tourists from non-European countries make up only 15 per cent of all visitors. The internal market is not only large, with an estimated 180 million tourists within the European Community alone; it is also very complex. An increase in real incomes and leisure time, a broadening of life-style expectations (Mitchell, 1984), and developments in mass tourism and package holidays (Pearce, 1987a) are all factors responsible for a growth in the market. Variations in these socio-economic factors between European countries account for the complexity of tourism demand, characterised by different types of holidays and by changing flows of tourists from one country to another.

There has also been an increased awareness of the economic importance of tourism during the 1980s, due to the increased size of the industry, and to a growing appreciation of how long-term structural changes have led to an expansion of the service economy. The increasing share of services in international trade is one trend that has focused attention on tourism as a major contributor to a country's balance of payments. Obviously, such contributions can be positive or negative ones, but either way many governments now see tourism as a major part of their economic policies (OECD, 1986). In extreme cases, such as Spain, receipts from tourism contribute 21 per cent of total receipts, compared with 5.1 per cent in France and 3.9 per cent in the UK (IMF, 1985; Williams and Shaw, 1988). However, even in those economies where tourism plays a smaller role in foreign earnings, there has been an increased appreciation of the tourist industry's potential to create new jobs. The growth of large-scale unemployment in many of the industrialised Western European countries has led to governments assessing tourism in a new light. Tourism's potential role as a creator of jobs is further strengthened because of the labour-intensive nature of the industry, and the apparent cheapness (to governments) with which employment can be generated (English Tourist Board, 1986).

Despite much debate on the subject (Young, 1973; de Kadt, 1979) there is still no clear consensus as to tourism's role in economic development. The picture is clouded not only by the different assessments of economic development, but also by its socio-cultural and environmental implications (Murphy, 1985). This chapter will provide a review of the relationships between tourism and development, and explore the changing nature of Western European tourism particularly within the context of state and private sector involvement. As a background to these issues, consideration will be given initially to the broad pattern of tourist movements.

Uneven demand for tourism

Tourist flows

During the post-war period there have been important shifts in the origins and destinations of both domestic and international tourists. Unfortunately, however, few reliable comparative data are available for the domestic scene, and geographical analysis is therefore only possible for international tourism (Williams and Shaw, 1988). Even here the situation is complicated since many tourist statistics are unavailable on a comparative basis for all Western European countries (Goodall, 1988, p. 19). There are different ways of enumerating tourists — both international frontiers and the place of accommodation can serve as census points, for example — and some countries undertake sample surveys rather than complete coverage (Williams and Shaw, 1988, p. 11). In these respects care is needed in comparing the different data.

In general terms, the relationship between tourist origins and their destinations within Western Europe is characterised by a movement from the northern, industrialised and urbanised countries to the warmer climates of southern Europe (Goodall, 1988). These countries have an obvious comparative advantage in terms of their climate and access to attractive coastlines. The European market is, however, highly segmented: major holiday destinations include beach locations, alpine areas, rural areas, urban/cultural attractions and business/conference-based centres. However, the mainstays of holiday tourism are beach locations, with an estimated 52 per cent of tourists from the European Community taking their main holidays at the seaside (Commission of the European Communities, 1987a, p. 29). Such preferences are highlighted by the flows of tourists into Mediterranean states, as shown in Figure 15.1. From such evidence it can be seen that France, West Germany and the UK are the origins of the majority of tourists.

As Figure 15.2 shows, most countries have experienced increases in foreign tourists since 1960, but it is Spain, Italy, Greece and France that have witnessed some of the greatest growth. The first three countries, but particularly Spain, have benefited from the development of inclusive, low-cost package tourism. In the case of Spain these packages are usually by air charter flights,

with the UK and West Germany providing the bulk of the tourists. For example, the flow of passengers using inclusive tours by charter flight (chartered by a tour organiser or travel agent) from the UK increased from 2.9 million in 1970 to 5 million in 1980 (Pearce, 1987a). The development of these inclusive tour charter flights has been one of the most significant factors contributing to the rise of mass tourism within the Mediterranean region although, as Pearce (1987b) has shown, such traffic flows are both complex and dynamic. In this respect, future technological changes in transport and government air-travel policies (such as deregulation) may play major roles in the flows of tourists.

Alpine holidays, developing in popularity since the late 1950s, are a second focus of mass tourism (Gilg, 1988). Alpine areas have been major beneficiaries of rising standards of living, which have allowed more people to enjoy second holidays, with a priority often given to skiing. For example, second holidays in France and Italy are predominantly in mountain areas (Commission, 1987a, p. 41).

Not all tourism in Mediterranean or Alpine areas is based on package holidays. Indeed, the characteristics of tour operators in the countries of origin play a major role in creating this type of holiday. In France, for example, the multiplicity of travel agents and a low proportion of direct sales by tour operators, together with close access to other Mediterranean countries, and a preference for domestic destinations, have resulted in only 5 per cent of French tourists taking organised packages (Pearce, 1987a; Commission 1987a, p. 45). This contrasts with a Community average of 13 per cent for all holidays in 1985.

Rural-based tourism is a more difficult market segment to assess, but it has been estimated that 25 per cent of all European holidaymakers spend their holidays in the countryside (Commission, 1987b). In many countries the limited evidence suggests that a gradual substitution of clientele is taking place, with low-income visitors being replaced by higher-income groups. These newer tourists are more demanding, wanting more structured tourist resorts within which to pursue their leisure interests. Such demands cause inevitable problems in many rural areas, since the original idea of many governments was to promote farm holidays to aid the rural economy, without fundamentally restructuring it. In West Germany

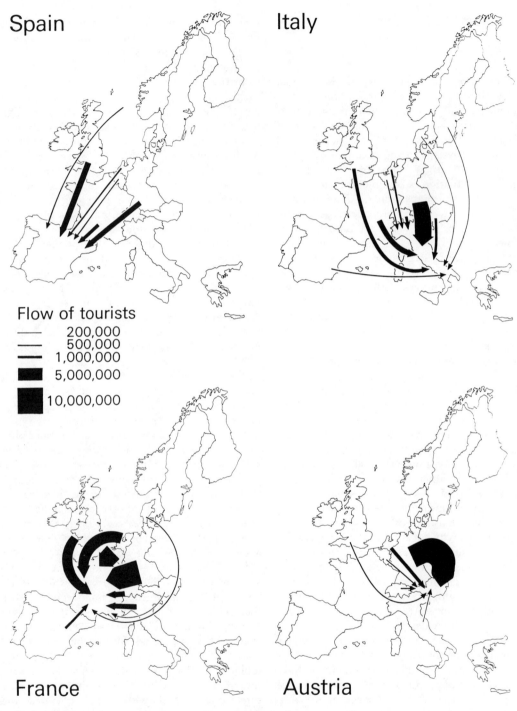

Only annual tourist flows larger than 200,000 are shown

Figure 15.1 *Annual tourist flows to selected countries, 1986*
Source: World Tourism Organisation (1987).

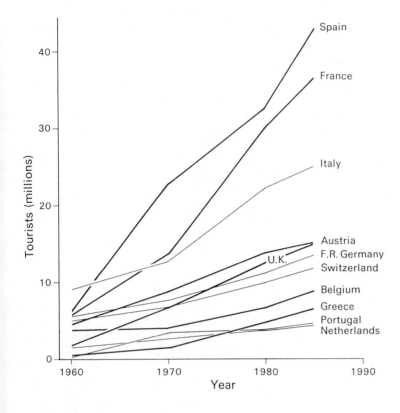

Figure 15.2 *Changes in the numbers of foreign tourists visiting Western European countries, 1960–85*
Source: OCED (1986).

such holidays were linked to the concept of 'soft' tourism, in that they did not require a sophisticated infrastructure and had a limited impact on the landscape (Schnell, 1988, p. 211). However, newer developments, in the form of integrated leisure centres, not only offer more organised holidays, but pose a threat to more traditional rural tourism schemes.

The final, major form of tourism is focused on urban areas and comprises two strands: cultural tourism and business/conference tourism. Every major European city benefits in some way from both types, with London (59.5 million overnight stays in 1985) and Paris (15.8 million) being major beneficiaries (Williams and Shaw, 1988, p. 18). Cultural tourism — although poorly defined — is a primary motive for much travel, and is important in both foreign and domestic markets. As Kosters (1981) points out, it is the second most important reason for international travel, after beach holidays. This type of tourism often links with other forms, with cities being visited from nearby coasts. In this context, and

also in respect of day recreational trips, it forms an important part of domestic tourism. Its significance for historic cities has been highlighted by Ashworth (1988), while others have stressed its growing appeal to a wide range of urban areas (Buckley and Witt, 1985). The development of urban tourism is very much related to the growth of short-break holidays, which in the UK provide one of the most rapidly expanding segments of the market, increasing by 100 per cent between 1975 and 1986 (English Tourist Board, 1987). Conference and business travel is also a major market: Law (1985) estimated that in 1984 there were some 14,000 international conferences. Within Europe, France plays the leading role in this market (Tuppen, 1985). Paris is the focal point for much of the business and conference tourism; in 1985 it was the world's leading conference venue with this activity attracting an estimated 650,000 visitors (Tuppen, 1988).

Much of the above discussion has concerned foreign tourism, but for many countries domestic tourists play an even more important role

Table 15.1 *Foreign and domestic tourist nights spent in hotels and similar establishments in Western Europe, 1985[1] (thousands)*

	Nights spent by foreign tourists	Nights spent by domestic tourists	Total nights
Austria	54,588	13,948	68,536
Belgium[2]	5,256	2,351	7,607
Denmark	4,591	4,153	8,744
Finland	2,097	6,692	8,789
France[3]	18,166	9,987	28,153
West Germany	23,895	104,614	128,509
Italy	64,760	106,998	171,758
Norway	3,713	8,193	11,906
Portugal	14,933	12,345	27,278
Spain	78,919	42,097	121,016
Sweden	3,549	11,984	15,533
Switzerland	20,320	13,013	33,333

Notes: 1. Excludes camping, caravans, rented houses, etc.
2. 1984 data.
3. Data for Ile de France only.
Source: OECD (1986, p. 172).

(Table 15.1). This is the situation in West Germany, Italy and the UK. Despite the favourable growth in overseas visitors, the tourist industries of these countries are very much dependent on domestic demand. This market can be extremely dynamic and has changed, in the case of the UK, very markedly since the early 1970s (Shaw *et al.*, 1988). UK holiday tourism (particularly seaside-based) declined from 62 per cent of all tourist spending in 1972, to 52 per cent in 1985; while over the same period business tourism increased from 17 to 28 per cent of spending. Meanwhile, the proportion of UK residents holidaying abroad increased from 10 to 24 per cent (British Tourist Authority, 1985).

Social constraints on access to tourism

A number of factors obviously influence people's decisions to participate in tourism, including stage in the family life cycle, cultural conditions, the amount of leisure time available, access to tourist areas and disposable income. In most studies economic factors have been given prominence. Thus, within Western Europe there is a strong correlation between GDP per capita and the proportion of people taking holidays (Commission, 1987a, p. 9). Indeed, tourism is characterised by a positive income elasticity of demand: the demand for holidays rises proportionately greater than increases in income. This is aided by the nature of the tourism product (there is an enormous variety of places to visit) and by the fact that the tourist is a collector of places (Waters, 1967). Another important factor is that, for many, the holiday is both a luxury and a basic right. Consequently, as Ashworth and Bergsma (1987) have shown from their work in the Netherlands, foreign holidays have been largely unaffected by stable or declining incomes.

Recent investigations within the European Community have shown that an estimated 56 per cent (140 million tourists) took at least one holiday away from home, for four days or more (Commission, 1987a, p. 7). In Scandinavia even higher proportions have been recorded, reaching 83 per cent in Sweden (1981) and 77 per cent in Norway (1978) (Sundelin, 1983). The Community average hides considerable variations, ranging from only 31 per cent of the population of Portugal to 65 per cent of that of the Netherlands. These differences raise the whole issue of access to tourism since, on average, 44 per cent of the Community's population did not go on holiday in 1985. As Table 15.2 shows, there is a variety of influences which condition access, but in every country except Denmark economic constraints were the most frequently cited. Most of these respondents probably comprise the 21 per cent of Europeans who habitually stay at home and who mostly have low-income occupations. Thus, in the EC survey, 49 per cent of manual workers stayed at home compared with only 18 per cent of those in professional occupations. It is possible, therefore, to conceive of a group who are 'the tourism poor', the size and features of which vary markedly among European countries.

The survey also showed that older people, those with large families and those living in rural areas are also less likely to take holidays away from home. Thus, 66 per cent of Europeans living in large towns went on holiday, compared with 45 per cent living in villages. Such differences appear to be greater in the less-developed economies, such as Italy, Spain, Greece and Portugal, than in the more developed ones. For example, in Italy

Table 15.2 *Reasons for not taking a holiday during 1985 (per cent)*

	(a)	(b)	(c)	(d)	(e)	(f)
Belgium	32	7	40	14	*	4
Denmark	38	8	23	16	*	16
West Germany	27	12	41	29	1	3
Greece	12	24	55	35	*	–
Spain	14	22	50	15	*	9
France	22	23	44	25	1	2
Ireland	14	10	61	10	1	6
Italy	30	22	31	21	2	2
Luxembourg	23	16	20	31	*	10
The Netherlands	22	9	32	27	1	14
Portugal	12	19	67	14	1	2
United Kingdom	14	6	50	21	1	16
EC12	22	16	44	22	1	6

(a) You preferred to stay at home.
(b) You were not able to get away from work.
(c) You couldn't afford it.
(d) Special reasons (your health, moving house, family reasons etc.).
(e) Worry about safety, terrorists etc.
(f) Other reasons.
Notes: In each country the sample comprised 100 people who did not take a holiday away from home in 1985. Because some respondents gave more than one answer, the total for each country generally exceeds 100.
* Less than 0.5%.
Source: Commission of the European Communities (1987a).

there is a generally low level of holiday taking and considerable divergence between different socio-economic groups (Travis, 1982). By comparison, in the UK and West Germany the proportion of people taking holidays shows very little variation related to age and socio-economic class (Mihovilovic, 1980). Clearly, there are important differences in people's access to tourism services within Western Europe, and it remains to be seen how quickly these non-holidaying groups will enter the market, and what impact they will have on both demand and the structure of provision.

New forms of tourism demand

The entry into the market of large numbers of new tourists from the developing economies of Western Europe will obviously change the size of the market. The initial impact of such changes will most probably be felt within the domestic market, although considerable scope exists for foreign holidays. Evidence of this latent demand was clear in the EC survey, which showed that even in those countries where foreign holidays are still only taken by a tiny minority, interest in visiting other European countries is high (for example, 96 per cent of respondents in Italy, 81 per cent in Spain and 78 per cent in Greece). Furthermore, such potential is still to be realised in countries like the UK, where in 1984 only 24 per cent of British adults experienced foreign holidays (Keynote Report, 1985). One of the major destinations of these new international tourists may well be Greece, which 36 per cent of all Europeans considered to be attractive for a future holiday (Commission, 1987a, p. 77). However, the demand for international tourism is notoriously fickle, and is determined by a combination of changes in relative costs, fashion and major tour companies' strategies. In the 1980s these all pointed to the European destinations losing market share (if not absolute numbers) to non-European destinations such as the USA, Thailand and Australia.

Other major changes will be concerned with the trend towards increasing market segmentation, and the growth of second or other holidays. In the EC survey, some 19 per cent of the sample had taken more than one holiday in 1985, with the greatest number of these excursions being short breaks. Such trends are most marked in the Netherlands, France, Denmark, the UK and West Germany. West Germany, for example, has since the early 1970s witnessed a growing trend towards short breaks (two to four days), and one-third of the population aged over 14 participated in this form of tourism in 1985 (Schnell, 1988). Similar changes are occurring in the UK, as detailed in Table 15.3, and have implications for the form of tourism provision. Present trends suggest that the growth will be in the two contrasting areas of urban and rural tourism. It is the latter, however, that may witness the greatest changes, with an increasing emphasis being placed on more structured multiple-activity holiday products, rather than on simple resources such as holiday cottages (Commission, 1987b).

Other changes are likely to be associated with a greater demand for quality tourism, especially from those entering the second-holiday market. Krippendorf (1986) believes that broader societal

Table 15.3 *Changes in the holiday market of British residents, 1976–86 (per cent)*

Type of holiday	Trips	Spending[1]
All tourism	+2	+15
All holidays	−8	−43
Long holidays (4+ nights)	−14	−48
Short holidays[2]	−3	−29
Short breaks[3]	+66	+106

[1] Spending adjusted for inflation, using the Retail Price Index.
[2] 1–3 nights in non-serviced accommodation.
[3] 1–3 nights in serviced hotels, guesthouses, etc.
Source: British Tourist Authority (1987).

changes related to the increased importance being given to the value of leisure will produce a new type of tourist. For many of these people the emphasis will be on leisure rather than work, producing what he has termed a more 'human form of tourism' (Krippendorf, 1984). This will result in different patterns of tourist behaviour, together with a new consciousness of holiday travel, producing 'critical consumer tourists'. These will be well-informed tourists who are very selective in their choice of holiday and will not be guided by the lowest price. As Zimmermann (1988) has shown for Austria, increased household income has led to a rise in the aspirations of tourists when on holiday. Such demands for quality, and an emphasis on environmental standards, could provide a new challenge for many European resorts.

According to Krippendorf's scenario, the future forms of tourist demand will produce a broad segmentation of the market, between the demands of two distinctive groups of tourists. One group will comprise those who are less demanding in their choice of tourist services. These may increasingly be people drawn from countries (Portugal, Spain and Greece) where holiday-taking is still developing, and where large numbers of new tourists will arise as household incomes improve. They will be added to the substantial traditional market drawn from the more-developed European countries. In contrast, the second group of tourists will probably be drawn from the industrialised nations, with a long history of holiday-taking. This group will form Krippendorf's 'new tourists', who will be more involved in foreign holidays and more critical in their demand for tourist

experiences. Unfortunately, they will also probably be shadowed by a third group, the 'tourism poor', who will continue to lack access to holidays, travel and recreation.

The delivery of tourist services

The role of the state and the private sector

Government involvement in the tourism sector has been of long standing in most Western European countries, dating back in many cases to the inter-war period. It was during this time that governments became aware of the economic importance of tourism, and the creation of many national tourist offices led also to the establishment of the first important transnational agency, the International Union of Official Tourist Publicity Organisations in 1925. The Second World War was an obvious disruption to such developments, but the immediate post-war years witnessed the setting up of the European Travel Commission in 1948. This international organisation was established to promote European tourism to North Americans, and was very much seen as another (although far smaller) spur to the rebuilding of the European economy alongside the Marshall Plan (European Travel Commission, 1988).

At a national level the state's role in tourism has moved through at least three phases (OECD, 1974). The first phase saw governments concerned with facilitating tourism by streamlining currency, health and customs regulations. This was quickly followed by a second phase, characterised by recognition of the boost to overseas earnings that tourism could give. During this phase, governments became very active in the promotion of tourism. The third phase, which began in the 1960s, has seen governments become more closely involved with the problems of tourist supply and the ways in which the state can enable its tourist industry to be successful.

Over time, therefore, most Western European countries have witnessed growing government involvement in tourism. This, however, must be seen in the perspective of the European tourist industry itself, which is still highly fragmented and very much a private sector activity. The role of the state in tourism will vary depending on the type of government and the importance of tourism to the country (Wolfson, 1964). The importance of the political character of government and its influence on tourism policy is well illustrated by Leontidou (1988) for Greece. Under the military dictatorship of 1967–74 tourism was used as a means of 'heating up' the economy, with special tax concessions being given to new hotels, while the emphasis was on boosting private enterprise. At this time regional imbalances were blatantly reinforced by the state's tourism policy. Since 1974 and the fall of the dictatorship, state involvement in tourism has increased, but major regional objectives have been formed relating to discouraging development in congested tourism areas.

Tourism policies are varied and include the promotion of a nation's tourist products (which features in all European countries), plus the planning and facilitating of tourism. Grant aid and the direct ownership of components of the tourist industry may well be elements in the strategy. Thus, in the UK, state aid was given to improve tourist accommodation following the Development of Tourism Act in 1969 (Shaw, Williams, Greenwood and Henessey, 1987). In many European countries tourism is identified in government structures with economic issues, as is the case in West Germany, the Netherlands and the UK (Airey, 1983). The situation in France is, however, very different. Here the government takes a much broader perspective on tourism, and is much less biased toward economic issues, especially in the more socially and culturally informed work of the Ministry of Leisure.

The main elements of state expenditure have usually been investment in infrastructure, tourist facilities and improvements to accommodation. The costs of infrastructure have been particularly important, especially through the construction of new roads and airports to open up tourism areas. The construction of Málaga airport, for example, was crucial to the development of Spain's Costa del Sol. Several governments have also set up companies to facilitate the construction of tourist facilities, with Italy's Cassa per il Mezzogiorno spending between 7 and 10 per cent of its budget on tourist facilities during the 1960s (White, 1976). For European Community members, the European Regional Development Fund has recently proved a further means of injecting public money into tourism; ERDF grants for the industry totalled 85 million ECU in 1985 (Pearce, 1988).

State involvement in tourism, therefore, has a wide variety of objectives and concerns, but such policy-making is strongly conditioned by the fact that much of the industry is under the direct control of private enterprise. The relationship between public concern and commercial activity is at best one of a loose co-operation between, on the one hand, national or regional tourist bodies and, on the other, private operators. Indeed, at the level of the local economy the relationships between public policy, usually in the form of land-use planning control, and tourist development are often in sharp conflict. This is usually because local bodies are often faced with controlling the excesses of tourist growth, and with attempting to reconcile environmental, social and economic problems that arise through tourism (Shaw, Williams, Greenwood and Henessey, 1987).

Company structures: family businesses and the multinationals

The tourist industry comprises a number of interlinked sectors, including tour operators, travel agents, accommodation and tourist facilities. While there are undoubtedly concentration tendencies within each of these sectors, as well as strengthening links between them, most tourist enterprises are small-scale. Inevitably, because of the very fragmented nature of much of the tourist industry, it is extremely difficult to gain any meaningful comparative statistics on ownership patterns. As a result the available evidence is largely based on individual case studies. In France, for example, 80 per cent of tourist firms employ less than ten people (Tuppen, 1985). Similarly, in Austria the tourism industry is characterised as small- to medium-scale, with a predominance of private capital and family-run businesses (Zimmermann, 1988, p. 54). The accommodation sector is, however, highly polarised, a trend that is increasing in countries such as the UK, as the share of larger hotels is growing. Thus, between 1951 and 1971 there was a 5–10 per cent increase in the number of hotel bedrooms, but a 40–50 per cent reduction in hotels and guesthouses (Stallinbrass, 1980). Similarly, in the Netherlands there was a fall of 55 per cent in the number of guesthouses between 1975 and 1984 (Pinder, 1988, p. 216).

There are marked trends towards both the concentration and internationalisation of hotel ownership, with the major organisations being

Table 15.4 *Transnational-associated hotel accommodation in selected Western European countries*

	Number of hotels	Rooms	Rooms as a percentage of all hotel rooms
Spain	58	14,883	1.8
West Germany	51	13,691	2.0
France	30	8,978	0.9
United Kingdom	28	8,631	n.a.
Italy	29	8,439	0.5
Switzerland	24	5,314	2.3
The Netherlands	27	5,271	5.6
Greece	15	4,630	1.6
Belgium	25	4,592	6.7

The percentage for each country is an estimated figure, since both 1982 and 1984 data have been used.
Sources: UNCTC (1982); OECD (1986).

dominated by North American companies. Thus, of the top ten major transnational hotel corporations, six are controlled by US companies, two by French, and two are under UK ownership (Williams and Shaw, 1988, p. 24). The distribution of these large hotel groups varies throughout Europe, but most countries have experienced such developments (Table 15.4). Such international hotels market more than just accommodation, since most sell a package of services related to leisure or business activities. Thus, according to one estimate, hotels obtain only 51 per cent of their income from providing accommodation (World Tourist Organisation, 1984, p. 58).

One of the most notable trends in the European tourist industry has been the growth of major tour companies since the 1960s. Many of these companies operate in the mass tourist market and reap the economies of scale accruing from catering for large-scale demand (Holloway, 1985) which marketing and advertising have rendered relatively uniform. These operators tend to concentrate on general air tours, and since the 1970s there has been a progressive integration with other sectors of the tourism industry, especially airlines (Goodall, 1988, p. 28). West German companies have witnessed the greatest degrees of horizontal integration, with almost 66 per cent of all tours being sold by two operators, the Touristik Union International and Neckermann. The former was established in 1968 by a merger of three other tour operators, while the latter emerged in 1963

from the merger of a department store group and other tour operators. In the UK the Thomson Group is the market leader (Heape, 1983), accounting for almost 30 per cent of the package holiday market in 1987. These large companies have considerable economic advantages, since they sell flights, accommodation, insurance and leisure activities at the holiday resorts. Scale economies also enable them to negotiate low-unit-cost airline and hotel contracts. There is also a number of specialist tour operators, which tend to be medium-sized firms or even subsidiaries of travel companies.

The third main element in the industry, that of travel agents, is also subject to concentration trends. In the UK this sector of the industry is dominated by a few large companies such as Thomas Cook and Lunn Poly which, in turn, are owned by the Midland Bank and the Thomson Group. It is estimated that these major UK agents probably control 20–25 per cent of all outlets and between 40–45 per cent of total sales (Buck, 1988, p. 172). The failure of the Association of British Travel Agents to bring stability to the market has resulted in fierce competition which has led to the closure of many of the smaller British travel agents, a trend that is likely to continue. Similar patterns are emerging in other European countries, where the process is also linked to the growing internationalisation of service provision. For example, in Spain during 1986 the Spanish travel agency, Melia, was taken over by a consortium comprising SASEA, the Swiss holiday company, and Interpart Holdings of Luxembourg.

Tourism and economic development

Development and dependency

Reliance on international tourism as a strategy for economic development has been criticised since it is often associated with a dependency upon external sources of growth (de Kadt, 1979). Such sources tend to be of a fickle nature, with the choice of tourist destination being susceptible to large fluctuations, either because of economic conditions in the tourist's country of origin, or perhaps because of the political situation in the holiday destinations. For example, the Libyan crisis and a fall in the dollar exchange rate in

1985 both had an impact on the European market, causing a decline in the number of North American visitors during 1985–6. Fluctuations in demand also arise from price changes in particular countries, especially in the undifferentiated market of mass tourism. Currency devaluations or revaluations can have a considerable impact on tourist numbers. In Spain, the devaluation of the peseta in 1959 was a key factor in the growth of the country's tourism industry from the 1960s. In Western Europe a variety of countries have tourism industries that are very dependent on the international market — in Austria foreign tourists account for 76 per cent of all overnight stays, and in Spain the corresponding proportion is 66 per cent.

Austria, in fact, has a greater level of international dependency than Spain. Not only do foreign tourists account for a greater share of stays, but 66 per cent of the visitors are from one country — West Germany. The holiday habits and trends of West Germans therefore play a dominant role in the development of Austrian tourism. One major impact has been the trend towards shorter holidays, and between 1981 and 1985 the number of overnight stays declined by more than 10 million (Zimmermann, 1988, p. 147).

Another dimension to the relationship between tourism and economic dependency is that of overseas investment, especially in the Mediterranean countries. In Greece, for example, foreign investment peaked in 1968, when it rose to 66.1 per cent of all investments in tourism (Leontidou, 1988, p. 85). This was mainly capital from the USA although, as Alexandrakis (1973) has shown, West German, Swiss and French investments were also important. Most of the foreign capital went into hotel complexes in prime coastal locations, creating a phase of speculative development with related environmental and sociopolitical problems (Spanoudis, 1982). Despite such difficulties, many European countries regard tourism as having an important economic role, through its impact on foreign earnings, employment creation and regional development (Heape, 1983).

International tourism and foreign earnings

The preoccupation of many governments with tourism is closely related to the fact that the industry is a major source of foreign earnings (Singh, 1984). As Figure 15.3 shows, the Mediterranean and Alpine countries all benefit from a positive balance in international tourism expenditure. Thus, with the exception of Italy, over 13 per cent of all exports derive from tourism in the Mediterranean countries, compared to less than 4 per cent in most northern European countries. The largest net earners in absolute terms are Italy and Spain, with West Germany having the largest deficit. Countries such as the UK have relatively small net balances which conceal large expenditures and receipts. However, the general north–south movement of wealth generated by tourism is seen by the European Community as very beneficial in contributing towards balance-of-payments stability within the member states (Commission, 1982, p. 13).

In terms of national income, tourism makes its greatest contribution in those countries supplying the mass demand, that is in Austria, Spain and Portugal (Table 15.5). The total income generated can be considerable when multiplier effects are taken into consideration. Within the European Community it has been estimated that direct income from tourism accounted for 4 per cent of GDP in 1979, and when multiplier effects were taken into account the figure rose to 10 per cent (Commission, 1985). The local multiplier effects are arguably the greatest in the larger or more specialised tourist economies, as these are better able to support linked specialist services and manufacturing. In Austria investments carried out by tourist enterprises have mainly benefited firms within the immediate local area, and it is estimated that 57 per cent of their capital expenditure is spent within 20 km — giving a highly localised multiplier (Zimmermann, 1988, p. 154).

There are, however, significant income leakages in tourism economies which are dominated by overseas firms, or which have weakly developed service and manufacturing sectors. Measurement of such leakages is extremely difficult, although Mathieson and Wall (1982) estimated that for Italy in 1975 leakages reduced tourist earnings to 59 per cent of receipts. Increased internationalisation of ownership of tourism facilities will probably increase the degree of economic leakage.

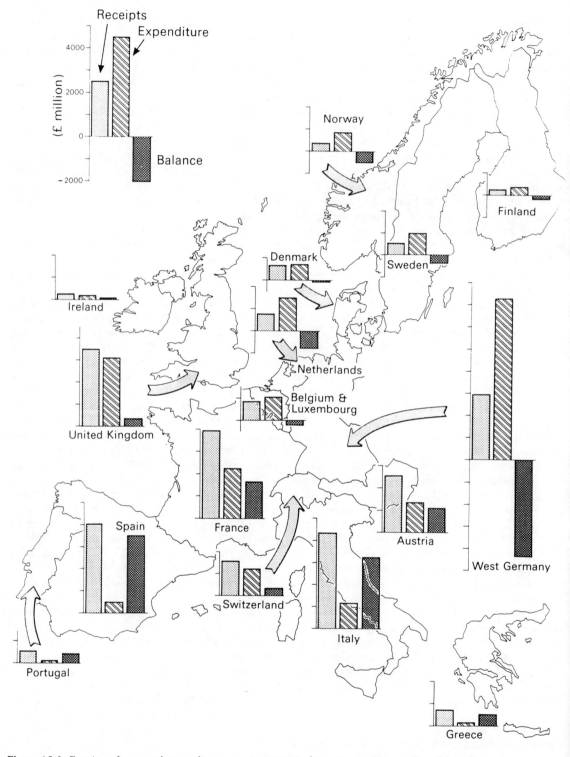

Figure 15.3 *Earnings from, and expenditures on, international tourism in Western Europe, 1985*
Source: OECD (1986).

Table 15.5 *International tourism and the national economies of Western Europe, 1984*

	Tourism receipts as a percentage of GDP	Exports of goods and services as a percentage of GDP	Tourism expenditure as a percentage of goods and services
Austria	8.4	17.7	9.2
Belgium and Luxembourg	2.2	2.1	2.5
Denmark	2.4	5.8	5.1
Finland	1.0	3.0	4.2
France	1.6	5.1	2.9
West Germany	0.9	2.6	7.4
Greece	3.9	17.9	3.1
Ireland	2.7	4.3	3.2
Italy	2.5	8.7	2.0
Netherlands	1.2	1.8	3.9
Norway	1.2	2.4	6.6
Portugal	5.0	13.5	2.3
Spain	4.8	21.1	2.3
Sweden	1.2	3.0	4.7
Switzerland	4.5	9.0	6.9
United Kingdom	1.4	3.2	3.3

Source: OECD (1986, p. 72).

Tourism and employment

Employment in tourism has become increasingly important: in 1984 the World Tourism Organisation estimated that tourism and tourism-related jobs account for about 15.5 per cent of total employment in Europe. This proportion has, however, remained fairly constant throughout the 1970s, reflecting the shift towards self-catering accommodation and attempts by some parts of the industry to resolve labour costs. Unfortunately, comparative employment data for individual European countries are only available for hotels and restaurants, which give only a partial picture (Table 15.6). For example, data from France show that employment in travel agencies, administration and other related sectors may equal or exceed that in hotels and restaurants (OECD, 1986, p. 182). Similarly, detailed analysis by Morrell (1985) for the UK suggests that hotels, other types of tourist accommodation and catering account for at most 57 per cent of all tourism-related jobs.

As Table 15.6 shows, throughout Europe there is a strong dependence on female labour in the hotel and restaurant sectors, with some of the highest levels being found in the more developed economies (for example, Norway, Sweden and the

UK). There are other significant characteristics of tourist employment, not shown in the comparative data, that relate to the quality of jobs created and whether the jobs are filled by nationals or immigrants. In terms of employment quality there is strong evidence that many jobs in the tourist industry are semi-skilled or unskilled. It has been estimated in Spain, for example, that these low-skilled jobs account for about 98 per cent of employment. Many of these jobs, and especially those in hotels, are of a seasonal nature, and in Spain about 44 per cent of all employment in hotels is seasonal. As Valenzuela (1988, p. 52) points out, this is the national figure and in the Costa del Sol and the Balearics seasonal employment accounts for almost 75 per cent of jobs. There is also a high proportion of tourism-related employment that is part-time, especially for women. In the UK, part-time employment accounts for at least 47 per cent of all hotel and catering jobs, while 31 per cent of all jobs in tourism, leisure and related services are part-time (Goodall, 1987, p. 116).

Levels of skill, the ratio of male/female employment, seasonality and the number of part-time workers are dependent on the nature of the tourism sector, the scale and ownership of enterprises and the nature of the local economy (Shaw,

Table 15.6 *Employment in tourism in Western European countries, 1985*

	% male	Total
Austria	36.2	117,028
Belgium[a]	49.4	56,850
Finland[b]	20.6	63,000
France,[a,b,e]	54.9	284,964
West Germany	40.3	690,000
Greece[b,c,d]	NK	39,600
Norway[b]	25.0	48,000
Netherlands	58.3	53,000
Portugal[e]	52.7	32,899
Sweden	36.2	80,000
Switzerland[a,b]	NK	178,200
United Kingdom[b]	38.4	476,100

Notes:
[a] 1983.
[b] Annual coverage.
[c] Considerable underestimate which does not include uninsured employees.
[d] Employers covered by the Hotel Employment Insurance Fund.
[e] 1984.
NK Not known.
Source: OECD (1986, p. 182).

Table 15.7 *Farms offering tourist accommodation, 1979–81 (estimated percentage)*

		Notes
Austria	9.8	Tyrol: 28% Vorarlberg: 15% Salzburg: 20%
England and Wales	8.0	
Finland	1.3	Aland: 32%
France	3.0	
West Germany	4.0	
Ireland	0.2	
Norway	2.9	In some municipalities up to 40%
Spain	0.4	
Sweden	20.0	
Scotland	5.8	Highlands and Islands: up to 16%

Source: Dernoi (1983).

Williams and Greenwood, 1987). Such factors also affect the geographical origin of the workforce, and the number of local people who are provided with tourism-related employment. Barker (1982) has, for example, shown that in many of the specialised ski resorts in France, the impact of corporate capital from outside the Alpine region has limited employment for local people to mainly unskilled, seasonal jobs. By contrast, in other countries low-skilled jobs are the domain of in-migrants. Thus, in Spain the most underdeveloped provinces of Andalusia (Granada and Jaen) provide the least-qualified labour force in the hotels of the Costa Brava (Valenzuela, 1988, p. 52).

International migration is another important dimension of tourism employment. The proportion of immigrants employed in hotels can be considerable, ranging from 18.3 per cent in Sweden, through 23.6 per cent in West Germany, to 39 per cent in Switzerland. In Switzerland only one-third of these immigrants have permanent jobs, and most Swiss will not work in the hotel and restaurant sectors because the jobs are poorly paid (Gilg, 1988, p. 134).

Set against these problems is the fact that small-scale tourism developments, especially in rural areas, appear to provide a much-needed boost to household incomes. In a survey of farm tourism enterprises in England, it was found that tourism supplemented farm income by between 13 and 19 per cent, while a quarter of the activities provided extra seasonal or part-time employment (Frater, 1983). The situation is similar in other parts of Europe; thus in Finland tourism has contributed 5–15 per cent extra income to farm-based enterprises, while in Austria the figure is around 4 per cent (Dernoi, 1983). Farm tourism is also fairly flexible, and unlike formal hotel structures it can expand or contract as the market demands. These types of benefits, together with a growing demand from tourists, have led to a growth in farm tourism in most countries (Table 15.7). In the 1960s and 1970s Spain, for example, witnessed an increase in both the scale and regional development of farm tourism. While the number of bed spaces increased from 1,152 in 1967 to 32,038 in 1977, the number of provinces involved grew from two to 41 over the same period (Dernoi, 1983).

Within the context of local economies, whether rural or urban based, it is difficult to evaluate the total level of employment created by tourism due to multiplier effects and external leakages. Furthermore, much of the evidence is somewhat fragmentary in nature, often being based on individual case studies (Vaughan, 1986). Thus, a review of tourist developments in the UK found estimates that the National Exhibition Centre in Birmingham provided almost 4,000 jobs from both direct and indirect employment (Law, 1987). Similarly, in Bath, domestic and international tourists help create jobs for some 15 per cent of the city's workforce and in 1986 they generated an income of £45 million. In the case of this city, however, it has been estimated that only a third of earnings are retained in the local economy (Bath City Council, 1987).

Even in those economies dominated by large-scale tourism enterprises such as Spain, tourism has made a contribution to improvements in living standards. Indeed, employment creation (both direct and indirect) in the major tourism regions of Spain provides powerful reasons why mass tourism, even with all its inherent problems, has been a major force. Thus, the industry is directly responsible for the Balearics and Gerona having the highest level of per capita income among Spain's 50 provinces (Valenzuela, 1988).

Tourism and regional development

Tourism is playing an increasing role in regional development policy, especially within the European Community. In 1984, tourism was specifically identified by the European Regional Development Fund (ERDF) as having much to offer in the creation of jobs, since it is a labour-intensive industry (Pearce, 1988). Grants have largely followed the general pattern of other regional assistance, although at the international scale a bias is evident towards the UK: between 1975 and 1985 61 per cent of the 251 million ECU invested in tourism by the ERDF was allocated to Britain. Italy accounted for a further 19 per cent. The general aims have been to improve infrastructure, which has absorbed almost a third of all investment, and to concentrate aid on the tourist attractions sector in the hope of drawing tourists to problem regions.

These EC schemes are, however, pre-dated by many other national attempts to use tourism as a tool for regional growth (Williams and Shaw, 1988). For example, in Switzerland a system of regional aid was introduced in 1974, with loans being made available to regions throughout the Alps and the Jura, and by 1985 some 52 development programmes had been approved (Gilg, 1988, p. 137). All types of policy initiative are to

some extent conditioned by the structure of the regional economy. In this context two factors are important. First, there is the question of whether tourist development is integrated into an existing settlement or is on a new non-integrated site. The scale of development will obviously determine the degree of integration, and a continuum can be recognised, ranging from small, highly integrated projects (such as farm-based initiatives) to large, non-integrated holiday complexes. In the Alps, for instance, there are major differences between high-level ski villages, built above the traditional settlement line, and low-level centres where tourism is more integrated with existing villages (Barker, 1982). In this example, a large high-level ski centre may well be developed by capital and labour from outside the local area. Similar problems have emerged in southern Italy, where large capital-intensive enterprises, unrelated to their local setting, have been developed. King (1988, p. 77) claims that these large hotels, many of them built with public money or regional aid, have done little to improve living standards in southern Italy, since almost a third of tourist revenue returns to the capitalists living mainly in Rome and Milan.

A second, and related, question is whether any new tourist developments are able to link with and utilise local resources. The initiatives most able to achieve such linkages are, again, small-scale developments and for this reason farm tourism has received increased attention from policy-makers. The arguments for farm-based tourism are discussed by Ilbery in Chapter 13, while an official perspective has been provided by the Commission of the European Communities (1987b). France has probably made most progress in integrating and utilising rural resources, and now has a long history of rural tourism planning through such schemes as *Logis de France* (established in 1949) and *Gîtes de France*. In other countries developments have been more recent and specialised. Thus, in Greece incentives offered by the Council for Equality have recently led to the establishment of eight women's agritourist co-operatives, offering tourist accommodation in rural areas (Commission, 1987b). However, as the popularity of rural-based tourism increases, there may well be a tendency for the marketing, and therefore some of the control over resources, to pass into the hands of larger companies. In some countries, such as the

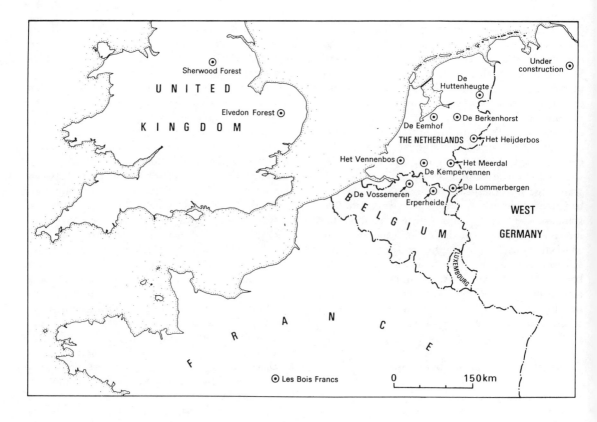

Figure 15.4 *Center Parcs holiday complexes*

Netherlands and the UK, such trends are already evident, with outside capital being invested in large-scale, self-contained holiday complexes such as those built by *Center Parcs*. This firm has now extended its activities from the Netherlands to Belgium and the UK (Figure 15.4).

It should also be stressed that tourism is no longer just the preserve of rural-based, regional policies; increasingly, it is also being used as an economic development instrument in many old industrial regions. A recent survey of 100 local authorities in England, including those controlling industrial areas, found that 94 thought tourism was either 'very important', or 'important' to their local area; moreover, 41 per cent had held these views for less than five years. This growing recognition of the importance of tourism in declining urban, industrial centres is based on both employment creation and the fact that tourism brings obvious environmental gains (Shaw, 1988).

The future of tourism development

All the signs point to tourism becoming an increasingly important force in most European economies. The growth in holiday-taking has yet to peak, as there are still substantial proportions of Europeans who do not take regular holidays. Furthermore, the increasing segmentation of the market, together with concentration tendencies within the industry itself, are producing a more complex arrangement of tourist flows and ownership patterns. Linked with these trends is the recognition of tourism's importance by most governments, particularly in the promotion of regional development and job creation. In this respect, there is likely to be an increasing involvement of public policy in tourism, both at a national level and through the European Community.

Previous models of tourism's impact on

development have concentrated on the spatial spread of influences from larger central places to remoter, rural communities (Kariel and Kariel, 1982). Such studies have viewed tourism as a vector or carrier of the urban-based culture into traditional, agrarian communities. The spiral of development associated with such changes involves an increase in tourism provision, leading to the growth of larger enterprises funded by external organisations. In the context of future tourism developments in Western Europe these past models are too limited in scope, since urban tourism is also of growing importance and tourism is being used as a spur to regeneration in many old industrial communities. In these cases, tourism is spreading different socio-economic values into some inner-city communities, and the spatial vectors of change are therefore more than simple urban–rural influences. This marketing of urban areas is also leading to a reassessment of traditional planning methodologies in such countries as the UK and the Netherlands (Ashworth and Voogd, 1988), as urban tourism increases its share of the market.

The intensification of the cultural and environmental problems associated with mass tourism will also dictate the future nature of development strategies. In Alpine countries there is already a growing demand for the more widespread introduction of 'soft' or 'green' tourism (Jungk, 1980). These terms encompass the use of existing, local facilities for tourism and the sale of locally produced goods. The idea is to limit the amount of technical infrastructure and thereby reduce tourism's adverse impact on the local community, while at the same time providing a boost to the local economy (Mose, 1988). Such smaller-scale developments more closely fit the demands of a growing number of 'critical consumer tourists' (Krippendorf, 1984). Future tourism developments will therefore become more polarised between the large-scale mass tourism market and those regions that pursue a more selective or 'green' approach to tourism.

References

Airey, D. (1983) European government approaches to tourism, *Tourism Management*, 4, 234–44.

Alexandrakis, N.E. (1973), *Tourism as a Leading Sector in Economic Development: a case study of Greece*, University Microfilms, Ann Arbor, MI.

Ashworth, G. (1988) Marketing the historic city for tourism. In B. Goodall and G. Ashworth (eds), *Marketing in the Tourism Industry*, Croom Helm, London, pp. 162–75.

Ashworth, G.J. and Bergsma, J. (1987) Policy for tourism: recent changes in the Netherlands, *Tijdschrift voor Economische en Sociale Geografie*, 78, 151–5.

Ashworth, G. and Voogd, M. (1988) Marketing the city: concepts, processes and Dutch applications, *Town Planning Review*, 59, 65–79.

Barker, M.L. (1982) Traditional landscape and mass tourism in the Alps, *Geographical Review*, 72, 395–415.

Bath City Council (1987) *Economics of Tourism in Bath*, Coopers and Lybrand Associates, London.

British Tourist Authority, (1985) *British Tourism Survey*, BTA, London.

British Tourist Authority, (1987) *British Tourism Survey*, BTA, London.

Buck, M. (1988) The role of travel agent and tour operator. In B. Goodall and G. Ashworth (eds), *Marketing in the Tourism Industry*, Croom Helm, London, pp. 67–74.

Buckley, P.J. and Witt, S.F. (1985) Tourism in difficult areas: case studies of Bradford, Bristol, Glasgow and Hamm, *Tourism Management*, 6, 205–13.

Commission of the European Communities (1982) *A Community Policy on Tourism*, Brussels.

Commission of the European Communities (1985) *Tourism and the European Community*, Brussels.

Commission of the European Communities (1987a) *Europeans and their Holidays*, vii/165/87-EN, Brussels.

Commission of the European Communities (1987b) *Rural Tourism in the 12 Member States of the European Community*, Directorate-General for Transport, Brussels.

Dernoi, L.A. (1983) Farm tourism in Europe, *Tourism Management*, 4, 155–66.

English Tourist Board (1986) *Jobs in Tourism and Leisure: an occupational review*, London.

English Tourist Board (1987) *The Short Break Market*, London.

European Travel Commission (1988) *40 Years of Joint Action 1948–1988*, Paris.

Frater, J. (1983) Farm tourism in England: planning, funding, promotion and some lessons from Europe, *Tourism Management*, 4, 167–79.

Gilg, A. (1988) Switzerland: structural changes within stability. In A.M. Williams and G. Shaw (eds), *Tourism and Economic Development*, Belhaven, London, pp. 123–44.

Goodall, B. (1987) Tourism policy and jobs in the UK, *Built Environment*, 13, 109–23.

Goodall, B. (1988) Changing patterns and structure of

European tourism. In B. Goddall and G. Ashworth (eds), *Marketing in the Tourism Industry: the promotion of destination regions*, Croom Helm, London.

Heape, R. (1983) Tour operating planning in Thomson Holidays UK, *Tourism Management*, 4, 245–53.

Holloway, J.C. (1985), *The Business of Tourism*, 2nd edn, Pitman, London.

IMF (1985) *Balance of Payments Statistics*, 36, International Monetary Fund, Washington, DC.

Jungk, R. (1980) Wieviel Touristen pro Hektar Strand?, *GEO*, 10, 154–6.

de Kadt, E. (ed.) (1979), *Tourism, Passport to Development*, Oxford University Press, Oxford.

Kariel, H.G. and Kariel, P. (1982) Socio-cultural impacts of tourism: an example from the Austrian Alps, *Geografiska Annaler*, 64B, 1–16.

Keynote Report (1985) *Travel Agents/Overseas Tour Operator: An Industry Sector Overview*, London.

King, R. (1988) Italy: multi-faceted tourism. In A.M. Williams and G. Shaw (eds), *Tourism and Economic Development*, Belhaven, London, pp. 58–79.

Kosters, M.J. (1981) *Focus op Toerisme*, VUGA, Amsterdam.

Krippendorf, J. (1984) *Die Ferienmenschen. Für ein neues Verständnis von Freizeit und Reisen*, Orell Füssli, Zurich.

Krippendorf, J. (1986) The new tourist-turning point for leisure and travel, *Tourism Management*, 7, 131–5.

Law, C.M. (1985) *The British Conference and Exhibition Business*, Manchester, University of Salford Department of Geography Working Paper No. 2.

Law, C.M. (1987) Conference and exhibition tourism, *Built Environment*, 13, 85–95.

Leontidou, L. (1988) Greece: prospects and contradictions of tourism in the 1980s. In A.M. Williams and G. Shaw (eds), *Tourism and Economic Development*, Belhaven, London, pp. 80–100.

Mathieson, A. and Wall, G. (1982) *Tourism: economic, physical and social impacts*, Longman, London.

Mihovilovic, M.A. (1980) Leisure and tourism in Europe, *International Social Science Journal*, 32, 45–52.

Mitchell, L. (1984) A geographical analysis of leisure activities: a life style case study. In J. Long and R. Hecock (eds), *Leisure, Tourism and Social Change*, University of Edinburgh, Edinburgh, pp. 67–74.

Morrell, J. (1985) *Employment in Tourism*, British Tourist Authority, London.

Mose, I. (1988) *Sanfter Tourismus im Nationalpark Hohe Tauern*, Vechta, Vechtaer Arbeiten zur Geographie und Regionalwissenschaft, Vol. 6.

Murphy, P.E. (1985) *Tourism: a community approach*, Methuen, New York.

OECD (1974) *Government Policy in the Development of Tourism*, Paris.

OECD (1986) *Tourism Policy and International Tourism in OECD Countries*, Paris.

Pearce, D.G. (1987a) Spatial patterns of package tourism in Europe, *Annals of Tourism Research*, 14, 183–201.

Pearce, D.G. (1987b) Mediterranean charters — a comparative geographic perspective, *Tourism Management*, 8, 291–305.

Pearce, D.G. (1988) Tourism and regional development in the European Community, *Tourism Management*, 9, 13–22.

Pinder, D. (1988) The Netherlands: tourist development in a crowded society. In A.M. Williams and G. Shaw (eds), *Tourism and Economic Development*, Belhaven, London, pp. 214–29.

Schnell, P. (1988) The Federal Republic of Germany: a growing international deficit? In A.M. Williams and G. Shaw (eds), *Tourism and Economic Development*, Belhaven, London, pp. 196–213.

Shaw, G. (1988) The role of tourism in the UK economy. In F. Zimmermann (ed), *Tourism and Development*, IGU Commission on Geography of Tourism and Leisure, Klagenfurt.

Shaw, G., Williams, A.M. and Greenwood, J. (1987) *Tourism and the Economy of Cornwall*, University of Exeter Tourism Research Group, Exeter.

Shaw, G., Williams, A.M., Greenwood, J. and Hennessy, S. (1987) *Public Policy and Tourism in England: a review of national and local trends*, University of Exeter Tourism Research Group, Working Paper no. 3, Exeter.

Shaw, G., Greenwood, J. and Williams, A.M. (1988) The United Kingdom: market responses and public policy. In A.M. Williams and G. Shaw (eds), *Tourism and Economic Development*, Belhaven, London, pp. 162–79.

Singh, B.P. (1984) *The Impact of Tourism on the Balance of Payments*, CPER, Athens.

Spanoudis, C. (1982) Trends in tourism planning and development, *Tourism Management*, 3, 314–18.

Stalinbrass, C. (1980) Seaside resorts and the hotel accommodation industry, *Progress in Planning*, 13, 103–74.

Sundelin, A. (1983) Tourism trends in Scandinavia, *Tourism Management*, 4, 262–8.

Travis, A.S. (1982) Leisure, recreation and tourism in Western Europe, *Tourism Management*, 3, 3–15.

Tuppen, J. (1985) *Urban Tourism in France*, Manchester, University of Salford, Department of Geography Working Paper no. 3.

Tuppen, J. (1988) France: the changing character of a key industry. In A.M. Williams and G. Shaw (eds), *Tourism and Economic Development*, Belhaven, London, pp. 180–95.

UNCTC (1982) *Transnational Corporations in International Tourism*, UNCTC, New York.

Valenzuela, M. (1988) Spain: the phenomenon of mass tourism. In A.M. Williams and G. Shaw (eds),

Tourism and Economic Development, Belhaven, London, pp. 39–57.

Vaughan, D.R. (1986) *Estimating the Level of Tourism-related Employment: an assessment of two non-survey techniques*, English Tourist Board, London.

Waters, S.R. (1967) Trends in international tourism, *Development Digest*, 5, 57–62.

White, P.E. (1976) Tourism and economic development in the rural environment. In R. Lee and P. Ogden (eds), *Economy and Society in the EEC: spatial perspectives*, Saxon House, Farnborough.

Williams, A.M. and Shaw, G. (1988) Western European tourism in perspective. In A.M. Williams and G. Shaw (eds), *Tourism and Economic Development: West European experiences*, Belhaven Press, London,

pp. 12–38.

Wolfson, M. (1964) Government's role in tourism development, *Development Digest*, 5, 50–6.

World Tourism Organisation (1984) *Economic Review of World Tourism*, World Tourism Organisation, Madrid.

World Tourism Organisation (1987) *Compendium of Tourism Statistics*, 8th edn, World Tourism Organisation, Madrid.

Young, G. (1973) *Tourism: blessing or blight?*, Penguin, Harmondsworth.

Zimmermann, F. (1988) Austria: contrasting tourist seasons and contrasting regions. In Williams, A.M. and Shaw, G. (eds), *Tourism and Economic Development*, Belhaven, London, pp. 145–61.

16

Conservation and the rural landscape

Brian J. Woodruffe

The European Campaign for the Countryside of 1987–8 emphasised the need for the protection and careful management of rural resources but also laid stress on the importance of keeping people resident in rural areas. This venture by the 21 member states of the Council of Europe was aimed at bringing to wider attention the point that conservation is not simply a means of protecting wildlife or landscape; it must also be involved with the economic and social functions of the countryside. The European Strategy of Conservation, established at the same time, stated that economy had to become synonymous with ecology, and added that a guiding principle was the integration of environmental policies within overall policies (Council of Europe, 1987a).

These proposed changes in approach to conservation as a process and as an exercise set out to counteract the myopic views of conservation and environmental protection held by some Western European states. There is no doubt that some countries, for reasons of prestige, have designated many parks and reserves but have executed little protection or conservation on the ground. The problems arising in the rural environment today demand solutions that cannot be implemented simply by the scheduling of land in one category or another. Instead, a broad-based and comprehensive approach with international co-operation is crucial for the survival of the countryside. Such an approach is in its infancy and must wait to be evaluated in the future. This chapter therefore looks back to appraise and assess some of the concepts and methods used in the recent past to conserve the rural landscapes of Western Europe. The issues discussed are of significance throughout the continent but they will be considered

primarily with reference to Alpine countries, where a number of very different national, regional and local frameworks have been devised to deal with the problems of continental proportions now threatening this region. These problems are widespread throughout the Alps, and they arise essentially from the new uses now being made of rural land resources — for non-traditional activities, such as specialised leisure pursuits, and for improved access to modern, functional settlements. But the problems are not unique to the Alps; they can also be recognised in some parts of lowland Europe, and especially along the Atlantic and Mediterranean seaboards where conflicts between tourism development, nature conservation and landscape protection are very evident. In all these areas, the principles and practices of conservation tend to be similar, but this chapter's focus on the Alpine environment allows more finely-tuned comparisons to be made. In making such appraisals, however, it needs to be noted that few formal evaluations have been undertaken of the effectiveness of past policies and plans, and moreover, very few methods exist to aid such evaluations.

In general, four clear themes dominate the planning framework for conservation in most of the countries concerned: the protection of scenic landscapes; the reservation of land for recreation and leisure purposes; the designation and management of wildlife reserves; and preservation and renovation of the cultural heritage. The four themes are not mutually exclusive and have not been treated separately in all the countries considered. Nevertheless, a common initial approach to conservation has been the categorisation of rural environments under a variety of labels, and it is

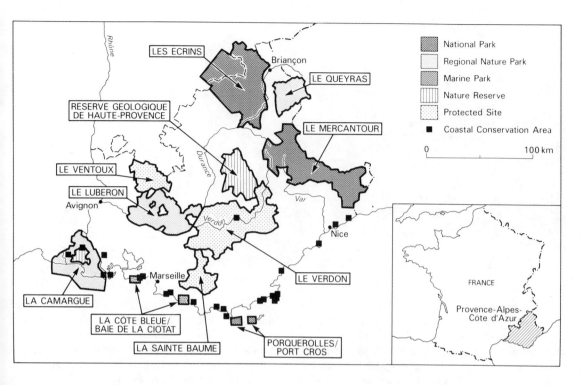

Figure 16.1 *Protected land in the region of Provence-Alpes-Côte d'Azur, France*

evident that in some regions extensive areas of land have been designated (Figure 16.1). The terminology used is not always a clear guide to the status of protection nor to the quality of management required, and efforts by the Council of Europe and by the International Union for the Conservation of Nature and Natural Resources (IUCN) to remedy this situation have not been entirely successful. It has been argued that a common terminology and comparable categories would aid the management of protected areas in Western Europe, and lead to a better understanding between governments and organisations involved with conservation across the continent.

Landscape conservation

The increasing pace and extent of urbanisation (Figure 16.2) has stimulated efforts to limit and direct development to the most suitable situations, and protection policies have been a notable part of this planning process. Many of these policies have involved landscape protection measures, so that no country in Western Europe is now without

them in one form or another. Paralleling this desire to keep attractive and heritage-rich landscapes free from pervasive development has come the need to set aside space for recreation and leisure-time activities. Many conserved landscapes can be used for informal recreational pursuits, and it is not unusual to find some protected areas, such as nature parks, being managed to meet the integrated objectives of scenic conservation and recreational enjoyment. Because of this duality of interest it is appropriate to discuss these two themes together.

National parks

At the top of the hierarchy of conserved landscapes in most European countries stand the national parks. Some are now very extensive and have been established for decades. Sweden was the first European country to designate national parks, and Table 16.1 shows how the system has evolved since 1909 when nine areas were set aside. The paramount aim during this time has been to attempt to keep the natural conditions of

259

Figure 16.2 *The urbanisation of villages within scenic landscapes. Urbanisation has stimulated demands for conservation measures to be integrated with development plans – Natters and Mutters, Tirol, Austria.*

these areas essentially intact and undisturbed. That they have outstanding aesthetic appeal or wildlife interest for the general public is a secondary consideration. Not all the parks are completely free of human use. Lapps still exercise their ancient rights to use the contiguous arctic wildernesses of Stora Sjöfallet, Sarek and Padjelanta as hunting and grazing grounds, although these rights are tolerated rather than sanctioned. Part of Stora Sjöfallet park contains a hydro-electric power scheme which has regulated river-flows and led to road construction (Curry-Lindahl, 1968). Compared to national parks in central Europe, these inspiring tracts of arctic scenery are under little threat of disturbance from tourists and are protected by their remoteness and difficulty of access. Another difference stems from the fact that, compared with most national parks elsewhere, all the Swedish parks are sited on state-owned land and thus problems with ownership or incompatible land uses do not arise. In many ways the Swedish system is a model for other countries to follow; it has succeeded in conserving vast areas of high landscape and natural value over a long period of time. Moreover, in designating not only wild countryside but also small islands (Ängsö and Bla Jungfrun), tracts of ancient forest (Norra Kvill), raised bogs (Store Mosse), and even traditionally farmed landscapes (Garphyttan), Sweden has demonstrated a flexibility in approach towards conserving its national heritage.

The mountain scenery of the Alpine region, extending almost 1,000 kilometres through seven countries, from the Julian Alps in Yugoslavia to the Maritime Alps of Mediterranean France, would seem to lend itself to conservation policies. In general it does, and yet in all this extent only eight national parks, covering little more than 5,600 square kilometres, have been established. Moreover, none of these parks is in state ownership, and current proposals envisage the creation of only one further park (in Austria). In West Germany, Austria and the South Tirol (Italy), a host of other landscape policies have been applied, but few of these are backed by national (as opposed to provincial) legislation. Furthermore, it has been argued that the implementation of existing policies has been quite ineffective in some national parks, most notably in the Stelvio NP in Italy, and in the Vanoise NP of France. Paradoxically, the lack of firm policy execution has been one reason advanced against the establishment of a national park in the Hohe Tauern range in Austria.

There does not appear to be a lack of good reasons to support further landscape conservation measures in the Alps. The region acts as a major recreational resource for the urban-oriented populations of Western Europe, and threats from even moderate levels of tourist infrastructure are penetrating way beyond the most accessible valleys and to heights far above normal residential altitudes. Expanding holiday resorts, purpose-built skiing centres and transportation networks are the sharp end of the economic and ecological changes infiltrating physical and social environments alike (Jülg, 1984; Knafou, 1987; Haimayer, 1987). The

Table 16.1 *National parks in Sweden*

Park	Date established	Area (ha)
Abisko	1909	7,700
Stora Sjöfallet	1909	127,800
Sarek	1909	197,000
Pieljekaise	1909	15,340
Sånfjället	1909	2,623
Hamra	1909	29
Ängsö	1909	73
Garphyttan	1909	111
Gotska Sandön	1909	3,700
Dalby Söderskog	1918	37
Vadvetjåkka	1920	2,630
Blå Jungfrun	1926	66
Norra Kvill	1927	28
Töfsingdalen	1930	1,590
Muddus	1942	49,340
Padjelanta	1962	198,400
Store Mosse	1982	7,750
Tiveden	1983	1,353
Skuleskogen	1984	2,500
Stenshuvud	1986	302
Total		618,372

Source: Statistical Abstract of Sweden (1989).

creation of more national parks or protective policies will not diminish the demand for facilities nor reduce visitor numbers but might, if combined with strategies to absorb the pressure, temper and deflect the impact on the sensitive Alpine landscape and prevent permanent ecological damage.

The existing national parks in the Alps differ greatly in concept and functions. The French parks are structured around a core/fringe concept; the Gran Paradiso NP in north-west Italy was formerly a royal hunting reserve presented to the nation in 1922; the Stelvio NP is centred on the 3900 m peak of the Ortler, but its perimeter encloses permanent settlements and farmed areas in neighbouring valleys (Figure 16.3). In this park summer skiing has penetrated into the glacial heartland and neither core nor periphery are said to justify national park status. By contrast the delimitation of Berchtesgaden NP in West Germany carefully avoided settled and pastoral land and, though visitors are not discouraged, it is being managed as a prime nature reserve, which it was before enlargement and upgrading in 1978.

In all the parks mentioned, scientific research does not figure highly as a function of the protected status, although some ecological investigation occurs in most parks. The Swiss NP, on the other hand, functions primarily as a reserve for scientific study (Schloeth, 1976). Covering 169 square kilometres of the Engadine, it encloses a series of valleys and peaks which, by Alpine standards, are only of modest aesthetic appeal. There are no glaciers, no permanent snowfields, no snow-capped mountains in summer. Conservation is effected by minimal management, and agriculture and forestry are now absent as the rights of local communities have been bought out. Ecological and physical changes are permitted to occur without interference, since one of the scientific objectives is to study and record them. A limited number of visitors is catered for, and visitor numbers have steadily increased to around 350,000 a year. But visitor access is largely restricted to a main road which bisects the park, where several small car parks that are the starting points for wardened trails have been sited. This restriction on visitor access, coupled with the size of the car parks (400 spaces in all), effectively limits the number of visitors at any one time as roadside parking is not possible. It undoubtedly makes a major contribution to the success that has been achieved in conserving the park. A visitor information centre has been developed in an attractive Engadine-styled building in Zernez, some 5 km outside the park. Here visitors are reminded of the park's specific aims, and it is made clear that it is not a playground or recreation area but a place where nature can be encountered on nature's terms. This strict strategy, it is important to note, has been devised and implemented without the introduction of state control. The park was established in 1914 through the joint efforts of the Swiss Society for Natural Sciences and the Swiss League for Nature Conservation, and in effect it is still run by them. Although they can be considered as semi-public organisations, neither is a government agency and the state provides only a part of their funding.

No national park in mainland Europe can be entirely free of external influences, and the Swiss NP is no exception. While no interference with wildlife occurs within the park, it is accepted that animals may be hunted should they wander beyond the boundaries. The red deer are culled when, in hard winters, they migrate to the surrounding valley floors, and the migrant birds

Figure 16.3 *The Stelvio National Park, Italy. The boundary of the park extends beyond the glacial heartland to enclose the living and working landscapes of its neighbouring communities.*

breeding in the park may clearly be affected by circumstances in their wintering sites (Schloeth, 1975). In recent years some tree species have been badly affected by acid rain and consequently the ecological quality of the park is changing; this has been accepted as a feature to be scientifically investigated. As the later discussion will demonstrate, compared to other national parks in Alpine Europe, the Swiss NP is therefore a very distinctive type with a particular concept underpinning its prime function.

The five national parks of mainland France are all sited in mountainous regions — the Cevennes in the Massif Central, the High Pyrenees, and the three Alpine ones of the Ecrins, Vanoise and Mercantour. All comprise a core or central zone set in a peripheral envelope (Table 16.2). The idea of a core/fringe structure is a very positive one in that it recognises that, because an attractive landscape will draw people and activities to it, necessary provision has to be made for them. In providing an outer zone where visitor pressure can be absorbed, the impact on the core is likely to be reduced and tighter controls can be exercised to conserve the natural environment that is the fundamental reason for the park's existence. In the central zone 'any action likely to harm the natural evolution of the fauna or flora and to alter the natural character of the park' is prohibited (INSEE, 1985). The concept also acknowledges

Table 16.2 *National parks, mainland France*

Park	Date established	Area (km^2) Core zone	Total	Population 1982
Cevennes	1970	844	3,209	41,272
Les Ecrins	1973	918	2,288	27,639
Mercantour	1979	685	2,150	16,445
Pyrenées Occidentales	1967	457	2,632	40,223
Vanoise	1963	528	1,993	27,973
Totals		3,432	12,272	153,552

Source: INSEE (1985).

that mountain areas possess resources for other uses — power generation, water supply, skiing terrain — and that, if rural people are to be retained in remote situations, development of new employment opportunities must be encouraged. However, for any given park, much depends on the way the concept is applied and on the manner in which policies and plans are implemented in both particular localities and the region as a whole.

In all French parks, enlargement of the tourist economy is highly significant for the rural communities as the traditional Alpine pastoral system of agriculture can no longer compete with the more productive lowland regions. Alternative forms of employment in the peripheral zones are therefore essential to sustain these communities (see also Chapter 13). At present, construction and service trades associated with tourism appear to be providing these alternatives and populations in all parks except the Pyrenees are steadily rising. Nevertheless signs pointing to future problems are appearing. In the Ecrins NP (Figure 16.1) and the Vanoise NP substantial tourist complexes have been constructed in the peripheral zones in recent years (Figure 16.4), and it is arguable that the core areas are being threatened (Préau, 1976; Knafou, 1987). Winter sports centres, often linked with one another by cable systems, abound on the northern slopes of the Vanoise — Tignes, Les Arcs, La Plagne, Courchevel, Méribel and Val Thorens. Here skiing pistes abut the perimeter of the central zone. Both parks contain high-level snowfields suitable for summer skiing and, as other locations become saturated, pressure to open these up is likely to increase. In the Ecrins, summer ski lifts from the resort of les Deux-Alpes

have already reached the snowfields adjacent to the central zone. Despite these environmental impacts in the peripheral zones, it can still be argued that these unique landscapes are being protected and the future of the mountain communities is being safeguarded. The core/fringe concept appears, therefore, to have the property of being able to blend development with conservation.

In Italy's Gran Paradiso NP, which adjoins the Vanoise NP, this same core/periphery concept has been used but in a more sophisticated form. To meet needs other than landscape conservation a four-zone system has been formulated for the central part: zone A comprises the wild heartland of mountain terrain (70.7 per cent); zone B is reserved for forestry and pastoral uses (28.5 per cent); zone C is reserved for siting recreational facilities (0.2 per cent); and zone D provides an area for the controlled development of rural settlements without prejudicing environmental quality (0.6 per cent). Around the perimeter of this zoned area, a buffer zone has been defined to protect the park from the urbanisation taking place in those settlements beyond the boundary. This approach is attempting, therefore, to avoid the problem of highly developed peripheral envelopes while allowing selective changes to occur (Gambino and Jaccod, 1985).

Other landscape protection measures

The zonation of land and zoning policies are features of land-use planning in many Western European countries and extend to the field of landscape conservation. Zoning is a versatile method of policy application; it has been utilised in areas of outstanding scenic importance, in regions of average landscape quality, in intensively used countryside, in situations where land has been abandoned, and where structural improvements have been planned. One region where a zoning approach has been in operation for almost two decades is the Bavarian Alps of West Germany. This is an area that is intensively used for recreation all year round, to the extent that by the early 1970s new roads, tracks, ski runs and cable systems were seen to be eroding the quality of the natural environment and threatening its stability. In an attempt to curb the extension of these networks, a set of landscape

Figure 16.4 *Val Claret in the Vanoise National Park, France. The French national parks have attracted new resorts to their peripheral zones with consequent impacts on the landscape.*

planning regulations was approved by the *Land* government of Bavaria in 1972. These sub-divided the whole 230 km stretch of the Bavarian Alps into three zones (Bay. SLU, 1973). Zone A covered land close to settlements and imposed restrictions which emphasised existing legal requirements. In zone B infrastructural developments were permitted only if they met very strict planning conditions. The tightest controls were in zone C, which covered 42 per cent of the region and was designated a 'quiet zone'. Here any extension to existing means of access was prohibited. Since its inception this scheme has controlled access very carefully and has contributed to the safeguarding from erosion of many sites, especially in the middle-altitude landscapes of zone B (Ruppert, 1984). Furthermore, this system of zonation complements other protection policies for other features of the landscape. Without specifying the details, it is worth recording that many forests and woodlands in Bavaria have been scheduled because they afford protection against soil erosion, noise penetration and air pollution, in addition to their timber-producing and wildlife conservation functions.

West Germany, Austria and the South Tirol province of Italy all employ a local form of landscape conservation area known as a *Landschaftsschutzgebiet*. Such zones are defined and backed by regulations conceived at provincial or *Land* level. The objectives for these areas, which are

often under productive agriculture or forestry, are to maintain the scenic beauty and to ensure that characteristic facets of the landscape are neither damaged nor destroyed. In pursuing these objectives it is recognised that human activities, particularly farming, may contribute towards the management of the landscape and that in some localities some form of economic support may be necessary to keep traditional systems intact. The Tirol regional government actively supports its farm families with social subsidies to retain this 'landscape gardening' aspect.

The size and composition of *Landschaftsschutzgebiet* exhibit great variety. In Austria, for example, the planning department of Carinthia has identified some 50 small landscape features such as lake-shores, woodland clusters, prominent viewpoints and knolls with churches (Kärntner Landesregierung, 1976). Salzburg province, by contrast, has focused on large tracts of valley floors and mountain ranges; in the absence of a Hohe Tauern national park, the northern side of the Grossvenediger massif has been given protection but the southern flanks, administered by the Tirol, still await some form of conservation measure (OROK, 1978).

Over 43 per cent (324,000 ha) of the South Tirol countryside of Alpine Italy has been designated as conserved landscape, and increasing amounts of finance are being made available to conserve its highly valued and unique landscapes

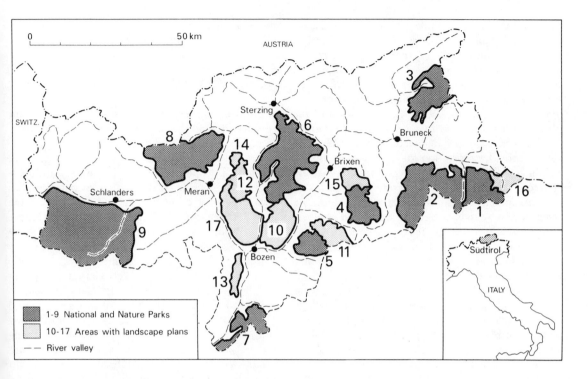

Figure 16.5 *Landscape protection in Südtirol, Italy.*
Key to areas: 1. Sextner Dolomites; 2. Fanes–Sennes; 3. Rieserferner; 4. Puez-Geisler; 5. Schlern; 6. Sarntal Alps;
7. Hornspitz; 8. Texel group; 9. Stelvio National Park; 10. Ritten; 11. Seiser Alm; 12. Meran 2000;
13. Montiggler Forest; 14. Hirzer; 15. Plose; 16. Sextner Valley; 17. Tschoggelberg.

(Südtiroler Landesregierung, 1983). This region includes some of the most spectacular scenery of the Dolomites and also possesses a rich cultural heritage of castles, religious buildings and vernacular architecture. These cultural features are protected not only as monuments but also in their setting. Surrounding many of them are *Bannzonen*, inviolate areas where change is only permissible if it respects traditional building styles and land uses. While it is admirable to have large acreages of countryside under protection, one of the problems facing the South Tirol planning authorities is that much modification and damaging use has occurred prior to designation. Efforts are now being made to restore sites and soften the impact of some of the new structures (Südtiroler Landesregierung, 1979a; 1988b). A more general point relating to the implementation of *Landschaftsschutzgebiet* policies is that they are very dependent on the seriousness with which conservation is viewed at the provincial level. Problems tend to arise when protected areas are also seen as public recreation sites and facilities begin to grow beyond informal car parks, picnic tables and footpaths.

There is another type of conservation measure operative in South Tirol which is not found in many other situations. In 1972 the government implemented a series of landscape plans (*landschaftliche Gebietspläne*) for selected areas (Figure 16.5) where the quality of the scenery was vulnerable to change and where new building needed to be sensitively sited (Autonome Provinz Bozen-Südtirol, 1973). The idea of such plans was to ensure that environmental assets were written into local (*Gemeinde*) plans and that further development of tourist accommodation and infrastructure respected landscape features and quality. Thus this type of measure was aimed at controlling additional change within a protective mesh and as a part of a planning process which envisaged continued development.

A similar format exists in the planning system of some Swiss cantons, but in these cases a distinction has been made between a landscape protection plan (*Landschaftsschutzplan*) and a landscape utilisation plan (*Landschafts-nutzungsplan*). This approach has been employed, for example, for the long-term planning of recreational land alongside the Rhine in the canton of Thurgau (Fingerhuth, 1975). An indication of the difference between the two plans may be gained from a selected list of items covered in each. The former provides protection for attractive groups of buildings; for farms or buildings retaining historic character; for stretches of riverside or small lakes; for viewpoints and panoramic stretches along the river; and for stands of trees or woodland clusters. Landscape utilisation, on the other hand, involves the re-siting or removal of camping sites from sensitive situations; the expansion of enclosed and well-laid-out camping grounds; improvements in the layout of marinas and associated car parks; the siting of picnic places with river and village views; the extension and proper signposting of the public footpath network; closure of footpaths which impinge on a nature reserve; and reduced tourist use of nature conservation sites.

While much has been done to safeguard Alpine landscapes, protection measures and plans alone will not solve the rural landscape issues currently requiring attention. Some problems are tied closely to economic changes, such as the concern over *Brachland*, ie. abandoned land, in Switzerland (Surber *et al.*, 1973). Thousands of hectares of once-productive land have gone out of use since the 1950s, with a consequent deterioration in landscape quality and an increasing danger of landslides, slope erosion and avalanches. This topic has been the source of much research and detailed mapping to assess the ecological effects and to identify the planning opportunities (Haefner and Günter, 1984; Aerni, 1986). The scale of the problem is now so great — 80,000 ha abandoned and a further 180,000 ha at risk — that protective measures to ensure effective management can be initiated only at federal level. Other problems centre on the physical consequences of forest clearance for the construction of ski runs (Figure 16.6; Baumgartner, 1983). Disturbance of terrain can lead to severe erosion and avalanche dangers (Simons, 1988), while schemes to remedy these situations need careful planning and are extremely costly (Südtiroler Landesregierung, 1987). In one way or another, these problems and others are all concerned with

methods of conserving either the landscape in its entirety or particular facets of it.

Nature parks and regional parks

As noted earlier, the conservation of scenic countryside and the use of such areas for informal recreation often merge in practice if not always in respect of aims and objectives. In many parts of Western Europe there are, however, landscape tracts that have been designated specifically with landscape protection and recreational use as the joint and primary objectives. In some cases nature conservation, socio-economic development and scientific research may be included as subsidiary objectives. 'Nature Park' is the most usual title given to these areas, as in West Germany, Austria and Belgium; in Spain the title 'Natural Park' is used; and in France they are known as 'Regional Nature Parks'. The degree of protection afforded and the level of effective conservation range widely, largely because many areas are administered on a regional basis rather than in accordance with a standardised national format. In many ways these areas are intermediate between national parks and purpose-built leisure complexes, and are comparable to the national parks and larger 'areas of outstanding natural beauty' in England and Wales. Most are designated for the distinctive quality of their cultural landscapes and many include villages and rural industries, as well as tracts of forest or woodland. Some countries have placed significant emphasis on establishing this type of conserved space: the 64 *Naturparke* of West Germany, for example, cover more than 22 per cent of the country and range in size from the 2900 square kilometre Altmühltal valley in central Bavaria to the compact Siebengebirge near Bonn, which covers only 42 square kilometres (Statistisches Bundesamt, 1988).

Where a specific system of protected landscapes is lacking, nature parks may form very adequate substitutes. This is the case in the Südtirol province of Italy where *Naturparke* are the most important feature of conservation planning and where they interlock with the landscape plan areas mentioned above and shown in Figure 16.5. Following the Landscape Protection Law (*Land-schaftsschutzgesetz*) of 1970, eight parks were proposed comprising around 139,000 ha, 18 per

Figure 16.6 *Winter sports and the environment. Newly created ski runs and access roads have altered the natural character of many Alpine landscapes — Axamer Lizum after the 1976 Winter Olympics, Tirol, Austria.*

cent of the provincial territory. While the landscape plans refer mainly to settled areas and tourist localities, the characteristic landscapes of the *Naturparke* embrace more rugged and wild terrain, largely undisturbed and free of permanent structures. High priority is given to the protection of the scenic environment, yet these areas are by no means unfrequented and herein lies the source of conflict within the landscapes. Five of the parks enclose the most spectacular scenery of the Dolomites, to which large numbers of visitors are attracted annually. The parks are intensively used for walking, climbing, and viewing flora and fauna, and the sheer volume of visitors has led to the widening and braiding of paths, together with roadside erosion through indiscriminate parking. This has occurred even though controls are effected through restricted road access, toll roads and the limited provision of facilities. Moreover, any form of control raises a quandary: visitors need to be encouraged because of their relevance to the economies of scores of local villages, yet the question of how to reconcile conservation of these unique landscapes with tourist numbers has yet to be answered (Südtiroler Landesregierung 1979b). Thus, while large areas of Südtirol have been delineated and designated, the practice of protection cannot claim to be completely satisfactory

267

Figure 16.7 *St. Leonhard in Abteital, Südtirol, Italy. Tourist transportation systems can facilitate access to remote parts of conserved landscapes, such as here on the fringe of the Fanes–Sennes Nature Park.*

(Figure 16.7). This situation is not very different from that in *Naturparke* in other countries, where management of visitors is critical to conserving the essential environmental character.

The regional parks (*parcs naturels régionaux*) of France differ slightly from the *Naturparke* model. Conservation of the cultural and natural heritage is again a prime aim, but this is coupled closely with a policy of attracting and welcoming visitors in order to strengthen the economic, and thereby the social, structures of local populations. New developments are required to be compatible with the environmental qualities of the parks and, although the interpretation of 'compatibility' varies a good deal from park to park, in most cases surprisingly little conflict or pressure has been caused by visitors and new facilities. Additional accommodation, for instance, has usually been contained within the existing fabric. Parks such as the Lubéron and the Queyras in Alpes-Provence (Figure 16.1) provide good examples. This combination of objectives, to stimulate economic change and execute landscape conservation, is unusual in Western Europe but it does appear to be working and, in many of the French parks, rural population figures are rising slowly and the capacity to accommodate visitors is expanding. Twenty-one regional parks now exist in mainland France (Table 16.3).

Yet a generally successful policy may have

significant exceptions as the Vercors massif, west of Grenoble, demonstrates. Here some very substantial and free-standing schemes for tourist accommodation have been completed and, in addition, several picturesque villages have sprouted estates of holiday villas on their peripheries. The visual and aesthetic impacts of these developments have stimulated the park's planning authority to issue guidelines in an attempt to safeguard distinctive elements of the Vercors landscape. These not only require architectural designs to conform to the region's traditional styles and materials, they also aim to influence the siting, orientation and setting of developments (Franconie, 1978). Nevertheless, such guidelines will not solve the issue of land-use change nor determine which uses are appropriate; as in other protected landscapes, EC and national economic support for agriculture may well be crucial in helping retain the mosaic of crops, pasture and woodland that is at the heart of most cultural landscapes. This point applies to regional parks in general, as much as to the Vercors specifically.

Conservation of nature and natural habitats

All the countries of Western Europe have defined

Table 16.3 *Regional natural parks, mainland France*

Park	Date established	Area (km²)	Population 1975	Population 1982	Visitor capacity
Armorique	1969	650	33,771	32,605	25,200
Brière	1970	414	26,276	33,877	7,500
Brotonne	1974	446	32,591	34,451	3,800
Camargue	1970	820	2,120	2,045	14,100
Forêt d'Orient	1970	669	17,293	18,349	5,200
Haut-Languedoc	1973	1,450	58,672	58,592	37,900
Haute-Vallée de Chevreuse	1983–4	256	n.a.	37,500	n.a.
Landes de Gascogne	1970	2,087	26,792	30,138	11,000
Livradois-Forez	1983–4	2,530	n.a.	99,000	n.a.
Lorraine	1974	2,059	73,784	74,138	12,900
Lubéron	1977	1,523	102,757	112,634	43,700
Marais Poitevin	1979	2,167	83,188	87,926	54,700
Montagne de Reims	1976	554	35,192	33,737	3,900
Morvan	1970	1,717	30,176	28,667	39,500
Nord-Pas de Calais	1983–4	1,656	n.a.	n.a.	n.a.
Normandie-Maine	1975	2,520	99,080	99,349	44,200
Pilat	1974	641	29,838	33,400	21,200
Queyras	1977	707	4,549	5,438	46,600
Vercors	1970	1,570	24,216	25,455	69,200
Volcans d'Auvergne	1977	3,355	85,413	86,937	130,800
Vosges du Nord	1975	1,167	73,915	73,730	20,800

Source: INSEE (1985).

nature reserves in one form or another. The key points centre on how reserves are financed and supported by legislation, on how well they are managed, and on whether other uses are excluded, tolerated or openly permitted. As with other protected designations, the status of nature reserves is diversified and ranges from sites of international recognition to those run by local organisations. All are important in their own way, but their general significance is that they should represent the wide range of habitats and eco-systems now recognised as vital to conserve European fauna and flora.

There are essentially two prongs to the legis-lative framework for nature conservation, and most Western European countries possess both: first, there is protection given to species wherever they are found within a country, and second, there is the creation of reserves or sites where nature has priority (Prieur, 1980). The total number of reserves is in itself no real indication of the effectiveness of conservation; but it can reflect the length of time a state has demonstrated an interest in its natural heritage and, superficially at least, its commitment to the nature-conservation process. Thus the 1.1 million hectares within Sweden's 1,215 nature reserves and conserved sites do indicate a long-standing responsibility towards the natural environment, even though the area in question is less than 3 per cent of the land surface of Sweden. Significant, too, is the fact that these reserves are centrally administered, recorded and documented. In Switzerland, because numerous public and private organisations hold nature reserves, the extent of protected land is not known and the status of protection afforded is incompletely understood (Council of Europe, 1987b). It is not known how many sites have been correctly mapped or marked on the ground. The Swiss League for the Protection of Nature has made an inventory of 4,000 sites said to be of importance, but variations in the degree of impor-tance have yet to be evaluated (Duffey, 1982). The League itself owns or leases only about 500 reserves, and much research and patient recording is required to assess the remainder.

By contrast, the provincial government of Carinthia in Austria holds a detailed register of all sites under its jurisdiction and has seen fit to publish this information in a series of regional guides (Kärntner Landesregierung, 1976). Landscape characteristics, rock types, hydrological conditions, vegetation types, fauna, and management details are summarised for each site. This approach illustrates a willingness to recognise that nature conservation is a public interest and that education is a necessary ingredient of a conservation policy.

Carinthia exemplifies the serious approach of some provincial governments towards conserving their natural environments. In Austria no national legislation exists, and responsibility for both forms of conservation falls to the nine provinces or *Länder*. The law to safeguard wildlife in Carinthia was passed in 1952 and aimed to protect single natural features (known as nature monuments or *Naturdenkmale*), selected species of wild plants and animals, and — in nature reserves or *Naturschutzgebiete* — spacially defined natural habitats. This wide-ranging legislation aimed to afford protection to relatively small pieces of natural heritage such as lakes, ponds, single trees and stands of trees, geological outcrops, waterfalls and springs. It also required lists to be made of endangered or rare plants and animals in threatened habitats or locations. Some 50 species of plants now have legal, if not *de facto*, protection against any site disturbance; limited picking of a further 33 flowering plants is allowed, but removal of the plant itself is forbidden. The list of animals includes deer, ants, woodmice, nine species of butterfly, amphibia and reptiles, and almost all native and migrant birds. Moreover, during nesting times each year it is forbidden to remove, cut down or burn shrubs, hedges and reed beds.

The mechanism for the establishment of nature reserves in Carinthia acknowledged that human uses might in some instances be an integral, or acceptable, characteristic of an area with conservation value. In other cases, infiltration by man might have disturbing and deleterious effects on either wildlife or habitat diversity. Accordingly, two types of reserve were defined — partial nature reserves (*Teilnaturschutzgebiete*) and strict reserves (*Vollnaturschutzgebiete*) (Poore and Gryn-Ambroes, 1980). Strict reserves were distinguished by 'total or highly natural qualities' such as are found in ancient woodlands, barren landscapes, rock or glacial zones, and bogs. Any interference within these natural habitats was prohibited. Partial reserves were defined as areas

Table 16.4 *Nature reserves (Naturschutzgebeite) in rural West Germany*

Land	Number of reserves 1980	Number of reserves 1987	Area (km²)	Land area (%)
Baden-Württemberg	286	512	343	0.9
Bavaria	181	335	1,037	1.5
Hessen	146	344	164	0.8
Niedersachsen	271	476	863	1.8
Nordrhein-Westfalen	247	469	244	0.7
Rheinland-Pfalz	110	266	168	0.8
Saarland	17	32	6	0.2
Schleswig-Holstein	97	117	177	1.1
Total	1,355	2,551	3,002	1.2

Source: Statistisches Bundesamt (1988).

in which rare species were located, or which were rich in small-scale features. In these areas human interference would be permitted as long as it was not contrary to the conservation objectives; in fact it could help to retain the diversity of plant or animal communities within the reserve, especially where traditional Alpine pastoral systems functioned (Kärntner Landesregierung, 1976).

Considering the scale of rural land-use change in recent years, much has been achieved in the Alps in the way of habitat retention under legislative frameworks similar to that in Carinthia. Nevertheless, in some parts of Western Europe, firm legal backing has not prevented the erosion of reserves, largely because protective measures have not been implemented as intended or have been overruled in favour of other uses. The increase in the number of reserves in West Germany looks impressive (Table 16.4) but a report by Fritz (1977) found that, in more than half of 900 reserves surveyed, recreational activities and traffic had been allowed to infiltrate; indeed, in 34 reserves the construction of weekend cottages had been permitted.

It has long been realised that national protective measures will not necessarily ensure the survival of species that migrate or depend on several habitats. Habitat modification outside reserves, and changes in farming practices, may unintentionally affect species numbers and composition (Nievergelt and Hess, 1984). A survey of animal species in Austria estimated that, of some 30,000 native species, 114 had disappeared in recent decades, 340 were threatened with extinction, and 2,200 were at risk. All amphibian species, 92 per cent of reptiles and 55 per cent of bird species were under severe threat of being lost in the near future. Agricultural changes were held responsible for this disturbing situation (Schacht, 1987), but recreational uses, too, were influential in damaging ecosystems, especially sensitive ones such as sand dunes and lakeshores (Satchell, 1982). However, there is also a small credit side to be noted, mainly in the form of a reintroduction of species into conserved areas. The most noteworthy examples are the return of griffon vultures to the Cevennes of France and the ibex to parts of the Swiss Alps (Terrasse, 1982).

Conservation of cultural and built environments

This review of rural conservation would be incomplete without some mention of efforts undertaken to retain man-made features in settlements and in the countryside. Conservation in these situations concerns living and working environments, and this fact has given rise to issues different from those involved with landscape and nature protection. In general, initial measures of protection began with the piecemeal recording of historic and cultural features, such as ancient monuments, ruins, and religious buildings. More recently, schemes have become more comprehensive in approach and scope, involving the conservation of historic centres of country towns, whole villages and sometimes broad tracts of farmed countryside. Many projects today are concerned with renewal and enhancement as well as conservation, while others are integrated with plans for residential expansion. This again recognises that development and conservation can be concordant and interdependent (Weiss, 1973).

Alpine countries are acutely aware of their historical heritage, and in most parts systems exist to provide valued features with some form of protection. In the German-speaking countries the system of *Denkmalschutz* applies. Taking the Südtirol as an example, this type of protection extends to castles, country houses, farmhouses and farm buildings, churches, field chapels and *Bildstöcke* or wayside shrines (Südtiroler Landesregierung, 1981). Particular attention has

been paid to the architectural heritage of buildings and villages in Ladin-speaking districts[1] because it is seen as important to preserve the traditions of this minority group. Determined efforts have been made in selected settlements to establish renewal areas (*Sanierungsgebiete*), whereby renovation and harmonious modernisation of farms can take place. This is an integrated approach combining socio-economic objectives with those for conservation, and it has met with some success. This is witnessed by the fact that the Ladin population is growing steadily and its employment base is broadening out beyond agriculture into service trades and tourism (Südtiroler Landesregierung, 1985).

Protection for individual features of the rural cultural landscape exists in many other parts of Western Europe. In France, for example, the system of *sites classés* and *sites inscrits* has been in operation since 1930 and extends throughout the entire country. By the mid-1980s some 4,900 and 2,350 sites, respectively, were protected (INSEE, 1985). In the Netherlands the *Monumentenlijst* is even more impressive. In the mid-1980s almost 43,000 features of the cultural landscape were covered and, although the majority were in urban areas, over 5,300 farm buildings, 1,000 mills and almost 300 castles were listed (Centraal Bureau voor de Statistiek, 1986).

Comprehensive renewal and enhancement within the built environments of rural areas has not been accepted as a mainstream conservation objective everywhere and, particularly outside West Germany, examples are not common. Even so, a number of small country towns, such as Hall-in-Tirol and Rattenberg in the Austrian Tirol, have benefited from pedestrianisation and townscape enhancement schemes (Figure 16.8). Similarly, the conservation sector scheme (*secteur sauvégarde*) in France is having far-reaching effects on some historic settlements. Uzès near Nîmes, for example, has had buildings restored to their former elegance, arcaded squares have been paved, trees and fountains have been added and commercial life has returned to the centre. In the Netherlands, too, the opportunities for wider protection schemes within settlements are well recognised. More than 300 town- and village-scapes which justify conservation measures have been identified, of which approximately half have been placed on the official list.

Many of these schemes are small in scale,

Figure 16.8 *Conservation of the built environment. Restoration of building façades, and enhancement of the market area of St Veit an der Glan, Carinthia, Austria.*

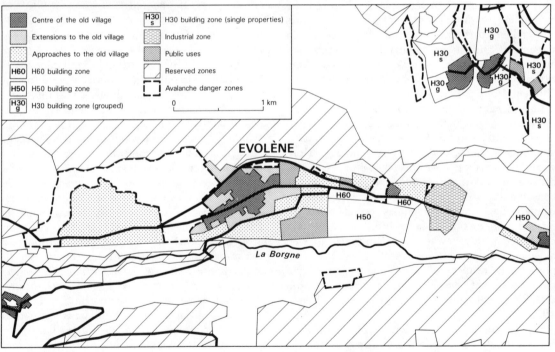

Centre of the old village

Extensions to the old village

Approaches to the old village

H60 H60 building zone

H50 H50 building zone

H30 g H30 building zone (grouped)

H30 s H30 building zone (single properties)

Industrial zone

Public uses

Reserved zones

Avalanche danger zones

0 1 km

EVOLÈNE

H60 H60 H50

H50

H30 g H30 s H30 g H30 g H30 s H30 s H30 s

La Borgne

Figure 16.9 *Zoning plan for Evolène, Valais, Switzerland.*

however, when compared to the widespread village renewal measures (*Dorferneuerung*) in West Germany (Henkel, 1984). Over 1,300 villages are presently involved, the aim being to modernise the fabric and yet retain the architectural heritage and aesthetic character. In some villages, buildings have been moved so that traffic can flow more easily, better access has been provided to farms, new uses have been found for redundant buildings, and landscaping of waste space has been carried out. Emphasis has been laid on improving community facilities to give villagers more social cohesion and an interest in their living environment. This process of *Dorferneuerung* may take place in conjunction with a local (*Gemeinde*) plan or as part of a land consolidation scheme. This stands in contrast to, for example, conservation area planning in the UK, where designation and enhancement are frequently carried out in isolation.

Zoning methods are employed also in village development plans, some of which incorporate conservation measures. The case of Switzerland demonstrates this point. Here the framework of these plans rests on a land-zoning system, and policies are set out for each zone. Figure 16.9 illustrates the zoning policy adopted around the village of Evolène in Valais (Commune de Evolène, 1975). Most zones are concerned with various types of residential expansion but the core

of the old village is acknowledged to be worthy of protection and a detailed set of restrictive policies has been formulated to control change (Table 16.5). This type of planning has enjoyed considerable success both here and elsewhere: old buildings have been renewed in traditional styles and control over new buildings has ensured that they harmonise with the existing fabric (Figure 16.10).

It seems from discussions in the planning sphere that conservation measures within settlements and their adjacent landscapes are likely to expand substantially in the future, and this should be to the benefit of many rural residents (Kuonen, 1986). In the Alpine countries and elsewhere, appreciation of the cultural heritage is growing rapidly, and visitor numbers to historic places and sites are increasing substantially. Grants and loans to renovate and maintain rural properties are more freely available now than in the past. Participation by rural people in this conservation process is important because, together with tourism, it will offer wider employment opportunities and broaden the economic base of these areas. Consequently, population numbers may be retained or even increased, thereby meeting one of the aims of the European strategy of conservation.

Figure 16.10 *Niederwald, Valais, Switzerland. The vernacular architecture of this village centre is protected by a zoning plan.*

Table 16.5 *Regulations for the centre of the old village, Evolène, Valais (translation)*

Aim: to encourage development in harmony with the regional styles. The appearance of small properties, arising from the high density of plots, and the contiguity of buildings of varying heights must be conserved. This zone includes old chalets, barns, *raccards** and stables.

(1) Without permission from the Commune, it is forbidden to
   a) demolish or remove beyond the perimeter of the zone, any of the above named buildings,
   b) install new bee-hives.

(2) Within the zone it is authorised to
   a) rebuild on stone bases old *raccards*, stables, barns and chalets as long as this integrates with the local character around the site,
   b) renovate existing buildings to make them inhabitable but without changing their dimensions and respecting strictly the local style.

In this zone uses which cause a nuisance are forbidden. Shops will be special cases to be studied by the Commune. Chicken-houses, wood-stores and rabbit enclosures are tolerated as long as no nuisance is caused and consideration is given to hygiene.

(3) Conditions for new buildings:
   a) Distance to edge of plot must be equal to one-third the height of each façade, but with a minimum of 3m.
   b) Maximum number of floors is 4, excluding inhabited roof space, up to maximum of 13m.
   c) Architectural details:
      i)   roof-shape: two surfaces with a slope of 40–50%.
      ii)   roofing materials: local stone slabs, black or grey artificial slates or cement tiles, wood shingles.
      iii) orientation: ridge must be parallel or perpendicular to the contour, or identical with the direction of existing properties.
      iv)   any outside aerial is forbidden.
      v)   materials and colours to be in harmony with the village.
      vi)   half of each side façade and of the west- or south-facing gabled side to be in wood.
      vii) stone work must be white, grey-white or grey; other colours considered in special cases.
      viii) painting of wood surfaces to be in varnish or clear paint.
      ix)   window dimensions must be specified sizes.
      x)   dimensions of the building must conform to specific ratios between height, width and depth.

* *Raccards* are wooden grain stores which, since they contain no stove, have little risk of fire. For this reason they have traditionally been used also as stores for family valuables (author's note).
Source: Translated from Commune de Evolène (1975).

Conclusions

This review has concentrated on the three main topics of landscape, wildlife and cultural conservation, but it needs to be remembered that other protective measures have been implemented throughout Western Europe. Little mention has been made of forestry practices and woodland management, of inland water and coastal conservation, or of schemes to control the physical forces shaping the landscape (Südtiroler Landesregierung, 1987), all of which fall under the umbrella of conservation processes. In focusing primarily on the countries of the Alpine chain, the intention has been to examine the various approaches that have been conceived to deal with common problems and issues within a fragile and highly valued environment. The amount of land in the Alpine region that has been scheduled for

protection of one kind or another is quite remarkable (Table 16.6), and it might be argued that designation and policy formulation have been the most successful parts of the wider conservation process. More success is likely to come from international co-operation through organisations like Arge-Alp and Arge-West which have brought together interested governments and organisations to implement programmes of action to mitigate common problems (Ruppert, 1984).

Publications from the Council of Europe (1981a; 1981b; 1987b) make it very clear that conflicts and issues found in the Alps are mirrored elsewhere in lowland and coastal Europe. Likewise, solutions to these problems are not always very dissimilar, despite the geographical differences: the imposition of protective categories, the formulation of planning policies, and the establishment of management objectives.

Table 16.6: *Protected land in Südtirol, Italy*

Designated areas (number)	Size (ha)	Total area (%)
National park (1)	53,447	7.2
Nature parks (6)	92,465	12.5
Nature reserves (103)	1,380	0.2
Nature monuments (479)	–	0.1*
Villages with landscape plans (78)	323,877	43.7
Other landscape protection plans (6)	43,590	5.9
Total	514,759	69.5

*Precise area has not been calculated.
Source: Südtiroler Landesregierung (1988b).

Not everywhere has firm control been exercised over development, and in only a few situations has development been truly integrated with environmental objectives on the lines of the European strategy (Council of Europe, 1987a). Reflecting on what has been achieved over the past 30 years, one is drawn to the conclusion that, while the foundations and frameworks for sound conservation are now firmly established, managing the future environment will depend on more enterprising and imaginative use being made of the opportunities, as they present themselves, in order to establish real and lasting practices of conservation.

Note

1. Ladin is a Romansh-type language spoken by some 15,000 inhabitants of the Gröden and Gadertal districts.

References

Aerni, K. (1986) Zur Nutzung der Flur im Binntal (Wallis), *Jahrbuch der Geographischen Gesellschaft von Bern*, 55, 211–34.

Autonome Provinz Bozen-Südtirol (1973) *Südtirol 1981 — Vorbereitendes Dokument für ein Landesentwicklungsprogramm*, Bozen.

Baumgartner, F. (1983) Zur raumplanerischen Beurteilung touristischer Transportlagen, *Raumplanung Informationshefte*, 3, 6–9.

Bay. SLU (Bayerisches Staatsministerium für Landesentwicklung und Umweltfragen) (1973) *Planung und Umwelt 2 — Erholungslandschaft Alpen*, Munich.

Central Bureau voor de Statistiek (1986) *Statistisch Zakboek, 1986*, The Hague, Staatsuitgeverij.

Commune de Evolène (1975) *Plan d'aménagement de la commune de Evolène: réglement sur la police des constructions*, Sion.

Council of Europe (1981a) *Behaviour of the Public in Protected Areas*, Strasbourg.

Council of Europe (1981b) *Birds in Need of Special Protection in Europe*, Strasbourg.

Council of Europe (1987a) Fifth ministerial conference on the environment, *Naturopa Newsletter — Nature and Environment*, 87, 1–4.

Council of Europe (1987b) *Management of Europe's Natural Heritage — Twenty-five Years of Activity*, Environment Protection and Management Division, Strasbourg.

Currey-Lindahl, K. (1968) *Sarek, Stora Sjöfallet, Padjelanta — Three National Parks in Swedish Lapland*, Raben and Sjögren, Stockholm.

Duffey, E. (1982) *National Parks and Reserves of Western Europe*, Macdonald, London.

Fingerhuth, C. (1975) *Erholungs- und Landschaftsschutzplanung Thurgauisches Bodensee und Rheinufer*, Büro für Orts- und Regionalplanung, Zürich.

Franconie, G. (1978) Construire dans le Vercors, *Courrier du parc naturel régional du Vecors*, 21, 1–36.

Fritz, G. (1977) Zur Inanspruchname von Naturschutzgebieten durch Freizeit und Erholung, *Natur und Landschaft*, 7, 10–20.

Gambino, R. and Jaccod P. (1985) Le Parc National du Grand Paradis — une proposition de plan, *Revue de géographie alpine*, 73, 217–46.

Haefner, H. and Günter, T. (1984) Land-use changes and ecological effects in the Swiss Alps. In E.A. Brugger, G. Furrer, B. Messerli and P. Messerli (eds), *The Transformation of Swiss Mountain Regions*, Verlag Paul Haupt, Bern, pp. 101–24.

Haimayer, P. (1987) Aspects écologiques, économiques et sociaux du ski sur glacier en Autriche, *Revue de géographie alpine*, 75, 141–56.

Henkel, G. (1984) Dorferneuerung in der Bundesrepublik Deutschland, *Geographische Rundschau*, 36, 170–6.

INSEE (1985) *Annuaire statistique de la France*, Institut national de la statistique et des études économiques, Paris.

Janin, B. (1985) Revitaliser un pays qui meurt: une nouvelle mission pour le Parc, *Revue de géographie alpine*, 73, 175–207.

Jülg, F. (1984) Le tourisme autrichien. In 25th International Geographical Congress, *Les Alpes*, International Geographical Union, Caen, pp. 217–26.

Kärntner Landesregierung (1976) *Die Natur- und Landschaftsschutzgebiete Kärntens 6 — Klagenfurt und seine Umgebung*, Klagenfurt.

Knafou, R. (1987) L'evolution récente de l'économie des sports d'hiver et de l'aménagement touristique de la montagne en France, *Revue de géographie alpine*, 75, 101–14.

Kuonen, J. (1986) Sachplan Ortsbild, *Raumplanung Informationshefte*, 2, 11–12.

Nievergelt, B. and R. Hess (1984) Changes in the fauna. In E.A. Brugger, G. Furrer, B. Messerli and P. Messerli: (eds), *The Transformation of Swiss Mountain Regions*, Bern, pp. 229–41.

OROK (1978) *Zweiter Raumordnungsbericht*, Österreichische Raumordnungskonferenz, no. 14, Vienna.

Poore, D. and Gryn-Ambroes, P. (1980) *Nature Conservation in Northern and Western Europe*, IUCN, Gland.

Préau, P. (1976) Le Parc national de la Vanoise, et l'aménagement de la montagne, *Revue de géographie de Lyon*, 51, 123–32.

Prieur, M. (1980) A more dynamic policy, *Naturopa*, 34–35, 14.

Ruppert, K. (1984) Der deutsche Alpenraum — aspekte raumwirksamer Staatstätigkeit. In 25th International Geographical Congress, *Les Alpes*, International Geographical Union, Caen, pp. 241–3.

Satchell, J. (1982) Recreation, ecology, landscape, *Naturopa*, 42, 26–8.

Schacht, H. (1987) The agricultural landscape, *Naturopa*, 56, 21–3.

Schloeth, R. (1975) Zur Veränderung der Fauna in den Schweizer Alpen. In Alpen-Institut, *Die Zukunft der Alpen*, Vol. 4, Alpen-Institut, Munich, pp. 161–7.

Schloeth, R. (1976) Nationalparks in der heutigen Zeit, *Terra Grischuna Bündnerland*, 35, 119–23.

Simons, P. (1988) Après ski le deluge, *New Scientist*, 14 January, 49–52.

Statistics Sweden (1989) *Statistical Abstract of Sweden*, Stockholm.

Statistisches Bundesamt (1988) *Statistisches Jahrbuch 1988*, Wiesbaden.

Südtiroler Landesregierung (1979a) Landschaftsschutz — ein ständig aktuelles Anliegen, *Informationsschrift*, 19, 8–30.

Südtiroler Landesregierung (1979b) Dolomitengebiet Fanes-Sennes-Prags — fünfter Naturpark in Südtirol, *Informationsschrift*, 20, 14–20.

Südtiroler Landesregierung (1981) Denkmalpflege in Südtirol 1974-1980, *Informationsschrift*, 27, 1–60.

Südtiroler Landesregierung (1983) Vernünftige Landschafts- und Umweltschutzpolitik, *Informationsschrift*, 35, 32–4.

Südtiroler Landesregierung (1985) Siedlungsgeschichte und Architektur Ladiniens, *Informationsschrift*, 40, 49–56.

Südtiroler Landesregierung (1987) Landschaftsplan Ridnauner Bach — Landschaftsschutz unter einer ganzheitlichen Sicht, *Informationsschrift*, 45, 29–36.

Südtiroler Landesregierung (1988a) Der Haushaltsplan des Landes für 1988, *Informationsschrift*, 47, 42–63.

Südtiroler Landesregierung (1988b) Zwei Drittel des Landes schon unter Schutz gestellt, *Informationsschrift*, 49, 32.

Surber, E., Amiet, R. and Kobert, H. (1973) *Das Brachlandproblem in der Schweiz*, Eidg. Anstalt für das forstliche Versuchswesen, Report 112, Birmensdorf.

Terrasse, M. (1982) Concrete measures, *Naturopa*, 40, 18–20.

Weiss, H. (1973) *Wo kann man bauen? Wie kann man bauen?*, Schweizerische Stiftung für Landschaftsschutz und Landschaftspflege, Bern.

17

Conclusion: Western Europe approaches the twenty-first century

David Pinder

Previous chapters have reviewed an extensive range of Western European development trends and problems. The pace and scale of change have been explored, the magnitude of the challenges to be confronted has been discussed, and responses to those challenges have been examined. A variety of recurrent themes has emerged, and in this brief conclusion it is appropriate to draw together the main areas of common ground. Although the discussion adopts a partly retrospective viewpoint, the primary intention is to take a prospective approach. Predictions are inevitably hazardous, but what are the major challenges likely to be faced by Western Europe as it enters the twenty-first century?

Uneven spatial development will remain a major feature of the economic scene. Existing development gaps, in terms of interregional and international comparisons, are too great for rapid progress to be made towards a more even development surface. At the international scale, Figure 17.1 emphasises the extent to which much of southern Europe continues to lag behind most other areas. Taking the extremes, per capita GDP in Switzerland is virtually three times the Portuguese level. If relatively prosperous countries are compared, significant variations are again evident. In Denmark and Sweden, for example, per capita GDP is almost 20 per cent higher than in Austria and Belgium. Similarly, regional data for the European Community draw out the entrenched nature of intranational prosperity contrasts (Figure 17.2). These are least marked for the poorer and smaller nations: in Greece the ratio between the least prosperous and most prosperous regions is 1:1.4, while for Belgium it is 1:1.6; in Spain, by contrast, the ratio is 1:2.2,

while in Italy it is as high as 1:2.5.

Several earlier chapters indicate that current influences are likely to intensify, rather than to dampen, geographical prosperity contrasts. Wise and Chalkley have underlined the frequent inability of regional policy to deliver growth at the scale originally anticipated and, most recently, its failure to cope with the unemployment consequences of industrial restructuring. While much of that restructuring has been associated with economic recession extending from the mid-1970s to the mid-1980s, Watts has demonstrated that it remains a significant force. The actual process of restructuring and its timing may vary markedly from industry to industry, but the consequences of the reorganisation of capital will continue to be felt throughout the 1990s and are unlikely to be confined to traditional 'sunset' industries. At the macro scale, compensation may come through the successful exploitation of technological change, yet Thwaites and Alderman suggest that it would be inadvisable to assume that this will be the common experience throughout Western Europe. The potential advantages of new technologies are being exploited most effectively by the more prosperous areas and, unless effective countervailing policies are devised, this trend is likely to continue. Similarly, although Mason and Harrison highlight a significant revival of small and medium-sized enterprises, very few of these have substantial long-term growth potential and, once again, they are most likely to cluster in regions that are already economically successful. The evidence offered by Daniels with respect to the rise of producer services reinforces still further the disequilibrium interpretation of the future. Despite the apparently liberating potential of new

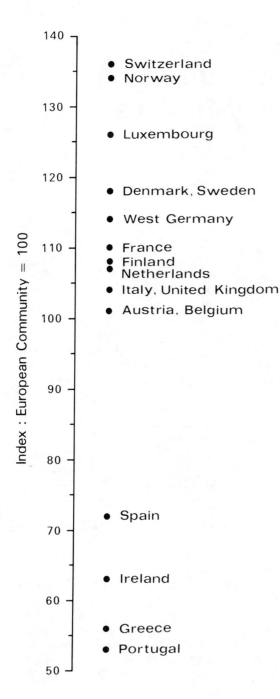

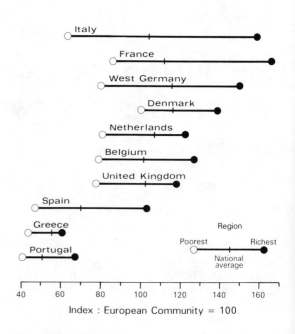

Figure 17.2 *Regional variations in per capita GDP in principal EC countries, 1985*
Source: Commission of the European Communities (1989) *Basic Statistics of the Communities*, Brussels.

telecommunications technologies, key producer services give every indication of maintaining their existing prefence for core areas.

To a great extent these conclusions are based on analyses of the operation of market forces. However, policy may also contribute to spatial disequilibrium, and it is widely anticipated that the European Community's policy of creating a unified market by the end of 1992 will perpetuate, and perhaps exacerbate, interregional prosperity gaps throughout much of Western Europe. In this context a tension can be observed between the goals of overall growth and those of regional policy. Movement towards the single market aims, through greater efficiency born of effective intra-Community competition, to strengthen the Western European economic system and thereby improve its competitive edge in the world economy. Taking a global perspective, it is based on the concept of attack as the best form of defence. Yet, ever since negotiations to form the original EEC began in the 1950s, it has been recognised that competition policy may sharpen interregional prosperity contrasts. For highly prosperous regions the prospects may be good;

Figure 17.1 *Per capita GDP in Western European countries, 1986*
Source: Commission of the European Communities (1989) *Basic Statistics of the Communities*, Brussels.

freer competition may well accelerate still further the process of economic growth as wider markets are captured. But the danger is that this will be to the detriment of less-competitive parts of Western Europe. Here, free competition may force industrial restructuring and employment losses while, because of deficiencies such as low incomes and low rates of saving and investment, market forces are unlikely to produce rapid progress towards healthier economic structures. Moreover, the single market is intended as a prerequisite for a related Community policy, monetary union, which may reinforce the geographical impact of market integration. Once monetary union is achieved, Community member states with weak economies and major problem regions will have relinquished currency devaluation as a strategy available to combat the deleterious regional economic effects of free competition. Against this background, the European Community's pursuit of a high-profile regional policy, detailed by Wise and Chalkley, assumes considerable political significance. Modest though the results may be, political commitment to the alleviation of regional problems is intended to encourage relatively vulnerable member states to accept the main thrust of integration policy.

It is also evident that Western Europe's links with the remainder of the global economy will continue to strengthen. In part this will reflect corporate restructuring. Watts's examination of a sample of activities indicates that, for many reasons, the pace of such restructuring is likely to vary considerably between industries. Yet the trend is well established and, in the case of some large (but perhaps academically neglected) industries such as the food sector, it is effectively producing a quiet revolution. But global integration will also be a function of intensified competition between major world regions, as Dicken has shown. In this highly competitive environment much is likely to depend on the ability of the manufacturing sector to exploit technological advances, both to raise productivity and to gain innovatory advantages. This is emphasised by several key economic indicators for the European Community (Figure 17.3). These do not combine to create a favourable scenario: the late 1980s were characterised by a modest but shrinking positive trade balance; by slow GDP/GNP growth relative to Japan and the USA; and by low rates of gross fixed capital formation. Such trends

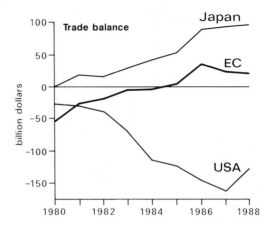

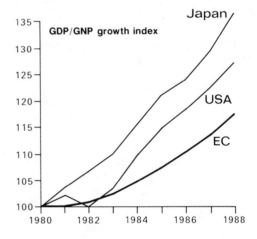

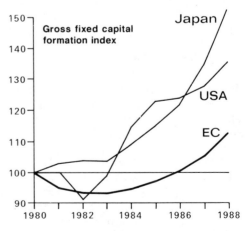

Figure 17.3 *Comparative economic indicators for the EC, the USA and Japan, 1980–88*
Source: European Investment Bank (1989) *Annual Report*, Luxembourg.

underlie the European Commission's determination to accelerate growth through market integration in the wake of the Single European Act.

Increasing integration may also be anticipated with respect to other sectors. Although many services will continue to cater for the needs of individuals and are likely to recognise strong contrasts in consumer preferences between countries, Burt and Dawson have highlighted the importance of mergers, takeovers and innovations as forces encouraging the rapid growth of international retail corporations and consortia. Similarly, Daniels's examination of producer services underlines the growing importance of international organisations and interaction, not least with respect to world financial markets. In both retailing and producer services, forces for change are leading to new international structures both within Western Europe and between this region and other segments of the global economy. This is also true of the energy sector, as Odell's analysis demonstrates. Within Western Europe the movement to break down national energy monopolies is expected to bring a significant increase in competition, controlling costs and therefore contributing to competitiveness. Externally, Odell argues, world reserves are such that supply-side competition will restrain prices and allow Western Europe to establish an attractive balance between internal and external energy sources. Moreover, this external integration should be possible without a return to too great a dependence on source areas that, potentially, are politically hostile. Improvements in resource availability in politically secure areas, coupled with the judicious exploitation of internal energy reserves, offer opportunities to reduce dramatically the dangers of energy imports.

The rise of 'green' issues was an impressive feature of the 1980s, and environmentalism will continue to figure prominently on the political agenda. This will be ensured partly by the continuing activities of a multiplicity of pressure groups and partly by growing political recognition of the importance of environmental protection. Political recognition reflects national political parties' increasing awareness of the power of the 'green' vote, but there is also an important international dimension. Above all, this is a consequence of the strong commitment to environmental improvement and protection developed by the European Commission during the 1980s, as

Williams's survey underlines. It is one thing, however, to establish environmentalism on the political agenda, and quite another to set in motion rapid progress towards successful protection. Part of Williams's argument is that, while it may be politically expedient for governments to acknowledge the importance of safeguarding the environment, all too often at the national level there are strong pressures obstructing progress towards environmental goals. The prevention or reduction of pollution requires investment which may be opposed by interest groups within the productive sector (particularly manufacturers responsible for emissions) or within the bureaucracy (such as Treasury ministers averse to public expenditure to achieve higher effluent standards).

Largely for these reasons, the European Commission's commitment to environmental protection is seen by many observers as a potentially crucial factor leading (and if necessary forcing) governments into firm environmental protection programmes. Yet factors militating against rapid progress towards environmental goals are not simply political. The range of challenges to be faced by the environmental protection movement is exceptionally wide, extending as it does from the greenhouse effect to local issues of effluent and waste disposal. Moreover, to aim at environmental protection is to aim at a moving target: year by year new problems are identified and, simultaneously, new pressures are imposed on the environment, largely as a result of rising prosperity. As a group, the contributions by Shaw and Williams, Woodruffe and Fielding underline the strength of these pressures: the economic attractions of tourism and its environmental implications; recreational impact in areas of high landscape and ecological value; and the widespread effects of counterurbanisation in rural Europe. Thus, while the rise of environmentalism must be welcomed, and while it may be seen as a major new dimension of political thinking, the breadth and scale of the tasks ahead must be fully recognised.

This conclusion gains added emphasis if environmental issues are viewed from an urban perspective. Urban Western Europe is largely responsible for energy consumption by economic activity, by the domestic sector and by transportation. Although this consumption contributes to global environmental problems, on a local scale it also severely degrades the environments of urban areas. The reduction of urban air pollution can be seen as a priority target yet, even if it were attained, many urban environmental problems would remain. Above all, these problems relate to the decline of inner-urban residential areas and to dereliction caused by industrial restructuring and the redundancy of previously vital transport facilities (such as railway marshalling yards and outmoded port areas). Shurmer-Smith and Burtenshaw demonstrate that concern to deal with deteriorating urban environments has deep historical roots, but that particular stress has been placed on the development of effective policies in the post-war period. In the public eye, the fruits are seen in showpiece redevelopments such as the Forum les Halles, Covent Garden, the London Docklands and the rehabilitation of the centre of Bologna. More generally, Burt and Dawson point to the widespread improvements brought to urban cores through the application of policies such as pedestrianisation. But such progress has by no means eliminated the problems of urban decay. The priority awarded to urban redevelopment varies substantially from city to city and from country to country. Moreover, the passage of time entails further decay and, in addition, substandard areas expand as the standards set by society rise. Once again, the environmental target is not stationary, and the improvement of the urban environment will pose major challenges well into the twenty-first century.

Major socio-economic problems will be interwoven with these challenges. Although in the 1980s it was demonstrated that intensive investment in decaying urban environments can result in striking physical improvements, it also became evident that such investment does not necessarily deal with the deprivation that is frequently endemic among inner-urban communities. Indeed, physical regeneration may serve to reinforce that deprivation. New jobs brought into declining areas may be unsuited to the skills of local people, while an influx of employment may draw in new residents to gentrify housing and displace original communities. Meanwhile, the widespread concern of governments to rein in state expenditure may result in failure to invest in the construction or renovation of public housing and other social overhead capital. Such trends are of great significance for inner-urban populations as a whole, populations whose handicaps are typically compounded by problems of age, poverty, low educational attainment and ill health. But beyond these general difficulties lie others linked with specific groups in inner-urban society and, most particularly, with guestworker communities. King has demonstrated that these can no longer be regarded as temporary additions to the social scene: there is still demand for their labour in uncogenial activities and many guestworkers have with them families whose roots are becoming well established. The reality is, therefore, that the cities which in the past drew on this convenient source of labour to sustain their economic expansion will in future face challenges posed by the needs and aspirations of the newcomers. Given the strong cultural identities of many of the immigrants, is it desirable — and is it feasible — to aim at integration with indigenous populations? What are the wishes of the immigrants themselves? Are those wishes identical for all guestworker communities? How may education levels for the younger generation be improved to offer able individuals access to more fulfilling and productive employment? And how can employers be encouraged to recruit employees with immigrant backgrounds for jobs that are other than mundane?

Urban social problems may predominate, but rural areas will also continue to experience social pressures. The need for cutbacks in farming, detailed by Ilbery, suggests that in future rural societies will be even less firmly rooted in agriculture than is at present the case. Many farming families will either have to leave the sector or earn a much higher proportion of their income from supplementary activities such as leisure and tourism. For new entrants to the labour market the prospects of employment in agriculture will continue to dwindle. Conversely, rural communities may find that employment opportunities outside agriculture improve through continued counterurbanisation, but this is also likely to bring pressures. As Fielding has demonstrated, the economic benefits of counterurbanisation are spread more and more unevenly between regions and, where benefits are felt, the population influx may have far-reaching effects on local communities. Thus the breakdown of traditional social structures may be accelerated, while pressures on the housing market may sustain the flow of young people from rural to urban areas.

A considerable range of challenges that are likely to face Western Europe in the twenty-first century can therefore be identified. Yet it is evident that effective strategies to deal with these challenges are in most instances still awaited. In this important sense even the immediate future must be considered highly uncertain. Moreover, uncertainty has been amplified by unanticipated political developments. Until the very end of the 1980s it appeared that Western Europe would continue to develop as an increasingly coherent economic and political entity, primarily within the context of the European Community, and largely separate from Eastern Europe. Since then the demise of hardline Eastern European regimes, the opening of the Berlin Wall and the emergence of East–West negotiations relating to economic assistance and co-operation have severely under-mined long-held assumptions on the division of East and West. If, as seems possible, this division continues to crumble, a new Europe will emerge. This may be typified by integration rather than separatism, yet it will also be characterised by far greater developmental difficulties than those of Western Europe alone. Deficiencies in, for example, Eastern Europe's agricultural, industrial and distribution systems, in its standards of urban development and in its concern for environmental protection will in many cases dwarf the equivalent problems in Western Europe. Thus the welcome spread of more democratic government is likely to be accompanied by still greater challenges to which both Western and Eastern Europe must rise in the coming decades.

Index*

Abandoned land 213, 217, 219, 266
Acid rain 195, 202, 262
Acquisitions 60, 64, 68, 69
Action Programmes on the Environment 201, 203–4
Aerospace industry 2, 3, 5, 11, 65–7
Agricultural
  concentration 211
  conversion 216, 223
  extensification 214, 216, 219, 223
  Extensification Regulation 214, 216
  intensification 211
  modernisation 211
  policies 213–17
  restructuring 127
  specialisation 211
  Structures Regulation 214
  surpluses 212
Agriculture 13–14, 163, 211–23, 226–7
Ahold 150, 152
Airbus, European 66–7
Algeria 27
ALLEGRO 155
Alps, the 258, 260–75
Alternative energy 21, 32
Altmühltal Valley *Naturpark* (Bavaria) 266
Amsterdam 131, 170, 174, 176
Animals, protection for 270–71
Anti-urbanisation 230
Appreciation of oil and gas reserves 24

Arabian Gulf 27
Arab–Israeli War 27
Arctic 25
Asda 147
Asko 147
Athens 164
Atomic power, *see* Nuclear power
Australian firms 68, 69
Austria 15, 167, 182, 196, 277
Automation 64–5, 75–6, 98–101
  see also Technological change
Automobile
  industry 3, 5, 11, 38, 47–50, 52, 53, 63–5, 172
  transporters, ocean-going 5

Bagnoli (Italy) 62
Balance of payments 240
Banking, *see* European Investment Bank, Financial services, Producer services
Barcelona 63, 114, 129–31
Basilicata 97
Basingstoke 113
Bavarian Alps 263–4
Beach holidays 241
Belgium 3, 30, 31, 34, 109, 158, 162, 163, 167, 168, 170, 182, 185, 187, 193, 198
Benelux Economic Union 3
Benneton 145, 150, 151
Benthamism 139
Berchtesgaden National Park (West Germany) 261
Bercy (Paris) 136–7
Berlin 169, 170
Billingsgate (London) 132
Blois (France) 69
Blokker 150
Bochum (West Germany) 128

Bofill, R. 131
Bologna 129, 134–5, 281
Bolton Committee 93
Bonn 127
Brachland 266
Bradford 130
Branch plant economy 100
  see also Spatial division of labour
Bremen 128
Bristol 119, 175
British Aerospace 64, 65, 66, 67
British Isles 25
Brussels 116, 169, 174
BSN 68–9
Built environment, conservation of 128–9, 133, 134–6, 139, 271–3
Burton Group 159
Buying groups 147–8

C & A 150
Caen (France) 128
Calabria 97
Cambridge 79, 85, 86, 88
Canals 171
Capital accumulation 164, 165
Cardiff 130–31
Carinthia (Austria) 264, 270, 272
Cassa per il Mezzogiorno 97, 180, 184
Centre Parcs 254
Centre-periphery, *see* Core regions, Peripherality
Centro storico (Bologna) 129
Cessation of Farming Regulation 215–16
Cevennes National Park (France) 262, 271
Charing Cross (London) 132

*Prepared with the assistance of Claire Pinder.

Chernobyl (USSR) 33
City-centre redevelopment
  general 125, 127–8
  Nord (Hamburg) 134
  retail 155, 156, 159
  see also Inner-city
City of London 118
Civic Amenities Act (UK) 128
Clandestine migration 166
Climatic change, *see*
  Environmental degradation
Closures, industrial and
  commercial 13, 62, 64, 65,
  70, 75, 83, 135, 136
  see also Industrial restructuring
Coal 19, 20, 25, 30, 31–2, 33
  imports 30
  mine closures 30
  protection 30, 33
  subsidies 25
  underground gasification 31, 34
Coal-based electricity 19, 31, 33
Cologne 131, 172
Colonialism 163
Combined heat and power systems
  31–2
Common Agricultural Policy
  (CAP) 203, 211–23
  consequences of 212
Common Market 19, 29
  see also Single European Market
Comparative advantage 91, 94,
  95, 103
Competitive advantage 152
Component industry (automobile)
  65
Compressed natural gas (CNG) 33
Computerised numerical control
  (CNC) 98–9
Computer services 94
Concentration ratio,
  manufacturing 60, 61, 63, 66,
  68
Concorde, Anglo-French 67
Conservation 125, 128–9, 133,
  134–5, 139, 156–7, 236,
  238, 264, 271–3
Construction industry 171, 172,
  173
Consumer
  behaviour 143, 159
  electronics 50, 51–2
  expenditure 145
Consumption economy 126
Containerisation 5, 181
Convenience stores 153, 159
Copenhagen 127

Core-fringe concept (National
  Parks) 262–3
Core regions 79, 95, 96, 97, 98,
  99, 100, 110–11, 113, 115,
  116, 118, 119, 121, 185,
  187, 193
  see also Peripherality
Co-responsibility levy (CAP) 213,
  216
Corporate diversification 149–50
Corporations, industrial 59–70
  see also Large-scale
  manufacturing, Transnational
  corporations
Council of Europe (CoE) 13, 195,
  196, 197, 203, 258, 273
Council of Ministers (EC) 203
Counterurbanisation 14, 126–7,
  144, 228–39, 280, 281
Covent Garden, London 135–6,
  281
Coventry 64
Cultural
  features, conservation of 271–3
  heritage 271, 273
  tourism 243
Cyprus 196, 197

Daewoo Corporation 45
DATAR 130
Davignon Plan 62
Decentralisation 80, 125–6, 187,
  188, 193, 228
Defence industries 65–7, 95
Deindustrialisation 6, 45–6, 54,
  56–7, 58, 62–3, 75, 108,
  125–6, 179, 181, 192, 237
Demographic change, *see*
  Population
Denain (France) 62
Denationalisation, *see* Privatisation
Denkmalschutz 271
Denmark 15, 109, 180, 182, 185,
  187, 188, 217, 218, 277
Department stores 143, 150, 158
Dependency 86, 165, 248–9
Deregulation 30, 31, 109, 114,
  118
Designer community 129
Design guidelines 268, 274
De-urbanisation 126–7
  see also Counterurbanisation
Diffusion, *see* Spatial diffusion
Dijon 69
Discount stores 153
Distrigaz (Belgium) 31
Diversification, corporate 62,

63–4, 68, 147, 149–50
Dixons 159
DIY retailing 149, 152, 153
Docks de France 147
Dolomites, the 265, 267
Dorferneuerung schemes (West
  Germany) 273
Dunkerque 60, 136
Dusseldorf 158, 173

Early retirement, *see* Retirement
East Anglia 233–4
Eastern Europe 2, 5, 29, 282
Economic development models
  108
Economic restructuring 11, 12,
  37–9, 41–55, 56–8, 60–70,
  72–4, 81–3, 91, 93–4,
  103–5, 108–14, 125–7, 142,
  144, 171, 232, 279, 281
Economies of scale 32, 61, 63,
  66, 69, 147–8, 152, 153
Ecrins National Park (France)
  259, 262–3
Edeka 148
Education 3, 6, 78, 92, 94, 102,
  136, 163, 171, 174, 281
EFTA 15, 182
Egypt 27, 29
Elderly, the 143, 144, 227
  migration of 230, 233, 236–7
  see also Population
Electoral motives 184
Electricité de France 31
Electricity 13, 19, 31, 32, 33, 34
Electronic data interchange 111,
  114–15, 117–18, 131–3, 155
Electronics industry 2, 3, 5, 11,
  38, 50–2, 53
Emilia Romagna 97
Employment 6, 56–8, 65, 66, 68,
  72–4, 81–4, 85, 87, 88, 94,
  101, 109–10, 111–13, 115,
  163, 165, 166, 171–2, 175,
  176, 180, 181, 188, 189,
  190, 191, 231, 232, 233–4,
  235, 237, 240, 251–3
  see also Female, Part-time,
  Seasonal
Energy 10, 19–34, 280–81
  competition 20, 22, 27, 30, 31,
  33, 34
  conservation 22
  efficient use of 22, 31
  demand 3, 22, 27, 30
  geo-economics 22, 27, 29, 30
  geo-politics 22, 27, 29, 30

imports 20, 22, 25, 27, 29, 32, 33, 34
indigenous 22, 25, 27, 32, 33, 34
-intensive industry 19, 25, 34
market 19, 20, 21, 25, 27, 30, 31, 32-3, 34
policies 25, 30
resources 19, 22-5, 27, 29, 32
trade 27, 29, 31
transmission systems 31
transport costs 27
Engadine (Switzerland) 261
Enlargement, of EC 9, 29, 196
Enterprise
  attitudes to 77
  Zones 8-9
Entrepreneurship 4, 77, 185, 188, 192
  female 74
Environmental
  Assessment Directive 197, 198, 200, 203-6
  degradation 13, 22, 25, 31, 32, 33, 34, 195-6, 202, 238, 255, 259, 260-62, 263, 266, 268, 280-81
  attractions 79, 120, 128-9, 134-5, 241, 259-68
  improvement 133-9, 196-206, 281
  protection 10, 196-206, 213-14, 220, 236, 255, 258-75, 280
Environmentally Sensitive Areas 213, 214, 220
Essen (West Germany) 128
Ethnicity 164, 172
Ethnic minorities 74
  see also Guestworkers
Euratom 19
European
  Agricultural Guidance and Guarantee Fund (EAGGF) 189, 203
  Architectural Heritage Year 197
  Campaign for the Countryside 258
  Campaign for Urban Resources 137, 197
  Coal and Steel Community (ECSC) 3, 19, 189
  Commission 10, 11, 13, 91, 94, 96, 103-5, 196, 199, 200, 201, 203, 205, 213, 216
  Community 3, 4, 9, 10, 11, 13, 19, 25, 27, 29, 31, 34, 46,

54, 58, 59, 62, 65, 69, 179, 180, 182, 183, 185, 187, 188, 189, 190, 191, 192, 193, 195, 196-8, 199, 201, 217, 218, 223, 240, 280
  Community Council of Ministers 216
  Free Trade Association (EFTA) 46
  Investment Bank (EIB) 203, 204
  monetary union 179, 279
  Parliament 198, 203
  Regional Development Fund (ERDF) 203, 205, 247, 253
  Regional/Spatial Planning Charter (Torremolinos Charter) 197
  Social Fund (ESF) 203
  spatial strategy 195, 203
  Travel Commission 246
  Year of the Environment (EYE) 196, 197, 203
Evolène (Valais, Switzerland) 272-4
Exploration for oil and gas 22, 24, 25, 32
Exports 58, 65, 66
Extensification, agricultural 214, 216, 219, 223
Extensification Regulation (EC) 214, 216

Fachmarkt 153
Fallow land
  grazed 221
  green fallow 221
  permanent 219, 220, 221
  rotational 219, 220, 221
Family
  life cycle 145, 237, 244-5
  reunion 162, 163, 169, 170, 175
Farm
  diversification 213, 214, 220, 222
  Diversification Scheme 222
  tourism 241-3, 252, 253
  woodland 222
  Woodland Scheme 214, 222
Federal Republic of Germany, *see* West Germany
Female employment 2, 84, 172, 235, 251
Fertility, of population 76, 143, 163, 170, 175
  see also Population
Fiat 48

Financial services 117-19
Finland 15, 162, 182, 252
First World War 2
Flexible
  accumulation 109
  manufacturing systems 100
  specialisation 80, 84
Florence 135
Food and beverage industry 11, 67-9
Ford 48, 63, 64
Foreign direct investment (FDI) 41-7, 60, 249
  competitive bidding for, 46-7
  policies towards 46-7
Forestry Commission 222
Forests, protected 264
Forum-les-Halles, Paris, *see* Les Halles
Fossil fuels 32-3, 34
Fragmentation strategies 75, 109
France 3, 7, 22, 30, 34, 39, 48, 50, 54, 73, 78, 79, 81, 94, 96, 103, 109, 111, 113, 128, 130, 131, 143, 145, 147, 148, 155, 157, 158, 162, 163, 166, 167, 168, 169, 170, 173, 174, 175, 180, 181, 182, 183, 184, 185, 187, 188, 193, 198, 217, 218, 228, 241, 243, 245, 251, 261, 262-3, 268, 271
Franchises 75, 151
Frankfurt 111, 128, 130, 170
Free-market economy 181
Fuel oil 33
Fujisawa 53
Fujitsu 51
Fusion power 34

Gare d'Austerlitz, Paris 132
Gas-based electricity 31-3
GATT 46
Gastarbeiter, *see* Guestworkers
Gaz de France 31
GEAR (Glasgow Eastern Area Renewal Project) 139
GEC (General Electric Company) 51
General Motors 48, 63, 64
Geneva 169
Gentrification 126, 129, 133, 173, 176
GIB 152
Glasgow 130, 136, 139, 144, 158
Global economy 2, 4-5, 10, 37, 38, 54, 129, 185, 278, 279

'Golden-goose' phenomenon 25
Government intervention, *see* Planning and policy
Grand Travaux, Paris 125
Gran Paradiso National Park (Italy) 261, 263
Grants, *see* Subsidies
Great Depression 2
Greece 5, 15, 27, 94, 96, 103, 109, 162, 163, 164, 185, 187, 188, 190, 191, 277
Greenhouse effect 31, 195, *see also* Environmental degradation
Green politics 198, 280
Green rates of exchange (CAP) 211
Grenoble 127, 169
Gropius, W. 128
Gross fixed capital formation 279
Grossvenegiger massif (Austria) 264
Guaranteed
  prices (CAP) 211, 212
  thresholds (CAP) 213
Guestworkers 12–13, 162–75, 281
Gulf Co-operation Council 27
Habitat et Vie Sociale 139
Hague, The 136
Hamburg 131, 134
Harmonisation, of environmental regulations 199, 200
  see also Single European Market
Health 3, 6, 143, 174, 281
Higher education, *see* Education
High Pyrenees National Park (France) 262–3
High technology 2, 95, 102–3, 180, 181, 183, 184, 188, 189
  see also Technological change
Heritage 128–9, 134–5, 254, 271
Hitachi 51
Hohe Tavern (Austria) 260, 264
Honda 49
Hong Kong 38, 45, 50
Housing 12, 128, 129, 131, 136, 138, 139, 165, 172–4, 176, 236, 281
Hydrocarbons, *see* oil and natural gas
Hydro-electricity 19, 32, 33
Hypermarkets 143, 148, 150, 151, 152, 153, 156, 158
Hyundai 49

Iceland 197

ICI
Ideology, political 127–31, 140
Ikea 151
Immigrants
  Algerian 167, 168, 169, 170, 171, 175
  Asian 172, 173, 174, 175
  Caribbean 173, 174, 175
  Cypriot 172
  Filipino 164
  Finnish 167, 170
  Greek 168, 170, 173
  Irish 167, 171, 175
  Italian 166, 167, 168, 169, 170, 171, 173, 174
  Maltese 165
  Moroccan 164, 169, 170, 172, 174
  North African 164, 167, 168, 173, 175
  Portuguese 167, 168, 169, 170, 173
  Spanish 167, 168, 169, 171
  Third World 164
  Turkish 167, 168, 169, 170, 172, 173, 174, 175
  Yugoslav 167, 168, 169, 170, 172
Imports 20, 22, 25, 27, 29, 32, 33, 34, 48, 49, 58, 61, 65
Incentives, *see* Subsidies
Income levels 144
Incomes 164, 176
Indigenous
  development 180, 184, 186, 187, 188, 191, 192
  oil and gas 22–5, 34
Industrial cities 81, 126–7, 140
Industrialisation 1, 2, 12, 38–55, 56–70, 72–4, 81–7, 93–4
Industrial restructuring 38–9, 41–54, 56–8, 61–9, 72–4, 189, 231–2, 281
Industry
  aerospace 2, 3, 65–7
  automobile 3, 48–50, 63–5
  electronics 2, 3, 50–51
  food and beverages 67–9
  ownership 59–60, 61, 63, 65, 68
  pharmaceuticals 52–4
  steel 61–3
Information technology 114–15, 155
Infrastructure 4, 12, 185, 187, 189, 191
Inner-city areas 8–9, 12, 131–40,

  172–4, 184, 185
Innovation 3, 76, 84–6, 92–102, 277, 280
Institutions of Higher Education, *see* Education
Integrated
  energy systems 32
  Plan for Mediterranean Regions 213
  Rural Development Programmes 213
Integration 279–80
  see also Single European Market
Intermarché 148, 150
Internationalisation 41–7, 60, 65–70, 109, 150–52, 159, 279–80
International migration 162–75
Inward direct investment 41–7, 60, 249
Iran 27, 29
Ireland, Republic of 3, 25, 94, 109, 162, 163, 181, 182, 187, 190, 191
Isle de France 113
Italie, Paris 128
Italy 3, 19, 22, 27, 39, 50, 96–100, 109, 162, 163, 164, 166, 180, 184, 185, 186, 187, 188, 189, 190, 229–30, 277

Japan 5, 32, 34, 37, 38, 39, 41, 43, 46, 47, 48, 50, 52, 54, 65, 69, 94, 100, 279
Job creation 72–3, 81–3, 109–10, 180, 184, 185, 188, 189, 191, 192, 193
Joint projects 66–7
Joint ventures 102, 150, 152
Julian Alps (Yugoslavia) 260
Just-in-time system 65

Keynesianism 180, 181
Kondratieff cycles 3, 61, 76
Konjunkturpuffer 163, 165
Kuwait 27

Labour
  costs 63, 155
  markets 163, 164, 171, 173, 174, 176
  productivity 56–7
La Defense, Paris 132
Latin-speaking population 271
Land
  abandonment 23, 217, 219

Act (France) 131
development 235, 237
erosion 217, 266
redundancy 14, 211–23, 281
retirement 213, 216–17, 219, 222
Landscape
conservation 259–66
plans 264–66
Protection Plan
(*Landschaftsschutzplan*) 266
quality, deterioration in 266
Utilisation Plan
(*Landschaftsnutzungsplan*) 266
Landschaftsschutzgebeit 264–5
Land-use planning 205, 206
Lapps, the 260
Large-scale manufacturing industry 59–70, 184, 188, 189, 190, 192
Laura Ashley 151
La Villette, Paris 132, 136–7
Le Corbusier, C-E. 128
Le Havre 128
Les Halles, Paris 135, 137, 281
Less Favoured
Areas (LFAs) 213, 219, 222
Areas Directive (EC) 213
Regions (LFRs) 91, 101, 102–3
Libya 27
Life styles 129, 130, 145
Lignite 22
Lille 131, 134
Linkages 65
Linwood 64
Liverpool 127, 130, 136, 139
Llanelli 65
Loans 189–90
Local content 47, 50, 51
Location decision 60, 63
Lombardia 97
Lome convention 46
London 111, 131–2, 135–6, 138, 172, 175, 232
City of 118
Covent Garden 158
Docklands 131, 138, 281
Longitudinal Study of England and Wales 233–4
Longwy 62
Lower Saxony 221
Luberon Regional Nature Park (France) 259, 268
Luxembourg 3, 61, 109, 120, 182
Lyon 169, 174

Mall merchandising 148, 152
Malraux Act (France) 128
Malta 196
Manchester 114, 127
Manufacturing 2, 3, 4–5, 6, 11, 37–55, 56–70, 72–88, 91–105, 109, 171, 172
Marais, Paris 128–9, 137
Maritime Alps 260
Market
concentration 158
dominance strategies 147–8
segmentation 145–6
Marks and Spencer 159
Marseille 156, 173
Marshall Plan 2
Mass tourism 241, 255
Matorell (Spain) 63
Mazda 49
Mechanical engineering 93, 99
Mediterranean, the 27, 29, 30, 182, 191, 195, 197
Basin 162, 163, 164, 167
Package 213
see also Peripherality
Mercantor National Park (France) 259, 262
Mergers and takeovers 60, 63–4, 67–9, 119, 147, 149, 152, 280
Metropolitan regions 79, 96
see also Core regions
Mezzogiorno, the 100, 184
see also Italy
Middle East 22, 27, 29, 30, 34
Migration 12–13, 162–75, 228–39, 281
Rural–urban 127–8
Urban–rural 14, 226–38
see also Immigrants
Milan 133
Milk quotas 213, 223
Mining 172
see also Coal
Minority groups 74, 162–75
Mixed marriages 170
Monetarism 181, 182
Monetary Compensatory
Allowances (CAP) 211
Monetary union 179, 279
Monopolies 31
Monopsonies 31, 32
Montpellier 131
Monumentalijst (The Netherlands) 271
Monumentenwet (The Netherlands) 128

Multinational companies, *see*
Transnational corporations
Munich 111, 127, 130, 169, 170

Nationalisation 3
National Parks 259–63
Natural
gas 21, 22–5, 29–30, 31, 32, 33, 34
habitats, conservation of 270
monopolies 31
Natural Park (Spain) 266
Naturdenkmale 270
Nature
conservation 268–71
legalisation for 269
parks 266–8
reserves 269–70
Naturparke
South Tirol 266–8
West Germany 266
Naturschutzgebeite 270
NEC 51
Nederlandse Gasunie 31
Nestlé 68, 69
Netherlands, The 3, 22, 27, 96, 103, 109, 162, 163, 166, 168, 182, 185, 187, 188, 200, 232, 271
New
Commonwealth 15
international division of labour 5, 10–11, 171
product development 78, 84–6, 148
Right 129, 130, 184
spatial division of labour 231–2
towns 128
New York
Newly industrialising countries (NICs) 5, 37, 38, 45, 46, 50, 181
see also Third World
Nissan 49
Noisiel (France) 69
Non-fossil fuels 33
Non-tariff barriers 46, 49, 52, 54, 199, 200
North Africa (Maghreb) 27, 34, 162, 164
North Sea 22, 24, 25, 195
Northern Ireland 81–3
Norway 7, 15, 25, 180, 182, 196, 232
Nuclear
power 19, 32, 33–4
accidents 33

Numerical control, in
    manufacturing 98–101, 105
Nuremburg 169

OAPEC 27
Offshore oil and gas 22–5
Oil 19, 22–5, 27–9, 30, 31–2,
    33, 34
  companies 27
  distribution 20
  embargo on supplies 27
  exporting countries 20, 27, 29
  imports 21, 27, 29
  markets 25, 29
  price controls 20, 21, 25, 30,
    32, 34
  price forecasts 21
  price shocks 4, 10, 20, 22
  production levels 24
  production licences 24
  production profitability 24
  recoverable reserves 24
  refining 3, 20, 21, 27
  transport 20, 29
  use 21, 22, 31
Oil and natural gas
  revenues 9, 21, 25
  scarcity 25, 30, 34
OPEC 22, 181
OECD 13, 195, 196, 197–8
Other gainful activities (OGAs) 222
Oxford 65

Pacific Rim 181
Package holidays 241, 248
Padjelanta National Park (Sweden)
    260
Paridoc Group 148
Paris 111, 125, 128–9, 131–2,
    135, 136–7, 169, 171, 232
  East Plan 136–7
Part Dieu, Lille 134
Part-time workers 84, 144, 155,
    251
Patents 96–7
Pedestrianisation 157, 159, 281
Peripherality 3, 10, 57, 79,
    101–5, 111, 165, 182, 189,
    191, 277
Perpignan 69
Persian or Arabian Gulf 27, 29
Personal mobility 145, 156
Petite Bourgeoisie, see Self-
    employed
Petro-chemical industry 5, 20, 27
Pharmaceuticals industry 10, 38,
    52–4

Philips 51, 52
Piedmonte 97
Pipelines 27, 29, 31
Pisa 135
Planning, *see* Policy and planning
Plants, protection for 270
Plessey 51
Plymouth 128
Poland 5
Policy and planning 3, 5, 8, 10,
    13–14, 19–20, 25, 27, 30,
    31, 32, 34, 46–7, 50, 52, 54,
    62–3, 65–7, 69–70, 77–8,
    102–5, 119–20, 121, 127–8,
    130–31, 134–40, 163, 172,
    174, 179–81, 182, 184–93,
    197, 199–206, 211–23, 240,
    246–7, 249, 253–4, 258–75,
    281
Political coalition, urban 130–31,
    133
Pollution, *see* Environmental
    degradation
Pollution havens 13, 200, 205
Population 143, 169, 170–71,
    173, 181
  census 155
  growth 19, 20
  life cycle 145, 220, 237, 244
  mobility 145, 156
  suburbanisation 143
  see also Fertility, Immigrants,
    Migration
Ports 13, 126–7
Portsmouth 130
Portugal 15, 27, 94, 96, 103,
    109, 162, 163, 164, 185,
    187, 188, 189, 190, 193, 277
Post-industrial society 76–7, 109,
    126, 142, 144
Post-modern culture 130
Post-war reconstruction 3, 12, 20,
    127–8
Poverty 163, 281
Power stations 33
Premises, industrial 78
Preservation, *see* Conservation
Pressurised water reactors 34
Price-support policies, agricultural
    14, 211, 212, 222, 223
Primary energy 19
Privatisation 9, 30, 31, 33, 65
Producer services 11, 108–21, 280
  and economic development 108,
    121
  and information technology
    114–15

and location 108, 109, 110–11,
    113, 114, 115, 116, 121
  and organisation 115, 116
  see also Financial services
Product ranges 143, 148–9
Professional and technical
    employment, *see* Service class
Public expenditure 6, 8, 9,
    130–31, 180, 181, 182, 184,
    185, 186, 187, 190, 191,
    192, 193, 211–12, 215, 216,
    217, 219, 221, 222
Public housing 128, 138, 139,
    172, 173–4, 281
Public policy, *see* Policy
Public sector firms 6, 61, 63, 66

Qatar 29
Queyras Regional Nature Park
    (France) 259, 268

Racial discrimination 171, 172,
    174
Railways 171
Randstad (The Netherlands) 130
Rathausmarkt, Hamburg 134
Reading 113
Recession 2, 9, 21, 38, 61–3, 75,
    76, 86, 163, 175, 179, 180,
    181, 182, 183, 185, 189, 277
Recreation 133, 220, 222,
    266–70, 280, 281
  see also Tourism
Redundancy 75
  see also Unemployment
Refugees 162, 163
Regional
  Nature Parks (France) 266, 268,
    269
  Parks 266–8
  problems and policy 4, 8, 10,
    11, 13, 103–5, 179–93, 203,
    205, 277, 278–9
  sectoral specialisation (RSS)
    231–2
  separatism 193
Rejuvenation, urban 125–40
Renault 48, 63, 65
Renewable energy 34
Rennes (France) 131
Research and development 2, 66,
    78, 85–6, 92, 94–6, 102, 108
Residential energy use 19, 20, 31
Restructuring, economic, *see*
    Economic restructuring
Retail
  floorspace 147

format 150, 153
location 142, 152, 155
management 147
sales 145, 147, 158
services 150
strategy 142, 146, 152
Retailing 12, 133, 142–59, 172, 280
Retirement 214, 215, 230, 233, 236, 237
Return migration 163, 165, 166, 175
Reurbanisation 126–7, 229
Reuters 117
Rewe 148
Rhein-Main 113
Rhine 195, 196
Ripa di Meana 23, 201
Robots 65, 99–101
Rome 164
Rotterdam 128, 138
Rover Group 49, 64, 65
Rural
   areas 79–81, 211–23, 226–38, 245, 252, 253, 258–75
   communities 219–20, 236, 237, 263, 273
   depopulation 165, 226–7, 232
Rural-based tourism 241–3, 252, 253
   see also Tourism

Salzburg, province of 264
Samsung Corporation 45
Sanierungsgebeite 271
Sanierungsgesetze 139
Sarcellitis 128
Sarek National Park (Sweden) 260
Saudi Arabia 5, 21, 27
Science parks 78, 188
Science Policy Research Unit (UK) 85, 97
Scottish Development Agency 139
Sears 159
Seasonal employment 84, 169, 235, 251
Second homes 235–6
Second World War 2, 13, 19, 32, 127
Secteur sauvegardé (France) 128–9, 271
Security of energy supplies 27, 29, 34
SEDAS 155
Segregation 173
Self-employed and small business owners 233–5, 237

SEMAH 135, *see also* Les Halles
Semi-conductors 49, 50–1
Service class 233–5, 237–8
Service sector 5, 6, 11, 12, 73, 74, 77, 87, 108–21, 126–7, 133, 142–59, 165, 171, 180, 188, 232, 235, 237, 240–55
Set-aside 214, 216–17, 217, 218–20, 223
   agricultural 218–19
   environmental 219
   Regulation 214, 215, 216, 223
Shanty towns 173, 174
Sheffield 128–30
Shift-share analysis 58
Shipbuilding 3, 5
Shopping centres 157–8
   see also Retailing
Sicily 164, 167
Siebengebirge *Naturpark* (West Germany) 266
Siemens 51
Singapore 38, 50
Single European Act (SEA) 1, 9, 22, 30, 159, 197, 199, 200, 201, 202, 203
Single European Market (EC) 1, 11, 37, 46, 47, 50, 51, 52, 54, 69–70, 119–20, 159, 179, 190, 191, 192, 199, 200, 205, 278–9
Sites classés (France) 271
Sites inscrits (France) 271
Skiing 261, 263
Skilled migration 164, 172, 176
Slochteren (The Netherlands) 22
Slum clearance 174
Small and medium-sized
   enterprises (SMEs) 11, 72–88, 93, 180, 148, 188, 189, 191, 192, 237, 277
   towns 144, 227–9, 232, 235–7
Sochaux 65
Socialism 181, 182, 188
Social
   -marked economy 182
   mobility 233–4
   movements 128
   problems 162, 166, 172, 174–5
   relations of production 226, 231
   structure 162, 171, 174
   well-being 164, 165, 166, 174
'Soft' tourism 242–3
Solar power 32
South East England 79, 233–4
South Korea 5, 38, 46, 49, 50, 52

South Tirol (Italy) 264–8, 271, 275
Soviet Union 27, 29, 30, 34
Spain 6, 27, 48, 96, 103, 151, 152, 158, 162, 163, 164, 182, 183, 185, 187, 189, 190, 193, 277
Spatial
   diffusion 219–20, 222
   division of labour 79, 81, 100, 110–1, 113, 114, 115, 119, 231–4
Stanley Clinton Davis 201, 205
State subsidies, *see* Subsidies
Steel
   industry 11, 19, 61–3
   quotas 62
Stefanel 145
Stelvio National Park (Italy) 260–1, 262
Stora Sjofallet National Park (Sweden) 260
Strategy of conservation 258, 273
Strathclyde, *see* Glasgow
Structural policies, agricultural 211, 213–16
Structural shift
Stuttgart 169, 170, 172, 173
Style shops 155
Sub-contracting 65, 66, 75, 76, 80, 86
Subsidies 14, 25, 61, 62, 65, 67, 130–31, 180, 182, 184, 185, 186, 187, 188, 190–92, 211, 212, 214, 215, 216, 219, 221, 222, 264, 273
Suburbanisation 126–7, 143, 156, 229
Supermarkets 143, 159
Super-normal profits 25
Sweden 7, 15, 162, 167, 168, 173, 180, 182, 187, 196, 198, 277
   National Parks in 259–61
Swiss
   League for Nature Conservation 261, 269
   National Park 261–2
   Society for Natural Sciences 261
Switzerland 15, 53, 54, 162, 166, 167, 168, 173, 175, 182, 196, 197, 265–6, 269, 273, 277

Taiwan 38, 45, 50, 52
Takeda 53
Takeovers, *see* Mergers and takeovers

Tariffs 46, 51, 54
Technological change 11, 15, 22, 25, 30, 33, 75–6, 84–6, 91–105, 114–15, 131–3, 155, 277
  and decision making 104
  and employment 94
  and local economic development 93–4
  and productivity growth 93
  definition 92
  input measures 94–6
  output measures 96–101
  spatial variation in 94–102
Telecommunications 5, 131–2, 114–15, 278
Telecommuting 144
Textile industry 3, 4, 172
Thatcherism, *see* Monetarism
Third Periodic Report (ERDF) 104
Third World 164, 171
  see also Newly industrialising countries
Thomson Corporation 51, 52
Thurgau, Canton (Switzerland) 266
Tidal power 32, 34
Tirol (Austria) 260, 267, 271
Tokyo 118
Torremolinos Charter 197
Toshiba 51
Toulouse 66
Tourism 14, 84, 133, 169, 222, 240–55, 260–61, 263, 266, 267, 268, 273, 280, 281
  constraints 244–5
  dependency 248–9
  multipliers 249, 252–3
  receipts 240, 249–50
  rural-based 222, 241–3, 252, 253, 260–61, 263, 266, 267, 268, 273
  urban 243, 255
Tour operators 247–8
Town centre
  pedestrianisation 157, 159, 281
  redevelopment 155, 156, 159
Town planning 128, 130–31, 134–9, 156–8
Toyota 48, 49
TRADANET 155
Trade
  global 39–41, 279
  intra-EC 10, 39, 59
  tensions 46

see also Tariffs, Non-tariff barriers
Trade unions 171, 181
Training, *see* Education
Trans-Mediterranean relations 27, 29, 30, 34, 162, 163, 164, 166–8, 170, 175
Transnational
  corporations 5, 38–55, 59–71, 115, 129, 181
  hotel chains 248
  see also Foreign direct investment
TRANSNET 155
Transport 4, 5, 12, 13, 15, 20, 29, 175, 235, 280
  sector, energy use 20
Treaty
  of Paris 3
  of Rome 3, 19, 163, 197, 199, 200, 201
Turin 63
Turkey 29, 162, 163, 196

Unemployment 6, 13, 75, 163, 164, 165, 179, 180, 182–93, 277
  female 6
  long-term 185, 192
  youth 6, 182, 183
United Kingdom 19, 22, 30, 31, 39, 41, 44, 45, 48, 49, 52, 54, 93, 94, 97, 99, 109, 162, 166, 167, 170, 171, 174, 175, 176, 179, 180, 182, 183, 184, 185, 186, 187, 188, 190, 193, 199, 217, 218, 220, 221, 222
United Nations Economic Commission for Europe (UNECE) 196
United States of America 2, 15, 19, 20, 27, 37, 38, 46, 48, 49, 50, 51, 52, 54, 72, 76, 93, 94, 99, 196, 204, 279
  and production 37, 38, 39
  and trade 39
  and foreign direct investment 37, 41–4, 45
Urban
  'boosterism' 133–4
  change 129–40, 155–8
  Development Corporations (UDCs) 8–9, 131
  -industrial regions 81
  life-cycle model 126

lifestyle 129–30
  problems 12, 126–7, 162, 173
  renewal 12, 125–40, 158
  tourism 243, 255
  see also Inner-urban
Urbanisation 228–30
USA, *see* United States
Uzès (France) 271

Vanoise National Park (France) 260, 262–3, 264
Van de Rohe, M. 128
Vehicles 99, 100
Venice 125, 135
Vercors Regional Nature Park (France) 268
Vendex International 150
Vienna 169, 170
Village renewal schemes 273
Vocational training 185
Volkswagen 48, 63

Wages 12, 84, 171, 172, 235, 236, 237, 252
Wall Street 120
Waterfront revitalisation 129, 136–8
Waterstad, Rotterdam 138
Wave power 32, 34
Welfare 165, 174
  city 140
  State 5, 12, 174, 180
Waterschelde (The Netherlands) 31
West Germany 2, 3, 12, 22, 30, 31, 33, 39, 44, 45, 48, 51, 52, 53, 54, 93, 94, 96, 98–9, 109, 162, 163, 165, 167, 168, 170, 171, 172, 173, 174, 175, 182, 183, 185, 187, 193, 198, 199, 217, 218, 221, 222, 229–30
Wholesaling 172
Wilhemshaven 60
Wind power 32, 34
Winter sports centres 263
World economy, *see* Global economy

Young people 13, 237
Yugoslavia 162, 163

Zones d'Aménagement Concerté (ZAC) 131
Zoning systems and policies 263–4, 272, 273–4